T0257292

Phylogeography of California

Phylogeography of California

AN INTRODUCTION

Kristina A. Schierenbeck

UNIVERSITY OF CALIFORNIA PRESS

University of California Press, one of the most distinguished university presses in the United States, enriches lives around the world by advancing scholarship in the humanities, social sciences, and natural sciences. Its activities are supported by the UC Press Foundation and by philanthropic contributions from individuals and institutions. For more information, visit www.ucpress.edu.

University of California Press
Oakland, California

Library of Congress Cataloging-in-Publication Data

Schierenbeck, Kristina A., 1956-
    Phylogeography of California : an introduction / Kristina A. Schierenbeck.
        pages cm
    Includes bibliographical references and index.
    ISBN 978-0-520-27887-5 (cloth : alk. paper)—
    ISBN 978-0-520-95924-8 (e-book)
    1. Phylogeography—California.   2. Geology—California.   I. Title.
    QH105.C2S35 2014
    576.8′809794—dc23   2014003530

23   22   21   20   19   18   17   16   15   14
10   9   8   7   6   5   4   3   2   1

For Jim, Angela, and Celia

Contents

Acknowledgments   ix

PART I   GEOLOGIC AND ORGANISMAL HISTORY

1. Introduction   3

2. Historical Processes That Shaped California   21

3. The Cenozoic Era: Paleogene and Neogene Periods
    (65–2.6 Ma)   37

4. Quaternary Geologic and Climatic Changes   55

PART II   PHYLOGEOGRAPHIC PATTERNS IN
          VARIOUS TAXA

5. Conifers   67

6. Flowering Plants   77

7. Insects   103

8. Fishes   117

9. Amphibians   135

10. Reptiles   147

11. Birds   159

12. Mammals   171

13. Marine Mammals   197

PART III   SUMMARY

14. Consistent Phylogeographic Patterns across Taxa
     and Major Evolutionary Events   209

15. Conservation Implications and Recommendations   233

     Bibliography   257
     Index   317

Acknowledgments

I am honored to have had many fine mentors who have nurtured my passion for evolution, ecology, conservation, and the biological diversity of California. In no particular order these friends and mentors include George Corson, Doug Alexander, J. P. Smith, Bob Patterson, Ledyard Stebbins, Richard Mack, Rebecca Sharitz, Steve Edwards, Frank McKnight, Howard Latimer, Norm Ellstrand, Phyllis Faber, Jim Hamrick, Colin Hughes, Roger Lederer, and Wilma Follette. The friendships of Lily and Kader Aïnouche, Carla D'Antonio, Debra Ayres, Sue Jensen-Pollard, Kate McDonald, Charli Danielsen, Barbara Leitner, Deborah Jensen, Laura Morelli, Ann Bernadette, Chris Lozano, Marilyn Tierney, Tim Messick, Adrienne Edwards, Dawn Wilson, Doug Kain, Cindy Phelps, Ray Carruthers, Ellen Clark and the Third Avenue and Bunco gangs have benefited me greatly at various points in my life and in one way or another helped me to finish this book. Nothing I do would be possible without the unending support of my husband, Jim Eckert. Thanks go to my many family members for supporting my academic endeavors but especially Mary Huff and Nellie Huff, without whom I would have not been able to attend college. William G. Huff instilled in me an appreciation for California's rich history at a young age and provides some of the illustrations here. I thank the many enthusiastic and insightful students with whom I have been fortunate to interact over the past twenty years but particularly the dedicated graduate students; all have enriched my life and understanding of the biology of California.

Special thanks to those brave souls who provided comments on earlier versions of this work: Jay Bogiatto, Don Miller, Bruce Baldwin, Karen Burow, Andy Simpson, and especially John Avise, Peter Raven, and Arthur Shapiro. Chuck Crumly and Blake Edgar provided encouragement to complete this project and are much appreciated. California State University, Chico provided support to complete this work via a sabbatical.

PART I Geologic and
Organismal History

1 Introduction

What can we do with the western coast, a coast of 3,000
miles, rockbound, cheerless, uninviting, and not a harbor
on it? What use have we for such a country?

Daniel Webster, 1845

The geographic province we now call California was and in some places
remains every bit as rugged and inhospitable as Webster described. The
geographic parameters preventing extensive European expansion before
the nineteenth century are also the landscape on which the diverse flora
and fauna of this region have evolved. The goal of this book is to examine
and interpret the evolutionary history of the biota in California in a geologic context, as well as subsequent patterns in regional diversity that have
emerged across combined phylogenies. A number of phylogeographic patterns have indeed emerged; some previously identified are expanded here
in depth, and some new patterns are recognized. A survey of the phylogeography of the flora and fauna of California's diverse biota is organized by
major organismal groups, and these patterns provide the context in which
to ask further questions about evolutionary diversification in an area
defined by both physical and political boundaries. Comparing patterns of
many organisms provides the evidence needed to construct questions that
are narrower than those previously posed about the colonization of
taxa extant in California. Ultimately, this review provides a context for
landscape-level conservation efforts throughout the biogeographic provinces that roughly define the state of California.

Table 1.1   Number of species native and endemic to California (~411 k km²) compared to number of species native to the United States (9.83 M km²)
All 50 states included

Number of Species	Taxonomic Group		
	NATIVE TO CA	ENDEMIC TO CA	NATIVE TO UNITED STATES
Plants	4,844	1,416	18,743
Mammals	206	17†	428
Aves	640**	118‡	990^
Fishes	141*	67*	1,154
Reptiles	78	20#	311
Amphibians	64	35#	295

* native, resident and anadromous taxa, including 5 extinct species, 2 extirpated species (Moyle et al. 1995, 2000)
** www.californiabirds.org/ca_list.asp
† www.dfg.ca.gov/biogeodata/atlas/pdf/Mam_38b_web.pdf
‡ includes endemic and near-endemic species, from www.dfg.ca.gov/wildlife/nongame/ssc/docs/bird/BSSC-Overview.pdf
# www.californiaherps.com/info/endemicslist.html
^ Alaska included but not Hawaii

There are few places in the world that rival California in both topographic and biological diversity over similarly sized geographic areas. The state of California encompasses 411,015 km² and is 1,326 km long from corner to corner (Donley et al. 1979; Kreissman 1991). Plant communities range from Mesozoic coniferous forests along the north coast with rainfall of over 2,500 mm/year to halophytic communities in Death Valley with less than 3 mm/year of rain. In the northern Coast Ranges, an additional 200 mm/year of precipitation is added in the form of fog drip (Azevedo and Morgan 1974). California contains the highest peak in the conterminous United States (Mount Whitney, 4,406 m) and the lowest elevation in North America at Death Valley (-86 m), not more than 129 km apart. Each biogeographic region is remarkably heterogeneous in terms of topography, climate, and biological and geologic history. The diversity of species occurring within the political boundary of California cannot be rivaled in North America (Table 1.1), and Conservation International has recognized California as a globally important biodiversity hotspot (Myers et al. 2000).

Biogeographic studies of California are often focused on the California Floristic Province; however, for the purposes of colonization history from the south, north, east, and west, the biogeographic boundaries that define the California Floristic Province are expanded here to include areas beyond the state's political boundaries: the Mojave Desert to the south and southeast; the western Sonoran (Colorado) Desert and northern Baja California; the Transverse and Peninsular Ranges and Channel Islands to the southwest; the Sierra Nevada and western Great Basin to the east; the Cascade and Klamath-Siskiyou Ranges to the north and northwest, including southern parts of Oregon; the Modoc Plateau to the northeast; and the Coast Ranges, Pacific Ocean, and continental shelf to the west and northwest. Because of migration patterns, marine mammals and migratory birds and fishes are included if the breeding portion of their life cycle occurs in the California area. Somewhat arbitrarily, groups not included in this monograph are molluscs, noninsect arthropods, algae, fungi, and members of the Archaea and Bacteria domains. Although there is some literature on these groups, in particular arthropods other than insects and other marine invertebrates, it was not feasible to include the entire biota. Organization within each chapter is roughly from north to south and old to young, dependent on the literature available for each clade.

The relative geologic youth of the landforms that compose much of California combined with dynamic changes to the landscape during the late Cenozoic resulted in a variety of vicariant events over space and time. Whether species were residents of western North America since the Paleozoic or Cenozoic, geologic and climatic fluctuation had dramatic effects on their ranges and population structure and resulted in survival via refugia and range changes or extinction. Life history characteristics such as dispersal ability at each life stage, generation time, reproductive ability, and ecological characteristics such as degree of habitat specialization, competition, predation, mode of propagule dispersal, and availability of habitat or migration corridors all play an important role in the various outcomes for respective clades. The challenge of phylogeographic studies is to assess changes in population structure of once largely distributed populations or expansion from ancestral propagules into present-day population structures shaped by geologic and geographic processes.

There have been tremendous geomorphologic changes to the physical location that was to become California since the Paleozoic when the area was underwater, then moved gradually north and east and accreted to North America. These processes, combined with climatic fluctuation in the Miocene and, most recently, the Pleistocene, resulted in strong selection for both long-term residents and migrants. Species groups have radiated and resulted in incredibly diverse, ancient assemblages, as in the Klamath-Siskiyou Ranges, and the association of both paleoendemics and neoendemics along the Central Coast and in the Mojave Desert. Vicariant events such as mountain uplift and dry or arctic deserts were strong drivers of allopatric speciation throughout this process. California, because of its relative evolutionary youth, also has many examples of peripatric species, hybrid zones, and sympatric speciation. Dispersal events across very different habitats at different temporal scales have resulted in a mosaic of evolutionary scenarios.

Although California was initially explored by Europeans in the sixteenth century, there were no significant settlements until the late eighteenth century, and biological exploration of the area did not begin in earnest until the nineteenth century. Surprisingly, even *Sequoia sempervirens* (coastal redwood, Cupressaceae) was not formally described until 1795. The conservation of some the remoter areas of California can likely be ascribed to its relatively late European colonization. Early important contributions to the natural history of California were made by David Douglas (1799–1834), botanist; John C. Fremont (1813–90), botanist and geologist; William Brewer (1828–1910), botanist; James Cooper (1830–1902), geologist and naturalist; Willis L. Jepson (1867–1946), professor of botany at University of California, Berkeley, for over forty years; and Mary Brandegee (1844–1920), first curator of the California Academy of Sciences. A detailed account of the early, pioneering naturalists of California can be found in Beidleman's *California's Frontier Naturalists* (2006). Important contributions on the nature of California's diversity and the role of evolution were made by Joseph Grinnell (1877–1939), first director of the Museum of Vertebrate Zoology; Edward O. Essig (1884–1964), entomologist and founder of the entomology collection at UC Berkeley; Clinton Hart Merriam

(1855–1942), mammalogist, ornithologist, and entomologist; Lincoln Constance (1909–2001), director of the University of California Herbarium; and G. Ledyard Stebbins, evolutionary botanist and founding member of the Department of Genetics at UC Davis. More contemporary contributions to understanding the evolution of the biota of California have been made by many individuals but in particular by Daniel Axelrod, Herbert Baker, Bernie LeBeouf, Robert Haller, Lloyd Ingles, Arthur Kruckeberg, Harlan Lewis, Elizabeth McClintock, Jack Major, Robert Ornduff, James Patton, Peter Raven, and Robert Stebbins. Significant groundwork in phylogeography and evolution of the biota of California since the 1980s has been laid in the laboratories of David Wake, George Roderick, Brad Shaffer, Michael Caterino, Bruce Baldwin, Brett Riddle, Victoria Sork, Arthur Shapiro, Peter Moyle, Susan Harrison, Craig Mortiz, Greg Spicer, Wayne Savage, and Robert Zink.

Phylogeography of California summarizes and synthesizes the literature of the past fifty years, beginning roughly with the insightful and pioneering work of Axelrod in the 1960s. Raven and Axelrod's pathbreaking work in 1978 provides the foundation on which California phylogeographers have built an examination of the evolution of ancient, recent, native, and some migratory taxa to elucidate the major and minor evolutionary events that shaped the distribution, radiation, and speciation of the biota of this special place. Because phylogeography is a synthetic field drawing primarily from systematics, population genetics, geography, paleontology, and ecology, integration of these fields will enable us to predict and prioritize conservation areas during a time of rapid climate change, human disturbance, and invasive species. This work is not meant to be comprehensive but to provide a summary of the published literature on the evolution and diversification of some of the biota of California. It provides trends, examples, and, it is hoped, the generation of new hypotheses.

This book begins with a brief geologic history of the formation of the landscape on which California's species have evolved, followed by an evolutionary journey from the arrival of the ancestors of California's biota through their subsequent divergence within each major taxonomic group. Although the geologic and fossil records are not the province of phylogeography per se, the formation of California throughout the Paleozoic and Mesozoic provides the context for the development of the rich geology on which the biota gradually colonized and diversified. For example, the ancient substrates of the Klamath-Siskiyou Ranges originated at different latitudes from a variety of geologic processes that resulted in the formation of Paleozoic ophiolites and Mesozoic sedimentary rocks that underlay one of the most biodiverse regions in California. Fossil records prior to the Cenozoic help provide the visualization of the gradual colonization of the region prior to the dramatic climatic and geologic events of the late Cenozoic.

The writing is directed to the informed natural historian and is appropriate for an upper-division or graduate course on phylogeography or the evolution and natural history of California. Chapters are organized variously, according to available literature and clade distributions. Readers are provided with an evolutionary perspective of the basis of regional conservation and a context for how the California biota may respond to a rapidly changing environment due to global climate change.

CALIFORNIA TODAY

On the western edge of the North American Plate much of present-day California did not exist or was underwater until the Mesozoic. Plate tectonics were responsible for gradually accreting landforms to the North American Plate that became California. Land areas that were present prior to this time, which include the Mojave Desert and the Klamath Plate, have experienced dramatic changes in orientation, latitude, and climate. Following is a brief description of major contemporary landforms.

The Klamath-Siskiyou region includes the Siskiyou Mountains, Trinity Alps, Marble Mountains, Salmon Mountains, Scott Bar Mountains, and North Yollo Bolly Mountains, covering about 50,300 km² (Miles and Gouday 1997). The current substrate is a complex array of formations from the Paleozoic and Mesozoic, primarily composed of metamorphic rocks differentiated from the generally younger geology of the Coast Ranges and the Cascade Range (Sawyer and Thornburgh 1977). This area of northwestern and north central California and adjacent Oregon contains some of the most interesting taxonomic assemblages in western North America. Although there is some evidence of glaciation about 30,000 years ago (ka), this region largely escaped Pleistocene glaciation events, although the biota was certainly affected by climatic variability during this time. The large number of relictual plant species found in this region provides evidence that a number of high-elevation sites served as refugia (Soltis et al. 1997; Sawyer 2006).

North of the Sierra Nevada and beginning east and southeast of the Klamath-Siskiyou region is the Cascade Range, which consists of a chain of large volcanoes and dissected lava flows of early Paleogene through Holocene origin. California is home to the two southernmost volcanoes of the Cascade Range, Mount Shasta (4,319 m), a stratovolcano created by a series of eruptions over the past 600,000 to 100,000 years, and Mount Lassen (3,188 m), part of a volcanic center that began erupting 825 ka and most recently erupted during the period from 1914 to 1921. East of the Cascade Range in extreme northeastern California, the Modoc Plateau is a lava plain with an average elevation of 1,350 m, estimated to have formed from lava flows between 25 and 3 million years ago (Ma) (Schoenherr 1992; DeCourten 2008). The Modoc Plateau is also volcanically young, with activity as recent as 700 ka in the Medicine Lake area and eruptions as recently as 200 to 300 years ago (Harden 1998).

The Sierra Nevada stretches for roughly 650 km from southern Lassen County south to central Kern County, with an east-west breadth of 70–90 km (Storer and Usinger 1964; Howard 1979). Most of the base rock is a complex array of granitic plutons formed during the Mesozoic, although there are some weakly metamorphosed sedimentary and volcanic rocks of Paleozoic age that have been intruded by the granite batholiths (DeCourten 2008). Middle and upper Cenozoic rocks crop out in the northern parts of the Sierra Nevada. Geologists debate the temporal distribution of the uplift of the Sierra Nevada, but there is recent consensus that initial uplift occurred as long as 160 Ma (Cassel et al. 2009), with significant uplift from about 1,000 to 1,500 m (25 Ma) to 2,500 m (10 Ma) (Xue and Allen 2010). Pliocene uplift approximately 6–3 Ma is supported by tilted strata; however, isotope data suggest that the primary uplift occurred prior to 12 Ma (Mulch et al. 2008). The Sierra Nevada and its resident biota have been further shaped by at least nine major glaciations since the Pliocene (Gillespie and Zehfuss 2004).

The Great Valley of California lies between the Sierra Nevada and the Coast Ranges and continues about 680 km from the Klamath Mountains in the north to the Tehachapi Mountains at the southern end. It is a mostly flat plain, 60–120 km wide, undergoing deposition for as long as 100 Ma, with soils that are largely alluvial and lacustrine sediments originating from former inland seas and the nascent Sierra Nevada (Farrar and Bertoldi 1988). The Sacramento Valley occupies the northern one-third of the valley. The San Joaquin Valley occupies the southern two-thirds of the valley and comprises the San Joaquin Basin in the north and the interior-draining Tulare Basin in the south. Water flow in the valley is dispensed by the San Joaquin River in the south and the Sacramento River in the north; these rivers join approximately midway to form a delta composed of a westward series of freshwater, brackish, and salt marshes that flow into the San Francisco Bay. The San Francisco Bay estuary is one of California's most important ecological habitats, draining approximately 40 percent of the water in the state.

On the Pacific coast, south of the Klamath Range and extending to Santa Barbara County, are the Coast Ranges, a series of north-south ranges primarily of sedimentary origin that with few exceptions are less than 2,000 m in elevation (Harden 1998). Formation of the Coast Ranges is the result of a number of mechanisms. The northern ranges were formed by the movement of the North American and Pacific Plates (Atwater 1970). During the Oligocene, the Salinian terrane was located west of the present Central and Southern California coast. Beginning in the Miocene, northward movement and fragmentation of the granitic and metamorphic Salinian terrane resulted in the formation of islands that eventually became part of the Coast Ranges (Hall 2002; Kuchta and Tan 2009). Before the Pliocene, the central Coast Ranges existed as islands (Yanev 1980). About 5–3 Ma, the Santa Ynez Mountains became connected to the Gabilan and Santa Lucia Mountains via the uplift of the Temblor Range (Hall 2002). A seaway south of Monterey and present from about 8–2 Ma, was closed via continued uplift of the Coast Ranges and consequently allowed the Central Valley to fill with freshwater from the surrounding Coast Ranges and Sierra Nevada. Drainage of the Central Valley continued to occur via the Salinas and Pajaro Rivers until further uplift of the Coast Ranges about 600 ka diverted the drainage through the Carquinez Strait and into the San Francisco Bay (Sarna-Wojcicki et al. 1985).

Southern California is intersected by a number of small ranges that extend variously in either an east-west or north-south orientation. The Transverse Ranges lie between the southern end of the Coast Ranges and the Los Angeles Basin and extend from the coast eastward to the western and southern edges of the Mojave Desert (Jaeger and Smith 1966). The Transverse Ranges are geologically complex, consisting of igneous and metamorphic rocks in the eastern part and sedimentary material in the western part, and range in elevation from sea level to 3,506 m (Mount San Gorgonio) (Harden 1998). The largely granitic Peninsular Ranges extend roughly in a north-south direction and separate California from the Colorado Desert to the east. These ranges include high peaks such as Mount San Jacinto (3,286 m) and Santa Rosa Mountain (2,452 m) and low-elevation passes (Jaeger and Smith 1966).

California's offshore islands include the Channel Islands, which occur off the coast of Southern California and comprise four northern islands (Anacapa, San Miguel, Santa Cruz, and Santa Rosa) and four southern islands (San Clemente, San Nicholas, Santa Barbara, and Santa Catalina). Although steep cliffs provide dramatic relief on some of the islands, the topography is generally low, with elevations reaching no more than a few hundred meters (Harden 1998). The first emergence of islands in this area occurred in the Miocene (17–13 Ma). Pacific Plate movements resulted in the rotation of these islands away from the San Diego area during the Miocene (Jacobs et al. 2004). Most of the islands arose during the Pliocene (Hall 2002) or Pleistocene and were resubmerged about 500 ka. During the last glacial maximum (LGM), the northern Channel Islands comprised one landmass, approximately 6 km from the mainland (Hall 2002). At 260 ha, Santa Barbara Island is the smallest and is estimated to have been isolated for about 10,000 years (Rubinoff and Powell 2004).

The Mojave Desert is bordered by the Transverse Ranges (Tehachapi Mountains) to the west; the Sierra Pelona, San Gabriel, and San Bernardino Mountain ranges to the southwest; the Sonoran Desert to the south; and the Inyo Mountains and Great Basin to the north. The Mojave Desert is intersected by a number of mountain ranges separated by undrained, alluvial basins. Landforms of the Mojave Desert historically have been influenced by oceanic sedimentation and ancient volcanic activity. Recent events have further shaped the modern landforms and include periods of

Pleistocene glaciation in the Sierra Nevada and, to a lesser extent, recent volcanic eruptions in the Mojave and Great Basin regions. Active fault zones include the San Gabriel fault along the Transverse Ranges and the Garlock fault at the southern end of the Sierra Nevada (Cox et al. 2003; Bell et al. 2010). Contemporary landforms include desert mountains and aeolian dunes. Important historical events that helped shape the Mojave Desert include the middle Miocene Salton Trough that severed connections to the Baja Peninsula and the Sonoran Desert and the Bouse Embayment, which consisted of a chain of lakes that flooded and linked the Colorado River and the Gulf of California by 5.3 Ma. The link between the Colorado River and the Gulf of California was filled and closed by 4 Ma, at which time the Mojave Desert arrived at its present form (Ingles 1965; Bell et al. 2010).

BRIEF HISTORY OF PHYLOGEOGRAPHY

In the eighteenth century Alexander von Humboldt formally recognized the value of multiple disciplines to explain the occurrence of organisms. Darwin expanded on these ideas by incorporating the role of geologic history in natural selection and integrated selective processes with reproductive isolation, competition, and dispersal to explain evolutionary diversity.

Alfred Wegener's famously dismissed theory of continental drift, published in 1915, was not broadly accepted by the scientific community until almost sixty years later, during the late 1960s. As a result, earlier comprehensive works on the evolution of the California biota did not have the advantage of an understanding of the geologic context in which it occurred. Whereas Raven and Axelrod (1978) were able to incorporate plate tectonics into their work on the origin of the California flora, Axelrod's extensive work from the 1940s through the 1960s, as well as those of earlier biogeographic theorists on the evolution of the mammalian (Orr 1960) and herpetological fauna (Savage 1960) of California, was handicapped by the recency of the acceptance of continental drift (Riddle et al. 2000a). Ideas about vicariance biogeography, developed from the 1960s through the 1980s, were importantly influenced by Croizat's work on panbiogeography in 1958. Hennig's 1966 phylogenetic systematics was the foundation on which modern phylogenetics would develop, and many methods that have grown out of the field of phylogenetics are critically important to the detection of phylogeographic patterns.

By the 1980s, the intellectual groundwork had been laid for combining the fields of geology, paleontology, phylogeny, and population genetics. Wiley's 1988 review of vicariance biogeography appeared just a year after Avise et al. coined the term *phylogeography*. However, the field of phylogeography arguably began with Avise et al.'s 1979 study on southeastern pocket gophers by demonstrating that mitochondrial DNA (mtDNA) haplotypes were correlated with geographic divergence. Avise was the first to provide a comprehensive correlation of genetic variation at the population level and geographic phenomenon at the regional level with his recognition that phylogeographic pattern exists across phyla and even kingdoms. It is now widely accepted that species diversity and the distribution of continental biotas reflect climatic change, range shifts over geologic time, and changes in the distribution and gene flow of populations (Riddle 1996). Hewitt (2000) further developed phylogeography to incorporate the consequences of hybridization and speciation patterns.

Today phylogeography serves as a link between population genetics and phylogenetics and the landscapes on which they occur. Despite some overlap in studying the geographic distribution of genetic variation, landscape genetics and phylogeography are different (Storfer et al. 2007): phylogeography looks at historical events that shaped patterns; landscape genetics provides a more contemporary view (Knowles 2009, Wang 2010). In this work systematic, landscape, and population genetic studies have been included to assist in forming a clearer picture of phylogeographic patterns in California. The field of phylogeography has grown significantly in the past decade, as reflected by Google Scholar searches for articles with the keywords "California" and "phylogeography," which found 429 articles in 2000, 885 in 2005, 1,790 in 2011, and 15,600 in 2013.

Cladogenesis can vary widely based on speciation mechanisms among lineages; however, multitaxonomic studies have provided insight into historical gene flow and can lead to an understanding of the geographic and ecological factors important to speciation. A summary and analysis of 55 taxa in the California Floristic Province identifies a strong correlation among phylogenies and vicariant events such as the formation of mountains and deserts (Calsbeek et al. 2003). For animal taxa, there is a split in the Transverse Ranges into north and south clades, and a separation of east and west clades by the Sierra Nevada and Coast Ranges. Birds provide exceptions to these patterns, likely due to their high dispersal rates. Molecular clock data and pairwise divergences between different geographically located taxa establish an average split time of 2.49 Ma, correlating with geographic features, and suggest that these geographic changes began approximately 7 Ma. The same analyses for plants found shorter divergence times, averaging 1.35 Ma among similar vicariant events. More recent analyses expand, refine, and generally support their conclusions in the Transverse Ranges, Coast Ranges, and Sierra Nevada. More in-depth studies reveal additional phylogeographic patterns in birds and plants and are reviewed here.

There have been recent, significant strides in the interpretive possibilities of phylogeographic studies with the discovery of new nuclear and cytoplasmic markers, developments in coalescent analyses, paleoclimatic models, and niche modeling (Avise 2009; Hickerson et al. 2010; Sinervo et al. 2010; Camargo et al. 2010). Molecular techniques and robust statistical methods provide tools and data that can verify or dispute hypotheses about the origins and evolution of the biota of California. Genetic data are useful in detecting pathways of migration or enigmatic refugia (Petit et al. 2005). Fossil data can provide conclusive evidence of the existence of a species in an area and can inform molecular data (Gugger et al. 2010). Fossil data can also provide a measure of migrational patterns that, when combined with molecular data, can clarify climatic patterns and properties of past ecosystems. Extant species that have survived via glacial refugia are likely to show genetic signatures of within-refugium genetic drift, selection, and increased diversity among refugia, which can further inform historical ecological patterns (Gugger et al. 2010).

Phylogeographic patterns can result from a range of deep to shallow levels of divergence. Deeper phylogenetic structures are primarily based on vicariant events (Avise 2000). Depending on the temporal scale, dispersal events, bottlenecks followed by population growth, and gene flow among populations will all leave signatures of genetic variation (Slatkin and Hudson 1991). When corrected for ancestral polymorphisms and selected for the correct tempo for the time period under evaluation, sequence data can be used to estimate divergence times among clades (Nei 1987; Avise and Walker 1998). Low variability due to recent divergence can be a problem, particularly in places like California where variance and evolutionary events are recent; however, this issue can be overcome with the use of additional markers or the use of markers more suitable for measuring population-level variation, such as microsatellites and network analyses (Posasda and Crandall 2001).

A shared evolutionary history results in the coalescence of genes. Coalescent theory is important particularly for older taxa in which the preponderance of nuclear loci are expected to be monophyletic, in comparison to younger taxa in which fewer nuclear loci would be expected to be monophyletic (Palumbi et al. 2001). For autosomal genes, the probability that any allele will be passed on to the next generation is four times greater than for nonautosomal genes and consequently they will be four times more ancient (Avise 2008a). Some of these alleles will be linked via a number of mechanisms, including natural selection, ecological behavioral, and loss of gene flow. Most phylogenetic and phylogeographic studies use maximum-likelihood and Bayesian methods to infer genetic divergence. One must be careful, however, in interpreting variation due to differing mutation rates and coalescence. Coalescence can lead to a lot of variation and has tended to result in overinterpretation of data (Edwards and Beerli 2000). Statistical phylogeography provides the framework in which to quantitatively interpret alternative hypotheses (Knowles and Maddison 2002). Another method, Nested Clade Analysis (NCA), uses a hierarchical method to establish geographic structure that can result from restricted gene flow, population fragmentation, range expansion, and colonization (Templeton 2004) but has come under intense criticism because of false identification of population events (Knowles and Maddison 2002; Petit et al. 2008).

Incomplete lineage sorting is often correlated with some life history characteristics such as long generation times, large effective population sizes, or evolutionarily recent divergence (Syring et al. 2007; Eckert and Carstens 2008). Powerful molecular tools and Bayesian analyses can be used to sort even recent population divergences, but ongoing gene flow or gene flow following divergence can obscure lineage sorting. Topology methods are used to find the trees that have the highest probability of being the true phylogeographic pattern and include maximum parsimony and maximum likelihood (very accurate but slow to operate). Bayesian methods are the most accurate method to infer phylogeographic pattern, with the most popular programs currently being BEAST (Bayesian Evolutionary Analysis Sampling Trees; Drummond and Rambaut 2007) and MrBayes (Ronquist and Huelsenbeck 2003), which employs the use of Monte Carlo Markov Chains. Monte Carlo coalescent simulations of methods of genetic isolation (e.g., allopatry, island, peripatric, and stepping-stone) with different levels of gene flow revealed that if five or more loci are used, correct topologies can be identified with greater than 75 percent accuracy (Eckert and Carstens 2008). Regardless, it is clear that multiple loci from different genomes, evolving at varying but well-established rates of evolution, are essential in obtaining a clear evolutionary history of any lineage. Also essential for an accurate phylogeny is the thorough sampling of populations or terminal cladistic units.

The optimal molecular marker to measure phylogeographic pattern is dependent on the time since divergence. A significant amount of genetic variation needs to have accumulated among populations that are within the evolutionary time period in question. For animals, mtDNA is often used because it is maternally inherited and does not undergo recombination. The mtDNA control region (mtDNA CR), usually about 1 kilobase (kb), is a noncoding sequence involved in the initiation and regulation of replication and transcription that has proven quite useful in phylogeographic and population genetic studies of animals. Additional mtDNA regions of choice in phylogeographic studies include the cytochrome oxidase subunit I (COI) and subunit II (COII) for measuring within-species genetic structure. Cytochrome *b* (cyt *b*) and the NADH genes (*ND1*, etc.) are also known to have rates of evolution that are appropriate from measuring fairly recent divergence among mammals (Smith and Patton 1993). Chloroplast DNA sequences used in plants for population studies include a number of spacer regions (e.g., *trn*D-*trn*T, *trn*T-F, etc.). Whether plant or animal, the ideal study will use a number of markers from both the nuclear and cytoplasmic genomes and find concordance among taxa through the analysis of multiple loci within each species (Avise et al. 1987).

Nuclear DNA (nucDNA) markers used in population studies can be problematic due to biparental inheritance, intraallelic recombination, gene families, and heterozygosity. Although their use is not without concerns, the ribosomal internal transcribed spacers (ITS1 and ITS2) are known to be intra- and interspecifically variable in plants (Baldwin 1992) and variable among higher-level clades in animals (Coleman and Vacquier 2002). Few nucDNA regions have been identified as useful because of their slow rate of evolution, but some introns have proven informative (Brito and Edwards 2008). Microsatellites are often developed from the nuclear genome but sometimes also from cytoplasmic genomes. They are developed for use in a single species but often can be applied to related species; the number of loci in common will decrease with increasing genetic distance. The more rapidly evolving, usually codominant microsatellite loci provide a different genetic signature in time and are often ideal for detecting divergences since the Pleistocene in both plants and animals (Avise et al. 1987).

Deviations from heterozygosity expectations can be estimated using coalescence or equilibrium techniques that were developed initially by Sewall Wright and improved by theoretical and computational developments since the 1990s. In addition to the coalescence methods described above to infer haplotype networks, a number of population statistics are used to assess pattern and process in population divergence. First, H_E simply estimates the extent of genetic variability in a population via the expected proportion of heterozygous loci per individual, whereas H_O is the direct count of the frequency of heterozygotes. Wright's F-statistics, in particular F_{ST}, are widely used to determine evolutionary processes among populations with values that range from 0 to 1. Small F_{ST} values indicate that allele frequencies are similar among populations, and increasingly larger values reflect increasing large differences in allele frequencies among populations. F_{IS}, or the inbreeding coefficient, measures the deviation from Hardy-Weinberg equilibrium in subpopulations. Modifications of F_{ST} include R_{ST}, which accounts for mutation rates at microsatellite loci (Slatkin 1995); G_{ST}, which is a multi-allele analog of F_{ST} where D_{ST} is the total genetic diversity distributed among populations (Nei 1973, 1986, 1987); Θ, which estimates F_{ST} but corrects for error associated with incomplete sampling of a population or, specifically, haplotype data (Weir and Cockerham 1984); and Φ_{ST}, which measures differences among populations using sequence data (Excoffier et al. 1992). Nucleotide diversity (π) measures the degree of nucleotide polymorphisms within a population (Nei and Li 1979), and haplotype diversity (h) measures the number and frequency of haplotypes in a population (Nei and Tajima 1981). Analysis of Molecular Variance (AMOVA) estimates population differentiation from molecular data (Excoffier et al. 1992). Comparison of F_{ST} and R_{ST} can reveal levels of N_m, migration among populations; if R_{ST} is larger than F_{ST}, historical N_m was low among populations; if R_{ST} is equal to F_{ST}, N_m has been played a large role in the population structure; and if R_{ST} is less than F_{ST}, drift was a predominant factor in creating differences among populations. STRUCTURE is a popular program that detects the differences among populations, assigns individuals to populations, and identifies admixture in individuals and populations using allelic combinations (Pritchard et al. 2000).

Techniques now allow estimates of demographic events from contemporary sampling. Genealogical data and coalescent theory can be used to estimate effective population size (N_e) but will be affected by bottlenecks, breeding behaviors, number of progeny, and selection and thus require an average estimate of an evolutionary N_e from a group of related taxa (Avise 2008a). Differential mutation rates among loci and varying migration rates will also affect estimates of N_e (Hickerson et al. 2006). Consistent patterns among species often reflect similar responses to vicariant events, but inconsistencies may be present due to differential gene flow and different effective population sizes. Tajima's D is one of the best measures of vicariance events in species with variable populations sizes because it differentiates between neutral mutations and other evolutionary processes (Tajima 1989; Hickerson et al. 2006).

Life history characteristics, gene flow, and population demographics all influence phylogeographic patterns. Species with low rates of dispersal are generally easier to resolve, whereas high gene flow can increase the difficulty of interpretation. High levels of gene flow will be reflected in low F_{ST} values and high N_m. Taxa residing within the same region with similar ecological requirements, life histories, and behaviors often share similar phylogeographic patterns (Bermingham and Moritz 1998). However, loci under selection can alter expected phylogeographic patterns based on demographic patterns or the historical distribution of a population. Thus precise identification of refugia and vicariant events require data from a number of species.

The phylogeographic patterns illustrated here offer the opportunity to not only conserve regions with the highest biological diversity but also conserve the evolutionary process in the face of climate change. Understanding how species have evolved on the landscape gives us some measure of the ability to predict how they may respond in the future. Of particular importance is the use of genotypes in restoration that are compatible with projected regional changes in climate, for example, the response of characteristics such as bud-burst to changes in temperature and precipitation (Savolainen et al. 2007). Similarly, models of historical data from past climatic events can be useful in predicting the ability of alpine taxa to survive warming temperatures. It remains to be seen whether we can estimate the ability of a species to migrate based on past migrational events complicated by human landscape disturbance. Regardless, a combination of ecological and genetic data will provide the best foundation for conserving the ability of the California biota to respond to rapid environmental change (Gugger et al. 2011; Eckert 2011).

2 Historical Processes That Shaped California

The geologic history of California provides the basis for its present geologic complexity. The position of California along the continental margin gives a physical context that began when the earth's crust was formed and has resulted in rapid geologic change and effective biotic isolation from much of North America.

When the first sedimentary rocks were forming 3.8 Ga (billion years ago), California was nothing but unformed crust at different latitudes and longitudes under the sea and the last common ancestor of Archaea and Bacteria was diverging. Stromatolites, mats of Cyanobacteria (Domain Bacteria), are present in the fossil record from 3.5 Ga and survive in a few places on earth today. By 3 Ga, photosynthetic cyanobacteria dominated life on earth, and by 2.5 Ga their overabundance resulted in the overoxidation of the atmosphere. Although there is a continued peak in cyanobacterial diversification until about 1.4 Ga, their diversity then begins to decline. Acritarchs, fossil bacteria or eukaryotes, which are difficult to classify, are present since about 3.5 Ga and begin to diversify about 2.0 Ga.

The land currently known as North America starts forming in the early Proterozoic (2.5–1.6 Ga) as part of the assembly of the supercontinent Nuna (Hoffman 1999; Whitmeyer and Karlstrom 2007). Much of North American formation is thought to be a result of crustal growth after the Archean (3.8–2.5 Ga). The western limit of North America during the Proterozoic is estimated to be along the western border of Idaho and the eastern border of Nevada. The oldest rocks in California are about 2.4–1.4 Ga and provide the crystalline basement on which the rest of the rocks were deposited or intersected. Sedimentation, subduction, and tectonism formed a mosaic of material that formed the crust of Laurentia, the physical core of North America, and led to the differentiation of the mantle rock into mafic and felsic materials 1.8–1.0 Ga, a process known as the accretion-assembly-stabilization model (Whitmeyer and Karlstrom 2007). Dinoflagettes appear about 1.1 Ga when Laurentia was part of the supercontinent Rodinia.

The glaciations at the end of the Proterozoic, combined with early multicellularity, provided the ecological opportunity for the evolution of early Paleozoic organisms. Sexual reproduction is estimated to have evolved about 1.2 Ga and led to more rapid diversification via recombination and increased genetic diversity. The evolution of protist herbivores was likely important in the population control of prokaryotic and eukaryotic algae and provided selection pressure that resulted in diversification and multicellularity. Multicellular eukaryotes that evolved in the Proterozoic and that were globally present would have included the eukaryotic Acritarchs, likely marine algae, and a fairly common fossil from this era.

Limited sediments were deposited on the crystalline basement in the late Proterozoic (Neoproterozoic, 900–543 Ma) along the continental margin, and sedimentation increased and continued through the early Paleozoic. Extensive algal mats formed the carbonate platforms that were the continental shelves and formed the passive margin of Laurentia. These formations can be seen in Death Valley as they were changed by geologic processes in the subsequent geologic eras (Stewart and Suczek 1977). These mats were so extensive that they are thought to have resulted in a decrease of CO_2 in the atmosphere, which reduced the earth's temperature due to a reduction in the greenhouse effect. Cool temperatures persisted over a period that spanned 850–600 Ma and eventually led to increased glaciation. In turn, this cooling led to a cessation of the carbonate platform formation. Combined with the previous evolution of multicellularity and sex there is an increase in biodiversity, and by 580 Ma the Cambrian explosion of biological diversity has begun. Evidence for the first fungi is

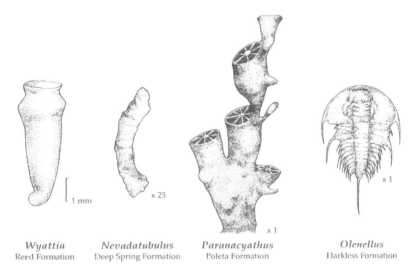

| Wyattia | Nevadatubulus | Paranacyathus | Olenellus |
| Reed Formation | Deep Spring Formation | Poleta Formation | Harkless Formation |

Figure 2.1. Examples of fossils from the Cambrian of the White-Inyo region. Used with permission from John Wiley and Sons. Original illustrations by Janice Fong from Mount and Signor 1989.

about 560 Ma, and this, with an increase in oxygen in the atmosphere, facilitated the movement of other life-forms to land 580–540 Ma. The rest of California, meanwhile, was being shaped as part of the accretionary belt that has formed since near the Proterozoic-Cambrian boundary.

The Proterozoic-Cambrian transition is well documented in the White Mountains of Inyo County and establishes the emergence of the paleofauna and the faunal provinces for western North America (Mount and Signor 1989) (Figure 2.1). This stratigraphic sequence, first documented by Charles Walcott (of Burgess Shale fame), contains stromatolites from the Proterozoic and is an excellent example of this important evolutionary transition. By the early Cambrian, Archaeocyatha (sponges), Trilobita (extinct marine arthropods), Brachiopoda (lamp shells), Mollusca (molluscs), Radiolaria (amoeboid protozoa), and Echinodermata (modern representatives of which include sea urchins and sea stars) all are present and have been described from Shasta and Inyo Counties (ucmpdb.berkeley. edu). The Marble Mountains are a small range in the driest part of the Mojave Desert that contain a Cambrian fossil bed consisting predominantly of Trilobita.

Around the Cambrian (542–488 Ma), Laurentia was an independent continent and still provided the core of North America. There is a fairly homogeneous fauna from this region in the Precambrian and early Cambrian, and fossils indicate that the area that was to become western North America resided in tropic regions (Mount and Signor 1989). The earliest members of Chordata appeared in the Cambrian and continued to diversify throughout the Paleozoic. Land plants or Embryophyta emerged from Streptophyta, a clade of green algae during the Ordovician (490–444 Ma) (Becker and Marin 2009). Around the Devonian (416–359 Ma), Laurentia collided against Baltica, forming the minor supercontinent Euramerica. Seed plants emerge in the Devonian, the now-extinct seed ferns were common in the Devonian 385 Ma, with conifer-like plants present by the Carboniferous (Taylor et al. 2009). However, through the Devonian most of California was still under the sea and fossils from this period continue to consist mostly of invertebrates.

The oldest known insect fossil, *Rhyniognatha hirsti* (Pterygota), found in the Rhynie Chert of the Devonian in what is now Scotland, is estimated to have occurred about 400 Ma (Engel and Grimaldi 2004). Pterygota includes the winged and secondarily wingless insects and is divided into the Paleoptera and Neoptera. The Endopterygota (Neoptera) radiated in the Carboniferous and comprises those insects that undergo radical metamorphosis and includes eleven major clades of insects, including Lepidoptera (moths and butterflies), Diptera (true flies), Hymenoptera (wasps, bees, ants), and Coleoptera (beetles), which account for 80 percent of all insects. Phytophagous members of the Coleoptera and Lepidopterans coradiated with flowering plants beginning about 130 Ma, and almost all rely on Anthophyta for food, with many species restricted to one or a few plant species. Also within the Endopterygota, the diverse Trichoptera (caddisflies) contains 45 families, 18 of which occur in California, that are exclusively aquatic and radiated in the Carboniferous and Permian (Resh and Carde 2009). Insects have been ubiquitous since their emergence and radiation, but many clades show genetic signatures from Cenozoic events in California.

The region immediately west of and joining Laurentia, or the North American Craton, is an area referred to as the "deformed craton." West of the deformed craton is the accretionary belt, which consists of the current continental belt that has been welded to the North American Craton through a series of diverse geologic processes since the Precambrian (Map 2.1). As part of the accretion process, during the Carboniferous, volcanic island arcs formed west of North America and were surrounded by extensive coral reefs and sand now found in the northern Sierra Nevada and eastern Klamath Mountains (Harden 1998). California was still underwater in the Carboniferous. Coastal waters were warm and inhabited by members of Chondrichthyes (cartilaginous fishes), Mollusca, Cnidaria (jellyfish, corals), and many algae. Estuarine environments were forming and dominated by large members of the Insecta and plants from the Equistaceae (horsetails), Pteridophyta (ferns), Coniferophyta (conifers), Lycophyta (club mosses), and Pteridospermatophyta (extinct seed ferns).

All major continents collide to form the supercontinent Pangaea around the Permian (299–251 Ma). Permian California rocks include both deep and shallow marine environments dominated by members of Brachiopoda (lamp shells), Echinodermata (sea anemones, sea stars), Mollusca, Chondrichthyes, and Cnidaria and are best represented in Shasta and Butte Counties (ucmpdb.berkeley.edu) (Map 2.2). The Permian period ended in the greatest extinction in the history of the earth, with over 90 percent of all marine and 70 percent of all terrestrial genera becoming extinct. Although the cause or causes of the Permian extinction are still under debate, recent work suggests massive volcanic activity was a major factor (Sahney and Benton 2008).

MESOZOIC ERA (245–65 MA)

At the beginning of the Mesozoic Era (245 Ma), California consisted of a landmass roughly at the present-day location of the Mojave Desert and was mostly a low landscape on the western margin of the westward-growing North American Plate into the Jurassic period. The climate of western North America was tropical, but the uplift of an ancient Sierra Nevada was occurring as long as 160 Ma, which may have created a tropical montane environment (Graham 2011). By the Cretaceous (120–65 Ma), the ancestral Sierra Nevada and the then adjoining Klamath Range to the north were located in latitudes now occupied by Southern California. Throughout the Mesozoic, rocks formed during the Paleozoic were capped by volcanoes,

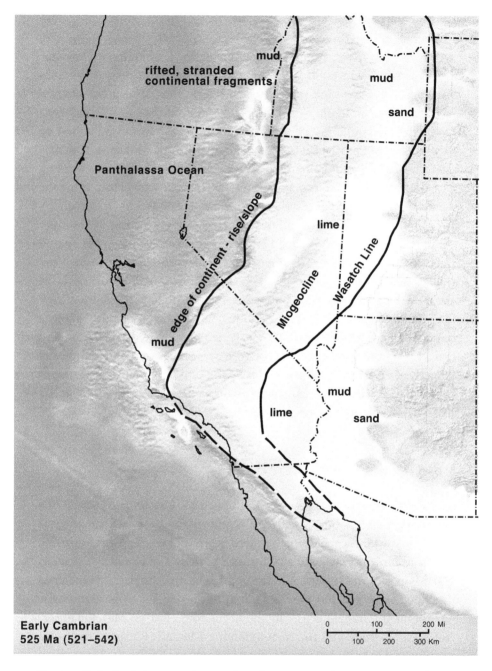

Map 2.1. Early Cambrian tectonic map of southwestern North America. Copyright, Ron Blakey, Colorado Plateau Geosystems. Used with permission.

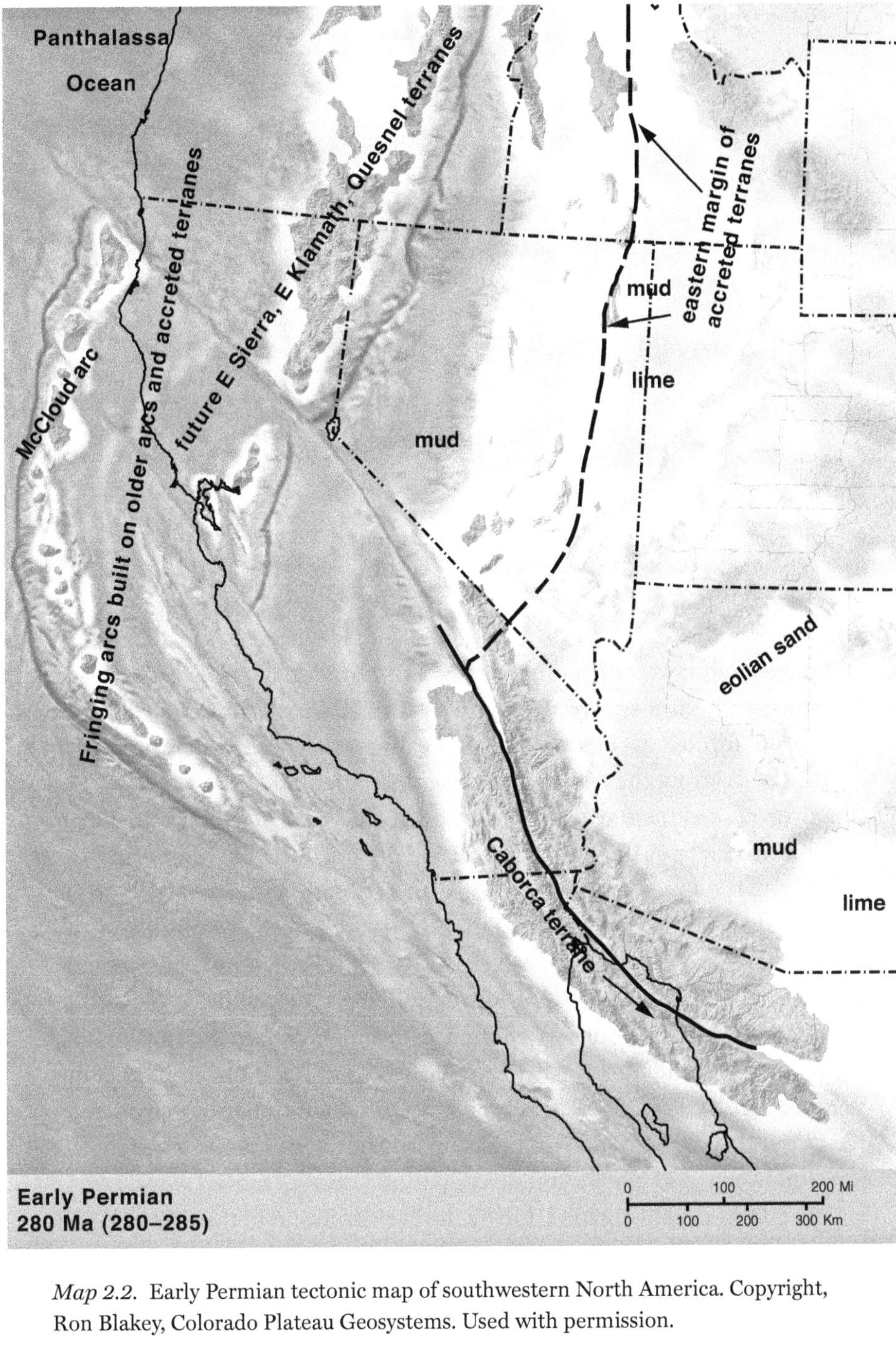
Panthalassa Ocean
Fringing arcs built on older arcs and accreted terranes
McCloud arc
future E Sierra, E Klamath, Quesnel terranes
eastern margin of accreted terranes
mud
lime
mud
eolian sand
Caborca terrane
mud
lime
Early Permian
280 Ma (280–285)
0 100 200 Mi
0 100 200 300 Km

Map 2.2. Early Permian tectonic map of southwestern North America. Copyright, Ron Blakey, Colorado Plateau Geosystems. Used with permission.

Figure 2.2. Artistic representation of marine Ammonita from Ordovician "California," by William G. Huff, from Camp 1952.

and associated lava and sediment flows and gradually moved to what is now the northern Sierra Nevada and eastern Klamath Mountains. These rocks were formed as a result of ocean volcanoes or via subduction zones that formed during the collisions, or a combination of both.

Paleozoic sedimentary and volcanic rocks were accreted to North America throughout the Mesozoic. During the Triassic, North America was still part of the supercontinent Pangaea and California was part of the continental shelf. The western edge of North America at the beginning of the Triassic was near the California-Nevada border. Limestones, shales, sandstones, and conglomerate were formed off the coast, and fossil beds in Shasta and Plumas Counties contain a diverse array of marine fossils, including Ammonoidea (extinct molluscs) (Figure 2.2), Brachiopoda, Echinodermata, Bivalvia (bivalve molluscs), and a number of marine reptiles. Over 100–50 Ma, Pangaea broke up into northern Laurasia and southern Gondwana, and North America as part of Laurasia moved westward.

A subduction zone formed during the Mesozoic and to the east resulted in a magmatic arc that was in roughly the same place as today's Sierra Nevada (Harden 1998) (Map 2.3). Intrusive igneous rocks formed during

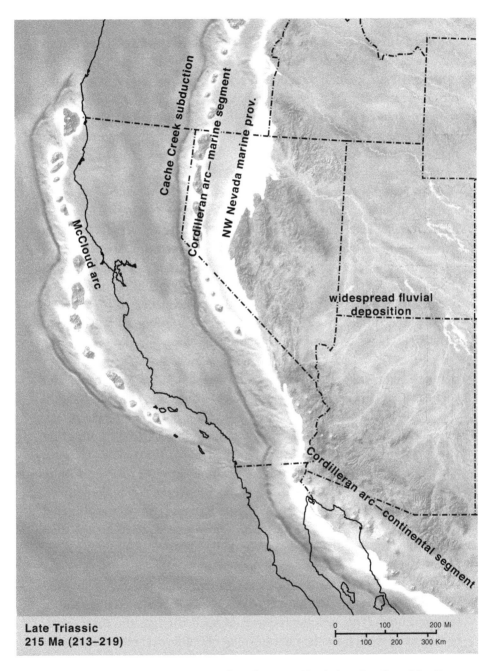

Cache Creek subduction

Cordilleran arc—marine segment

NW Nevada marine prov.

McCloud arc

widespread fluvial deposition

Cordilleran arc—continental segment

Late Triassic
215 Ma (213–219)

0		100		200 Mi
0	100	200	300 Km	

Map 2.3. Late Triassic tectonic map of southwestern North America. Copyright, Ron Blakey, Colorado Plateau Geosystems. Used with permission.

this period are found in the Klamath Mountains, Sierra Nevada, Mojave, and Peninsular Ranges. In the late Triassic the accretion process was continuing to interweave elements of the Pacific Plate and the North American Plate though subduction and volcanism. Accreted elements include remains from the Proterozoic and Paleozoic island arcs, which were surrounded by extensive coral reefs, sands, and muds. The western edge of Pangaea now remains in the eastern Klamath Mountains and northern Sierra Nevada where some of the most interesting and diverse geology in California is found today (Dickinson 2008).

As Pangaea began to break apart, the ancient Farallon Plate began, in the Jurassic, to subduct under the North American Plate as it moved eastward. As the Farallon Plate came into contact with North America, the subduction zone evolved into a transform boundary (Dickinson 2008). About 160 Ma, during the mid-Jurassic, the shoreline ran north-south essentially through the center of the state to Los Angeles and the Mojave Desert was a coastal desert, east of a western mountain range. This subduction began the creation of the ancestral Sierra Nevada and generated extensive volcanism. From 140–28 Ma, there was continual accretion of the North American Plate along the western margin and the subduction zone moved west. Exotic terranes composed of a variety of sediments were accreted to the North American continent. Eventually some of the Farallon Plate accreted to North America and is now part of a group of heterogeneous rocks known as the Franciscan complex in the Coastal Ranges (Hagstrum and Murchey 1993). Based on fossil Coniferophyta, Ginkgophyta, and Cycadophyta, the climate was warm and moderately mesic. The Monte de Oro formation of the Jurassic in Butte County consists of members of Ginkgophyta, Pteridophyta, and Cycadaceae.

The western Klamath Range and Coast Ranges result from continued subduction from the late Triassic to early Paleogene. As the Sierra Nevada was undergoing ancestral mountain building, sediments were accumulating to the west. Because everything west of the current Sierra Nevada was sea, the sediments in the Central Valley contain a number of fossil marine reptiles represented by Chelonia (turtles), Plesiosauria, and Ichthyosauria (Hilton 2003). This sedimentation continued through the Cretaceous. Sedimentary rocks from the Central Valley of California are now known as the Great Valley Group and range in age from the late Jurassic to the late Cretaceous and can be seen along the eastern edge of the inner Coast

Ranges. During the subduction process, the ophiolite assemblages that underlay the Coast Ranges have been uplifted and exposed and now form the serpentine component of the complex fabric of the Coast Ranges. The San Francisco Bay was formed during the Mesozoic and early Cenozoic through plate movement and erosion prior to and following its formation. Accretion stopped in the late Paleogene as the Farallon Plate was subducted and the Pacific Plate began to move laterally along the San Andreas.

The early Mesozoic terrestrial landscape was still dominated by Pteridophyta, Cycadophyta, Ginkgophyta, and Bennettitales (extinct seed plants). Modern Coniferophyta were recognizable by the early Triassic; by the middle Cretaceous, the Spermophyta emerged and would dominate all other land plants by the end of the Cretaceous. The only Mesozoic plant fossils in California are *Margeriella* sp. (Gymnospermopsida) from the Moreno formation in Fresno County (ucmpdb.berkeley.edu).

The terrestrial landscape of California during the Mesozoic was limited to what is now eastern California (Map 2.4). Southern California was primarily mountainous and did not provide conditions conducive to fossil formation; thus most terrestrial fossils are found in the seafloor sediment that formed throughout the Triassic, Jurassic, and early Cretaceous. The only reptilian Triassic fossils in California can be found in the Hosselkus limestone formation in the Klamath Mountains of Shasta County, which was then a warm sea with marine reptiles (Hilton 2003). A number of reptilian fossils have been described from the Cretaceous in Fresno County (ucmpdb.berkeley.edu).

Fossil Chelonians have been found in the Cretaceous marine rocks from a number of counties that were coastal during the Mesozoic. Virtually all of the marine Chelonii found in California were marine dwellers found in the clade Chelonioidea, which includes the Dermochelyidae (leatherback sea turtles), Cheloniidae (green sea turtles and relatives), and Toxochelyidae (extinct sea turtles). All are found in the Chico Formation of Butte County of the early Cretaceous (Hilton 2003). The Dermochelyid turtles are most similar to the modern-day leatherback turtles and are the oldest dermochelyids from the eastern Pacific (Parham and Stidham 1999). One possible exception to the marine nature of turtles in the California Mesozoic is the genus *Basilemys* (Nanhsiungchelyidae); however, this remains controversial (Hilton 2003).

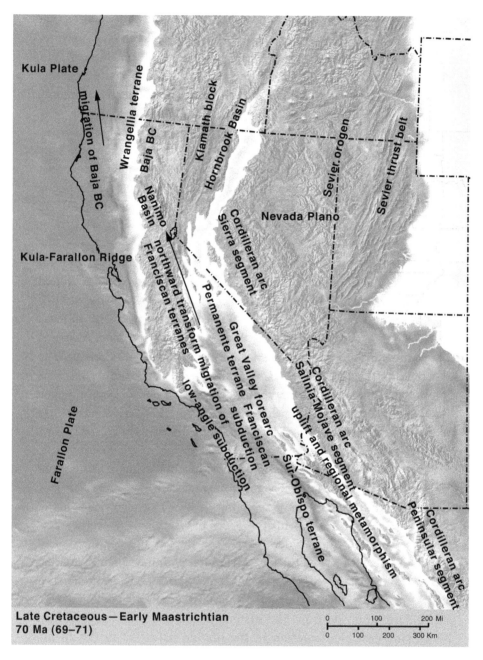

Kula Plate

migration of Baja BC

Wrangellia terrane

Baja BC

Nanimo Basin

Klamath block

Hornbrook Basin

Sevier orogen

Sevier thrust belt

Nevada Plano

Cordilleran arc Sierra segment

Kula-Farallon Ridge

northward transform migration of low-angle subduction

Franciscan terranes

Permanente terrane Franciscan subduction

Great Valley forearc

Salinia-Mojave segment Cordilleran arc

uplift and regional metamorphism

Farallon Plate

Sur-Obispo terrane

Cordilleran arc Peninsular segment

Late Cretaceous—Early Maastrichtian
70 Ma (69–71)

0		100		200 Mi
0	100	200	300 Km	

Map 2.4. Late Cretaceous tectonic map of southwestern North America. Copyright, Ron Blakey, Colorado Plateau Geosystems. Used with permission.

The late Jurassic forests of Northern California were influenced by moist summers and warm winters and were likely dominated by Coniferophyta, Ginkgophyta, Cycadophyta, and Pteridophyta. The late Cretaceous fossil record of terrestrial California is richer than the Triassic and Jurassic, and there is evidence of an increase in plant diversity.

Late Mesozoic Southern California and Baja California delta sediments include archosaurs and some early mammals. Plant fossils from this region are similar to those found in the Chico Formation in Northern California consisting of Ginkgophyta, Coniferophyta, and Cycadophyta. The sedimentary Point Loma Formation (San Diego County) from the late Cretaceous contains representatives from the Formaminifera (amoeboid protists), Mollusca, and Coccolithales (single-celled algal protists) (Kennedy and Moore 1971).

Bayesian and maximum likelihood analyses of fossil and DNA sequence data indicate that birds radiated in the late Cretaceous. Conservative estimates indicate that shorebird and seabird lineages likely diverged at least 74 Ma (Clarke et al. 2005; Slack et al. 2006). Independent lineages are respectively represented in western North America 95–85 Ma by the toothed birds, *Hesperornis*, a grebelike marine bird; *Ichthyornis*, considered by some to actually be a member of the galliform genus *Austinomis;* and members of the Enantiornithes (extinct primitive birds), but none survived the Cretaceous-Paleogene extinction (Clarke 2004; Longrich et al. 2011). Both *Ichthyornis* and *Hesperornis* have been found in the Chico Formation of Butte County. Early, modern crown birds (Neornithes) were present in the Cretaceous and underwent divergence into first the Neognathae and other, likely paraphyletic clades Palaeognathae (ratites, tinamous) and Ornithurae (extinct transitional birds) (Gauthier and de Queiroz 2001; Slack et al. 2006). *Vegavis iaai*, an extinct member of the clade Anseriformes from Cretaceous Antarctica, establishes that ducks, chickens, and ratitelike members were also present during the Cretaceous, but there are no fossils from California (Clarke et al. 2005).

Molecular and fossil evidence establishes the origin of Anthophyta (flowering plants) to about 160 Ma and by 70 Ma; all major plant groups had evolved and were diversifying (Doyle 2012). During 110–95 Ma of the Cretaceous and through the early Paleogene, globally there was one flora. Alaska was home to many members of the Anthophyta, and review data support their dispersal along the coasts and into the continent along riparian corridors (Herman 2002). There is little fossil evidence of early flowering plants in what is now California, primarily because California was still forming during the Mesozoic. Mexico is considered an area in which the conditions for rapid radiation of the Anthophyta may have occurred due to the fluctuating climate, proximity to the tropics, and high elevations, but fossil preservation would have been low under these conditions (Stebbins 1974; Doyle and Hickey 1976).

The inimitable Sequoioid clade (Cupressaceae) emerged in the early Cretaceous. Close relatives of *Sequoiadendron giganteum*, the modern giant sequoia, were present throughout North America during the late Mesozoic, although the oldest *Sequoiadendron* fossil is from the Miocene (Axelrod 1964; Harvey 1986). The Sequoioid clade continued to be abundant throughout the Cretaceous but underwent contractions in range beginning in the Paleocene, with *Metasequoia glyptostroboides* (dawn redwood) no longer present after the Pliocene and *Sequoiadendron* and *Sequoia* being reduced to narrow, relictual habitats in the southern Sierra Nevada and central and northern coast of California, respectively (Jagels and Equiza 2007; Pitterman et al. 2012).

By the Cretaceous, continental and oceanic plates were subducted at northeastward angles to the now independent North American Plate, and terranes, or regional assemblages of rocks, continued to become attached to the margin of the North American continent (Karlstrom et al. 1999). The Sierra Nevada and the Transverse Ranges and their drainages, in which many of California's endemic fish taxa occur, were still being formed during the Mesozoic. The Klamath province broke away from the Sierra Nevada and moved 161 km northwest, where it sat as a large island in an ocean that extended from Southern California to the Modoc Plateau. *Gila coerulea* (Cyprinidae), the Klamath blue chub, may have originated from the related Southern California species, *G. orcuttii*, the arroyo chub, as the Klamath region split off from the Sierra Nevada. Many other fish species may have dispersed throughout California similarly (Avise and Ayala 1976).

Milankovitch cycles are based on the Earth's orbit and axis tilt and direction, all of which vary over time and result in the amount of solar energy reaching the Earth's surface. Together these changes cycle about every 100,000 years and initiate changes to the Earth's temperature that greatly influences the landscapes on which species evolve. Generally, species survive 1–30 million years, but most survive 1–10 million years (Bennett 2004). Microevolutionary changes accumulate within species about 20 ka, but Milankovitch cycles force disruption to communities every 20,000–100,000 years. The isolation of allopatric species generally occurs over 1–10 million years, and every 26 million years there are major extinctions and sorting of species (Bennett 1997). This ecological sorting can be considered to have started in California during the Miocene, with subsequent climatic and rapid geologic change contributing to the evolution of its unique biota.

3 The Cenozoic Era: Paleogene and Neogene Periods (65–2.6 Ma)

The Cenozoic Era (65 Ma–present) comprises three periods, Paleogene (65–23 Ma), Neogene (23–2.6 Ma), and Quaternary (2.6 Ma–present). The Paleogene (Paleocene, Eocene, and Oligocene Epochs) and the Neogene (Miocene and Pliocene Epochs) were formerly included in the Tertiary Period, but the term *Tertiary* is no longer formally recognized by the International Commission on Stratigraphy. The Quaternary Period is discussed in separate chapters for each of the major clades considered here and includes the Pleistocene and Holocene Epochs.

PALEOGENE PERIOD (65–25 MA)

Paleocene Epoch (65–55 Ma)

During the Paleocene Epoch (65–58 Ma), the Klamath Range had separated from the Sierra Nevada, the terranes that would become the Coast and Transverse Ranges had become accreted to the continent, and the Cascade Range began to form in what would become its current position. Fossil evidence indicates there was climatic change from tropical to temperate and the ancestral Cretaceous Sierra Nevada had eroded extensively (Harden 1998).

Most of western California was still underwater during the Paleogene, as evidenced by marine fossils. Marine faunal diversity increases in the period but continues to contain Cnidaria, Bivalvia, Gastropoda, Scaphopoda, Echinodermata, and Foraminifera. San Francisco Bay was formed during the Mesozoic and early Cenozoic through plate movement and erosion prior to and following its formation and subsequently has been greatly influenced by climatic fluctuations. River and estuarine deposits are more common in the northern counties. Floristic elements of the biota in terrestrial sediments in the southern counties show a pattern of increased temperate climate from tropical (Benton and Harper 2009).

California did not escape the major extinction event that occurred 65 Ma and resulted in the extinction of approximately 50 percent of all plant and animal genera on the planet, although with varying impacts among clades. Fossil evidence indicates the first archaic mammals appeared in the Jurassic Period of the Mesozoic, approximately 225–210 Ma. Mammalia and Aves, evolutionarily constrained by predation and competition from nonavian Archosaurians, were already present in California and adaptively radiated throughout the Cenozoic across a diverse landscape. However, it was not until near the beginning of the Cenozoic that the great reptile die-off ushered in the new age of the mammals. Two archaic, placental orders were among the first to populate North America and included the Condylarthra and Creodonta, which ultimately gave rise to the modern herbivores and carnivores, respectively (Ingles 1965). Most modern major clades of Mammalia were present by the Eocene but were all less than 10 kg. Mammalian fossils in the Pacific states show three patterns that summarize the evolution of mammals in western North America, including the indigenous development of several families from the Paleocene to the Pleistocene, multiple immigrations from Asia throughout the Cenozoic, and a period of mass extinction during the Pleistocene (Ingles 1965).

Migration among the northern continents was extensive during the early Paleogene. Proboscidea, widespread in North America at this time, evolved from a Paleocene ancestor and diversified throughout the Oligocene and Miocene. *Mammut* (mastodons, Mammutidae) evolved in Africa but by the Miocene had migrated to North America, as had *Mammuthus* (mammoths, Elephantidae), but both became extinct in the Pleistocene, with a few exceptions in Alaska that survived until the Holocene. With generalist diets that make them remarkably adaptive, members of Rodentia were present in North America by the Paleocene (Janis et al. 2004).

During the Paleocene and Eocene, the Sierra Nevada was likely low hills with extensive marine sedimentation and the Pacific coast was the western edge of the early, hilly Sierra Nevada. In the Paleocene and Eocene, the climate became increasingly humid (Wolfe 1995) but also underwent major fluctuations in temperature, alternating between humid subtropical and tropical conditions, and the California vegetation was similar to the rest of temperate North America. Generally Paleogene California hosted a warm-humid climate and associated vegetation.

Eocene Epoch (55–35 Ma)

The Eocene Epoch included climatic drying and ended with a general cooling period in which there is fossil evidence of worldwide extinction (Harden 1998). The highest mean temperature in the entire Cenozoic of about 30°C is thought to have occurred in the early Eocene (55–48 Ma). The temperature gradients were not thought to be great, and in general there was little ice and high precipitation worldwide. North America and Europe were connected via Greenland, and North America and Asia were connected via the Bering Strait. North America continued to be invaded by modernized mammal orders originating in Asia, which, via the Bering Strait, replaced the more primitive resident Condylarthra and Creodonta (Rose 2006).

In the middle Eocene, global cooling increased due to the separation of Antarctica and Australis, and the creation of the circum-Antarctic current, a pattern that continued through the late Eocene and resulted in increased seasonality and lower mean annual temperatures worldwide (Lawver and Gahagan 2003). These late Eocene weather patterns and reduced predation resulted in an increase in body size of mammals that coincided with an increase in open savannahs.

Western North American deserts arose during the early Eocene, and the cooling and drying trend worldwide in the early Oligocene and middle Miocene saw the evolution of more xerically adapted species (Hafner and Riddle 2008; Riddle et al. 2008). The oldest known Artiodactyla, ancestor to all even-toed hoofed mammals, evolved in the early Eocene and were similar in size to modern lagomorphs.

During the Eocene there is very little presence of Coniferophyta in California, which likely resided in more appropriate climatic refugia, although *Larix* (larch) fossils have been identified from the Eocene (45 Ma) (Axelrod 1990; Schorn 1994; Gugger et al. 2010), and undisputed *Pseudotsuga* (Douglas fir) fossils have been identified from the early Oligocene (32 Ma) (Schorn 1994; Gugger et al. 2010). The Paleogene flora of California is documented through a number of fossil floras in the Sierra Nevada at La Porte (1,200 m, early Oligocene), Susanville (1,500 m, middle Eocene), and Colfax (1,000 m, early Eocene) (Millar 1996). *Zamia* (Cycadaceae) and *Taxus* (yew) are present in these floras, with some conifers such as *Pinus* (pine), *Abies* (fir), and *Picea* (spruce) identified via pollen. The La Porte and Colfax floras also contain *Magnolia* and *Ficus* (fig).

A drop in sea level about 50 Ma resulted in a lowering of the seaway in the Central Valley and subsequently resulted in an increase in erosion in the rivers flowing down the western slope of the Sierra Nevada. Sea levels rose at about 49 Ma and resulted in the transgression of the sea onto land. The region around the Yuba River drainage became known as the Domengine-Ione Seaway and sediments were deposited as the seaway was 274–305 m above current sea levels and land met sea around the towns of Lincoln (Placer County), Folsom (Sacramento County), and Ione (Amador County). Fossils deposited in the ancient Yuba River Basin indicate the weather of the early to middle Eocene was subtropical or tropical with a dominance of *Magnolia*, Lauraceae, Moraceae, and Menispermaceae, indicating a mean annual temperature (MAT) of about 16°C, in contrast to the MAT of 12°C today when the area is dominated by *Pinus ponderosa* (ponderosa pine, Pinaceae), *Salix* spp. (willow, Salicaceae), *Populus* sp. (cottonwood, Salicaceae), *Arbutus menziesii* (madrone, Ericaceae), and *Alnus* sp. (alder, Betulaceae) (Wolfe 1995).

Oligocene Epoch (35–25 Ma)

The Oligocene Epoch was a time of continued but gradual uplift of the Sierra Nevada, movement of the coastline west, and some volcanic activity. A major climatic transition occurred about 34 Ma in the late Eocene to early Oligocene with a temperature decline of 13°C over one million years in some areas. The drop of global mean temperatures 3–4°C in about 300,000 years during the early Oligocene was due, in part, to decreases in atmospheric CO_2 and resulted in cooler and drier climates and the expansion of xeric plant communities (Zanazzi et al. 2007). Ice sheets were forming across North America but not in California, and in general, climates became more seasonal and drier (Wolfe 1978). The cooling trend and marine regression through the Oligocene resulted in a reduction of marine plankton and the migration of marina biota south or retreat to warmer refugia. By the end of the Oligocene, the California landscape as we know it today was beginning to form. A low inland sea occupied what is now the Central Valley, the Coastal Ranges were forming, and the Sierra Nevada was beginning to rise out of the basin and range landscape of what is now the Great Basin (Harden 1998).

The warm-humid adapted vegetation from earlier epochs declined, and there was an increase of conifers and flowering plants adapted to cooler, drier climates. Environmental heterogeneity is reflected in a diversity of plant communities with taxa such as *Sequoia* and *Pinus* (Millar 2012). The ancestral condition in the Cupressaceae reflects adaptation to mesic environments, with two convergent adaptations to xeric environments for California within the clade. *Calocedrus* emerged in the late Cretaceous and diverged with *Juniperus* and *Cupressus* in the early Eocene (Pitterman et al. 2012). The *Juniperus/Cupressus* split is well supported for 39–32 Ma, and physiological adaptations involving vascular structure functions allowed expansion into increasingly xeric climates (Pitterman et al. 2012). Diversification in *Juniperus* was rapid during the Miocene. Remarkably, the morphology of the Cupressaceae has remained stable for 70 million years (Stockey et al. 2005).

The Oligocene is characterized by the appearance of extensive grasslands that were important to the expansion of grazing animals (Figure 3.1). Once grasses emerged as the primary vegetation throughout much of western North America, high-crowned, cement-covered teeth evolved to cope with grasses. Today, worldwide there are 170 species of Artiodactyla and 16 species of Perissodactyla; however, earlier, during the Cenozoic, perissodactyls were more abundant. Most Artiodactyla have a three- or four-chambered stomach to more efficiently process food and absorb nutrients. In contrast, perissodactyls do not have a complex digestive

Figure 3.1. Landscape in early Oligocene at Titus Canyon showing *Protitanops* (large titanotheres) and *Mesohippus* (three-toed horse). Illustration by William G. Huff, from Camp 1952.

system. Because artiodactyls are better adapted to thrive on grassland vegetation, it is thought that that is why they replaced perissodactyls ecologically. Extinct Artiodactyla common in the North American Paleogene include the Oreodonta (ruminating ungulates), which went extinct in the Pliocene. Camelidae (camels) and Oromerycidae (extinct camels) were common throughout North America during much of the Cenozoic, evolving from small four-toed animals to two-toed by the Oligocene. *Desmostylia*, an extinct clade of hippo-sized mammals from the late Oligocene through the late Miocene, is known from fossils in Alameda County (Domning et al. 1986). Carbon, oxygen, and strontium isotope analyses from *Desmostylia* enamel from the middle Miocene led to the conclusion that they were more similar to terrestrial than marine relatives (Clementz et al. 2003).

NEOGENE PERIOD (25–2.6 MA)

Miocene Epoch (25–5.3 Ma)

The Miocene Epoch was a time of major change throughout the California landscape. Around 23 Ma North America crashed into South America, forming the minor supercontinent America. The subduction of the minor Farallon Plate resulted in the abutment of the Pacific Plate to the North American Plate and the gradual development of the San Andreas fault from 24–12 Ma throughout California. Simultaneously, the Klamath Range moved westward, and volcanism was rampant in the Cascades. Sierra Nevadan uplift increased from about 1,000–1,500 m (25 Ma) to 2,500 m (10 Ma) (Xue and Allen 2010). The late Neogene uplift of the Sierra Nevada is estimated to have occurred 7.5–2.7 Ma, and by the end of the Pliocene, the Sierra Nevada was about 3,500 m. Throughout the Miocene and Pliocene, erosion of the Sierra Nevada and the sedimentation of the Central Valley continued and continue today. There was little change in elevation to the immediate east of the Sierra Nevada. The Coast Ranges remained low throughout this period, however, beginning in the Miocene; the north-south aligned southern Coast Ranges gradually became occupied by the western Transverse Ranges and the northern Peninsular Ranges as they were rotated by a series of faults into their current east-west alignment by 5.5 Ma. Modern animal taxa are suggested to have started diversifying about 7 Ma as the Sierra Nevada, Coast Ranges, and Transverse Ranges continued to form (Calsbeek et al. 2003).

The Miocene and Pliocene are well represented with a number of diverse fossil floras representing a fairly stable climatic period of 23 million years. Temperatures began to warm and dry, then cool and dry, and by the Miocene there is first evidence of the Mediterranean-type climate, which was fully developed by the Pliocene. The middle Miocene is well documented by paleofloras from the Gold and Webber Lake areas (eastern Butte County and Sierra County, respectively) of the Sierra Nevada and support deciduous broadleaf and conifer vegetation in a mesic temperate climate (Axelrod 1985; Wolfe and Schorn 1994).

Movement between the Pacific and North American Plates resulted in the right lateral rotation of a block in Southern California that is currently defined by the Santa Ynez Mountains in the north and by the northern

Channel Islands and Santa Monica Mountains in the south (Ingersoll and Rumelhart 1999) about 18 and 12 Ma. This movement was pivotal in delimiting the Los Angeles Basin and the Santa Maria, Santa Ynez, and Santa Clara Valleys.

Beginning about 10 Ma, Death Valley broke up the area of the Mojave Desert, which was previously a relatively flat plain about 1,000 m elevation. The White and Inyo Mountains were separated from the Sierra Nevada about 4–3 Ma. The Sierra Nevada continued to rise, and the rain shadow continued to form east. Saline basins in which there is no drainage outlet began to form in the southern Central Valley and in the Mojave Desert. Deserts that formed during the early Eocene experienced the cooling and drying trend that was occurring worldwide in the early Oligocene. By the middle Miocene the evolution of more xerically adapted species was occurring in North America (Hafner and Riddle 2008; Riddle et al. 2008).

During the Miocene to middle Pliocene there were large bays in Los Angeles, Santa Clara Valley, Santa Maria, Morro Bay, Santa Ynez, and Alexander Valley (Maps 3.1, 3.2). Monterey had a large bay that extended 250 km inland. These bays served as likely barriers to terrestrial amphibians like *Taricha torosa* (California newt, Salamandridae) and *Ambystoma* spp. (mole salamanders, Ambystomatidae) and small mammals (Wake 1997). All of these bays and estuaries were warmer but began cooling toward the end of the Pliocene (Hall 2002). Late Miocene diatomite formations are left as evidence in Orange County as interior waterways receded and ocean levels dropped.

Since it was drier in the Miocene, fauna adapted to warmer fresh waters were extirpated from Southern California. The San Joaquin River has only a few primary native fish, which include Cyprinidae (minnows) and Centrarchidae (freshwater sunfishes), and they would have had to persist in the remaining waters of the late Miocene and early Pliocene until it cooled again (Near et al. 2003).

Throughout the late Neogene, sea levels fluctuated with glaciations into the late Pleistocene. These fluctuations resulted in extinctions of estuary biota from the Miocene and Pliocene and the input of new biota from other areas. The reduction in precipitation also resulted in less stream input into the estuaries. Smaller estuaries tended to be cut off because the waves built berms at the mouths of the estuaries and isolated taxa that occurred within

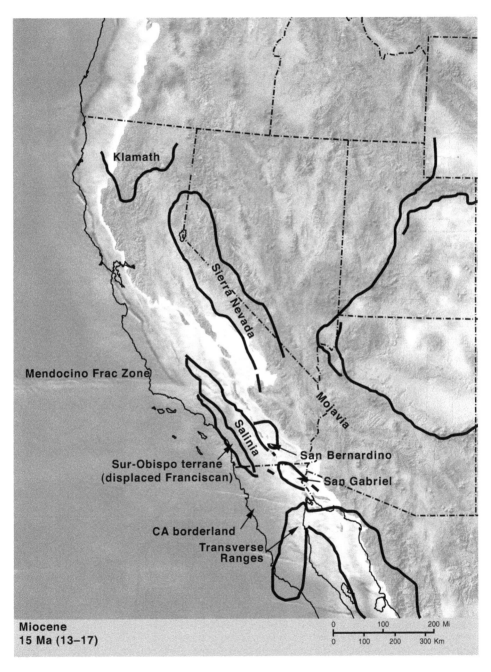

Map 3.1. Miocene tectonic map of southwestern North America. Copyright Ron Blakey, Colorado Plateau Geosystems. Used with permission.

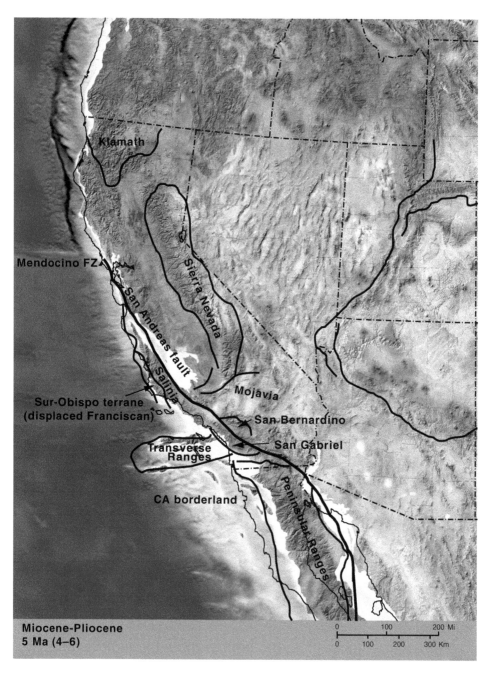

Klamath

Mendocino FZ

San Andreas fault

Sierra Nevada

Salinia

Sur-Obispo terrane
(displaced Franciscan)

Mojavia

San Bernardino

San Gabriel

Transverse
Ranges

CA borderland

Peninsular Ranges

Miocene-Pliocene
5 Ma (4–6)

| 0 | 100 | | 200 Mi |
| 0 | 100 | 200 | 300 Km |

Map 3.2. Miocene-Pliocene tectonic map of southwestern North America. Copyright Ron Blakey, Colorado Plateau Geosystems. Used with permission.

them from other populations. Late Miocene–early Pliocene genetic differentiation is reflected in such estuarine species as *Eucyclogobius newberryi* (tidewater goby, Gobiidae), nearshore Embiotocidae (surfperches) and *Pleuronectes* spp. (flatfish, Pleuronectidae) (Jacobs et al. 2004).

Oceanic global circulation patterns changed during the Miocene because of continental shifts, resulting in warmer climates. Marine diversity was greatly influenced by upwelling 15–12 Ma. Terrestrial diversity was also affected by the summer-dry climate in the middle Miocene (Jacobs et al. 2004). Marine biotic connections have fluctuated since about the late Miocene from the Arctic and the North Pacific with closure of Panama about 4–3 Ma (Marincovich and Gladenkov 1999) and influenced the biota of the entire west coast of North America (Jacobs et al. 2004).

The Monterey Formation of the Coast Ranges is a large sedimentary formation that accumulated from 23–12 Ma that is rich in fossil plankton. The range of accumulation tells a story of upwelling poor conditions at the beginning of the Miocene to upwelling rich conditions by 12 Ma. The maximum upwelling of 14 Ma probably was the beginning of the Mediterranean climate and summer fog banks. Upwelling had mainland effects due to the cool water stabilizing the onshore climate by preventing the movement of fronts onto land and resulting in a summer-dry Mediterranean-type climate (Axelrod 1986). Late Miocene fossils from the Monterey Formation include Mysticeti (baleen whales), Pinnipedia, Sirenia (manatees and dugongs), and Desmostylians. Additional California marine fossils from this period include *Morus* cf. *vagabundus* (a Gannett) and *Loxolithax stocktoni*, a middle Miocene toothed cetacean from the Valmonte Diatomite near Los Angeles, and a member of the Phocoenidae (porpoise) from the Duxbury Reef of Marin County (Hilton 2003).

Factors such as temporal stability and continued and less seasonality of upwelling contribute to the long-term equilibrium diversity of both marine and terrestrial environments. Kelp forests that need cold, nutrient-rich water are important for offshore ecosystem stability (Steneck et al. 2002). Kelps radiated in the late Miocene and continued to speciate in the eastern Pacific, which now has one of the most species-rich regions of kelp in the world. *Hydrodamalis* (sea cows, Dugongidae, Sirenia), dependent on kelp forests, were present in the late Miocene but were hunted to extinction in 1768 (Domning 1978). Members of the Sirenia do not presently occur in

California but are part of the fossil record from mid-Miocene Santa Barbara, San Luis Obispo, and San Diego Counties (Domning 1978; Domning 1987; Aranda-Manteca et al. 1994). Members of the Otariidae (eared seals) and Phocidae (true or earless seals) were present in California by the late Pliocene and Pleistocene (Kellogg 1922).

Consistent with Raven and Axelrod (1978), a detailed analysis of species ranges, climatic variables, and assembled percentages of floras establish that California vegetation originated from different areas and the Mediterranean climate and associated vegetation have been present only since the last 10–5 Ma (Ackerly 2009). North temperate components occur in the Sierra Nevada and northern mountains, and subtropical components are highest in the Coast Ranges and foothills of the Sierra Nevada, including chaparral. Desert floristic components are found in the San Joaquin Valley and southern coast. Sclerophyllous leaved taxa are suggested to have occurred prior to the Mediterranean climate but spread with the establishment of the climatic regime. Some *Arctostaphylos* (manzanita, Ericaceae) diversification was present by the Miocene as evidenced by fossil records, although the evidence is not strong (Edwards 2004). The flora of the Tehachapi Pass from the Miocene 16 Ma is a "live oak–pinyon–cypress woodland" with the additional presence of *Arbutus*, *Persea* (avocado, Lauraceae), *Umbellularia* (bay laurel, Lauraceae), *Ceanothus* (California lilac, Rhamnaceae), *Arctostaphylos*, *Cercocarpus* (mountain mahogany, Rosaceae) and *Rhus* (Anacardiacae). Precipitation was 65 cm or less with mild temperatures and no frost (Axelrod 1995). The current patchy distribution of *Sequoia* has been attributed to the dry period during the Miocene (Axelrod 1986).

By the end of the Miocene, most modern bird genera had evolved. The earliest records of fossil birds for California are, not surprisingly, seabirds, since most of California was underwater during the Miocene. Species present include six species in the Sulidae (gannets and boobies), five taxa from the Procellariiformes (tubenosed seabirds), and rare occurrences of Phalacrocoracidae (cormorants). Land birds from the Miocene include *Cyrtonyx cooki* (an ancestral quail) (Zink and Blackwell 1998), two extinct *Buteo* (hawks, Accipitridae) species, and *Miohierax stocki*, another raptor (Howard 1944; Olson 1985).

Fossils from Perognathinae (pocket mice) and Dipodominae (kangaroo rats and mice) are known from North America 20 Ma (Korth and Reynolds

1991; Wahlert 1993; Riddle et al. 2000). The *Chaetodipus baileyi* species group (pocket mice) is estimated to have originated during the middle Neogene (16–5.5 Ma), supported by the concomitant emergence of *C. hispidus*, an inhabitant of arid lands east of the Rocky Mountains, and *C. formosus*, an inhabitant of the desert scrub of the Mojave Desert and surrounding areas (Riddle et al. 2000). *Chaetodipus* likely emerged from less xeric environments in the late Neogene, as supported by fossil botanical data, as drying continued (Axelrod 1979).

By the Oligocene, all major extant clades of Mammalia were present. By the Miocene and, at the latest, the Pliocene, there was much diversification and mammals were superficially similar to those present today. A number of taxa now extinct are the grassland mammals, which included *Amebeledon*, a shovel-tusked mastodon; *Teleoceras*, a short-legged rhinoceros; *Cranioceras*, a horned, hoofed deer ancestor; *Merycodus*, an extinct pronghorn; and *Pliohippus*, a one-toed grazing horse (Janis et al. 2004). Within the Artiodactyla, the Bovidae emerged in the Miocene and diversified in the Pliocene. In California, only three subspecies of *Ovis canadensis* (bighorn sheep, Bovidae) remain from the vast herds that roamed the grasslands (Luikart and Allendorf 1996).

One of the best-documented fossil records of mammal evolution is present in the Equidae (horses) that evolved in North America. Evolutionary trends in this group are very well documented. Fifty species of members of the Equidae have been named from the North American Pleistocene; however, there is little agreement that this is accurate (Weinstock et al. 2005). Many North American forms of the horses are considered close relatives to Asian forms, but others, such as the stilt-legged forms, are thought to be endemic. *Hyracotherium* was the earliest member of the Equidae; it was about the size of a medium dog, had hoofs covering four toes on the front legs and three toes on the rear legs, and resided in forests. An extensive fossil record clearly links *Hyracotherium* to modern horses. The ancestor to modern horses with high-crowned teeth adapted for grazing was concomitant with the spread of grasses in the Miocene. Trends in horse evolution included increase in size, reduction in the number of toes, longer limbs, and further development of teeth adapted for grazing.

Felidae (cats) originated in the mid-Miocene of Africa but have been present in North America since the Miocene, with fossil specimens of

Pseudaelurus, ancestor to extant felines and pantherines, present since about 15 Ma. The Felinae clade (modern cats) has been present in North America for 10 million years, although *Smilodon* (saber-toothed cat) species (extinct Machairodontinae) did not emerge until about 2.5 Ma and became extinct 10 ka (Tedford 1987).

The Isthmus of Panama began forming about 19–15 Ma as either a peninsula of southern Central America or as a volcanic island chain; regardless, the land bridge was not completely formed until 3 Ma (Coates and Obando 1996; Kirby et al. 2008). Once formed, the land bridge provided an important conduit for biotic migration between North and South America and resulted in what is known as the Great American Biotic Interchange. Recent dated molecular phylogenies established different divergence dates for plants and animals between the two continents. Plants have been dispersing between the landmasses for 50 million years, with most divergences older than 20 Ma, whereas most animal divergence dates are less than 10 Ma (Cody et al. 2010).

Pliocene Epoch (5.3–2.6 Ma)

Cooling continued through the Pliocene Epoch and resulted in the development of polar ice caps. The late Miocene and early Pliocene generally experienced a firmer establishment of the Mediterranean-type climate and summer fog banks in California due to a strong California Current System (CCS). However, the CCS fluctuated throughout the Pliocene, and these fluctuations influenced weather patterns. Miocene and mid-Pliocene estuaries increased in the late Pliocene and Pleistocene during a time of increased cooling, filled rivers, and increased sedimentation during the winter (Jacobs et al. 2004). Upwelling started again in the late Pliocene and Pleistocene and was concomitant with the continued uplift of the Coast Ranges, Transverse Ranges, and Sierra Nevada. Tectonic movement along the coast resulted in an increase in rocky shores and fewer estuaries, but by the end of the Pleistocene, upwelling and sea levels rose (Jacobs et al. 2004). *Oncorhynchus* spp. (salmon, steelhead, Salmonidae) expanded from the sea into the mountain riverine systems during the Pliocene and Pleistocene (Waples et al. 2009). *Oncorhynchus rastrosus* (saber-toothed salmon), a large (2 m long) salmon with fanglike teeth, is known from

Figure 3.2. Artist representation (copyright Stanton F. Fink) of the Pleistocene, *Oncorhynchus rastrosus* (saber-toothed salmon). Used with permission under Creative Commons licensing from the artist.

Pliocene deposits in Oregon, the San Francisco Bay Area, and Stanislaus County (Cavender and Miller 1972) (Figure 3.2).

In addition to its importance to ecosystem structure and function and the building of fish and invertebrate faunal diversity (Vermeij 1993), upwelling provides an important source for organic material in estuaries (Smith and Hollibaugh 1997; Jacobs et al. 2004). The coast of California, like its mainland, has a higher diversity of marine animals, both vertebrate and invertebrate, compared to other coastal areas due to seasonal diversity that is stable through time, connections to other regional biota such as the North Pacific, and within the system geographic variation and separation that promote speciation. California also has reduced seasonality in nutrient blooms compared to other areas and provides a constant supply of plankton for egg production and larval development (Briggs

2003). Currently, upwelling due to the Coriolis Effect and Ekman Transport results in cool, nutrient-laden deeper waters in the summer off the coast of California (Jacobs et al. 2004). Upwelling waxed and waned due to glaciation and had climatic effects as far south as southern Washington and Idaho and in the Sierra Nevada during the Neogene and especially during the Pleistocene. Factors would have reduced the differential between the ocean and land. Upwelling stopped in the mid-Pliocene for about a million years, except for the extreme northern Pacific and returned in the late Pliocene and Pleistocene with some fluctuation (Marlow et al. 2000; Jacobs et al. 2004).

Sporadic cooling and drying that began in the late Pliocene continued throughout the Pleistocene. Early glaciations of the Sierra Nevada are evidenced by Searles Lake, now a dry lake in the Mojave Desert. Milankovitch cycles with minor cycles embedded resulted in cyclic glaciation periods, more prevalent moisture, and a generally cooler and less seasonal climate (Cane et al. 2006). Local climate and a heterogeneous environment are key to the diversification of the biota of California (Stebbins and Major 1965; Raven and Axelrod 1978). Range shift dynamics due to climate change in the Pliocene and Pleistocene combined with ecological processes can explain much of the diversification seen in younger taxa in California.

Late in the Pliocene (5–3 Ma), the compression of the Pacific and North American Plates resulted in the uplift of the Coast Ranges and the further development of the rain shadows created by them. In the area of the Transverse Ranges, the Pacific and North American Plates come together at the San Andreas fault and compress together. The Transverse Ranges run east-west and perpendicular to the primary zone of compression and began rapidly uplifting about 5 Ma, with the basins in between filling with sand over the last 4 million years (Wright 1991). The Los Angeles Basin and the Channel Islands are also part of this complex system of slip-and-thrust faults resulting from this compression. By the late Neogene, the development of the Coast and Transverse Ranges resulted in the elimination of the inland sea that occupied the Central Valley. Combined climate change and coastal mountain uplift resulted in the drying of the water in the Central Valley and the increase of climatic drying in Southern California. As the climate continued to dry, forests became grasslands and mammalian grazers increased in frequency over browsers. Throughout the Miocene and Pliocene, erosion of the Sierra Nevada and the sedimentation of the Great Central Valley increased, and they continue today.

Conifers now with a limited subalpine distribution were more widespread during the Pliocene for species such as *Pinus longaeva* (bristlecone pine), *P. monticola* (western white pine), *P. balfouriana* (foxtail pine), and *Tsuga heterophylla* (western hemlock, Pinaceae) (Millar 1996). In general, montane conifers were more widespread, with *Abies bracteata* (Santa Lucia fir, Pinaceae), *Sequoia sempervirens,* and *Quercus agrifolia* (coastal live oak) present in the Sierra Nevada. The Sierra Nevada flora also included the Salicaceae, Ericaceae, Juglandaceae, and Ulmaceae (Graham 1993). *Abies concolor* (white fir) was not present in the Sierra Nevada until 6–8 Ma, and *A. magnifica* (red fir) was widely present in the Pliocene Sierra Nevada. By the early Pliocene, oak woodlands, along with *Arctostaphylos viscida* ssp. *mariposa* (mariposa manzanita), were present on the western slope of the Sierra Nevada. In the Tule Lake (Siskiyou County) area 3–2 Ma, the Cupressaceae and Taxaceae were an important component of the vegetation (Adams et al. 1990).

Although diversification began during the Miocene, current genetic structure for animal species was estimated to be 3.4–1.2 Ma between pairwise genetic divergences for species that span mountain ranges and 0.66 Ma for those species that did not span major geologic features. For plants, pairwise genetic divergences were estimated to be 1.1–1.5 Ma for those taxa spanning mountain ranges and 0.986 Ma for those not spanning mountain ranges (Calsbeek et al. 2003).

Recent DNA and fossil evidence places *Hippidion* (Equidae) as probably arising during the late Pliocene (3 Ma), in contrast to previous views that it was present in North America in the late Miocene (Weinstock et al. 2005). New World stilt-legged horses are now confirmed as North American endemics and not immigrants from Eurasia and probably originated south of the Pleistocene ice sheets where fossils from about 0.5 Ma are found and where they persisted to about 13 ka based on Nevadan specimens (Weinstock et al. 2005).

The Pliocene Avifauna shares many similarities with the Miocene, with a high frequency of seabirds, *Mancalla*, a flightless Auk, Phalacrocoracidae (cormorants), Cathartidae (New World vultures), and two extinct *Gavia* (loons, Gaviidae); an extinct member of the Podicipedidae (grebes) that cannot be ascribed to a modern or extinct genus are present in the Pliocene fossil record (Murray 1967). Although members of the Passeriformes were likely present, their small, hollow bones are not conducive to fossilization.

In general terms, throughout the Paleogene and Neogene, California was part of a mostly contiguous forest that was present across both sides of the Bering Land Bridge that also served as a migratory corridor (Tiffney 1985). Taxa that survived the Cretaceous-Paleogene extinction experienced extensive migration and diversification, with gradual isolation and continued divergence as habitats became fragmented with climatic fluctuations and geologic change (Xiang et al. 1998; Sanmartín et al. 2008).

4 Quaternary Geologic and Climatic Changes

PLEISTOCENE (2.6–0.011 MA)

The Quaternary was a time of dramatic evolutionary change due to strong climatic latitudinal gradients (Bennett 2004). Oceanic biotic exchanges occurred between the Arctic, the eastern Pacific, and the equator, particularly in times of high sea levels. Prior to the eventual cutoff of the Pacific from the Caribbean by the Isthmus of Panama (3 Ma), there was some transfer of fauna such as fishes in the Gobiidae (gobies) and Labridae (wrasses). Fossil evidence supports the west coast as an area that supported speciation throughout the Cenozoic. Comparatively, there has been less extinction on the west coast than the Atlantic or South America in the Pliocene and Pleistocene (Landini et al. 2002).

The Transverse Ranges, particularly the San Bernardino and San Jacinto Mountains, experienced 2 km or more of uplift during the Pliocene and Pleistocene (Kendrick et al. 2002). Other significant uplift along the western edge of California from the late Pliocene through the Pleistocene include a 1 km or more uplift of the Santa Lucia Range (Ducea et al. 2003), the Palos Verdes Peninsula, and San Clemente and San Nicholas Islands (Ward and Valensise 1996; Lindberg and Lipps 1996).

Fewer mountains in the late Miocene translated to fewer rain shadows and less runoff (Axelrod 1986). In addition, an increased presence of ice in North America resulted in more precipitation and a cooler climate in California about 3 Ma. By the end of the Pleistocene the climate was much different from that of the late Miocene or early Pliocene, with increased rain and seasonality. These climatic changes were particularly felt in the Sierra Nevada and increased wetlands in the deserts and Central Valley. There was a lot of moisture during this period, even during the interglacial periods through the Holocene. Increased Sierra Nevada uplift in the Kern River Formation and in the Coast Ranges about 3–2 Ma combined with increased precipitation resulted in increased sedimentation and the filling of estuaries.

By the late Pleistocene, sea level changes were less than 25 m to over 125 m and resulted in flooding and draining cycles of bays and estuaries; in fact, during the LGM, the San Francisco Bay was empty. Fluctuations in the Pleistocene resulted in the extinctions of estuary biota from the Miocene and Pliocene and input of new biota from other areas. Smaller estuaries would have been isolated, and as result there are genetic signatures in some species from these events. Estuaries and streams were colonized following glaciations and then isolated from others, resulting in the development of unique genotypes of gobies from Monterey Bay to Bodega Head (Dawson et al. 2002). Similarly, climatic fluctuations resulted in the genetic isolation of resident fish species such as *Gila intermedia* (Gila chub) and *Cyprinodon macularius* (desert pupfishes, Cyprinodontidae) from the Bonneville and Lahontan systems. Similar patterns of genetic isolation also resulted for those taxa that migrated into these systems, such as Cottidae (sculpins) and *Oncorhynchus clarkii* (cutthroat trout) (Smith et al. 2002). Refugia during the Pleistocene were present for species like *Oncorhynchus mykiss* (steelhead) and *Oncorhynchus tshawytscha* (chinook salmon), but there were also probably extinctions prior to the Pleistocene. During melting periods, increased water would have allowed riverine systems to be connected to the ocean and colonized by anadromous fishes like *Gasterosteus aculeatus* (three-spined stickleback, Gasterosteidae), which are resident in postglacial lakes, and this appears to have occurred many times (Thompson et al. 1997).

The California Bight begins from Point Conception to the area around San Diego. It is different in its tectonic history and has interesting water circulation. The CCS flows south, but in the Bight and nearshore, the water flows north; however, due to the presence of shoals and islands, northward flow is moderated. The Bight favors species with limited dispersal ability and low fecundity (Dawson 2001). Phylogeographic studies of coastal Oregonian and Californian faunas have placed a barrier to gene flow in the middle of the Bight (Burton 1998; Dawson 2001). Although Point Conception is sometimes implicated as a biogeographic barrier, the California Bight was much reduced about 300 ka and led to a vicariant event in the Bight itself, not Point Conception (Jacobs et al. 2004). The Bight is important for terrestrial biota too. There are about 100 species of plants and five vertebrates endemic to the Channel Islands that likely colonized the islands as refugial populations during the late Neogene drying and cooling that extirpated them from the mainland. The endemic and rare *Pinus torreyana* (Torrey pine) has populations at La Jolla (San Diego County) and Santa Rosa Island (Ledig and Conkle 1983).

Pleistocene weather cycles have been well studied. The Pleistocene was a time of high oscillations, possibly explained by increased tectonic activity, continental drift, and changes in the oceanic currents. Glacial periods resulted in low sea levels with interglacial periods with higher sea levels and extensive terrace deposits. Based on deposits and mollusc fossils, waters were estimated to be warmer about 120 ka and cooler about 80 ka (Muhs et al. 2002). There were over twenty glacial cycles during the Pleistocene that averaged 60,000 years with 10,000-year warming intervals. Sea levels periodically dropped by 120 m, providing a 2,414-km-wide connection between Siberia and Alaska via the Bering Land Bridge. The peak of the last glaciation event was about 23 ka, and San Francisco Bay would have been 120 m below its current depth and the Pacific shoreline 32 km west of present. Inundation at its current level around San Francisco Bay likely was reached 4 ka (Harden 1998). For cold-tolerant species occurring in the area, there would have been more gene flow around the bay, contrary to the current isolation of populations north and east of San Francisco.

During times of high sea level, the coastline would have been farther inland, affecting gene flow. The Colorado River as an effective barrier to gene flow during the late Neogene and early Pleistocene has been long

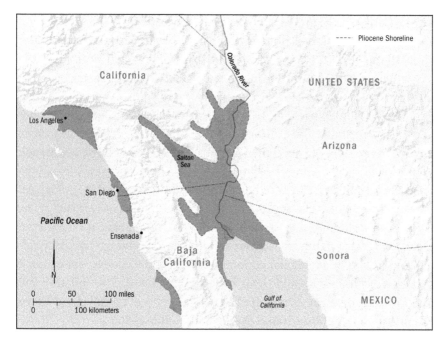

Map 4.1. Marine transgression into southeastern California in Colorado River channel showing the Salton Trough and Bouse Embayment during the Pliocene. Adapted from Durham and Allison 1960, used with permission from Oxford Journals.

recognized, and during times of high sea level this barrier would have been more formidable (Grinnell 1914) (Map 4.1). Genetic evidence supports the Colorado River as an important vicariant mechanism for species of *Peromyscus* (deer mice, Cricetidae), *Chaetodipus* (pocket mice, Perognathinae), *Thomomys* (pocket gopher, Geomyidae), and *Neotoma* (woodrats, Cricetidae) (Mascarello 1978; Lee et al. 1996).

During the LGM (26–19 ka) (Clark et al. 2009), ice sheets extended into the northwestern United States and had extensive climatic effects that were the impetus for much within-species diversification via the establishment of glacial barriers and stimulating the retreat of many taxa into more hospitable southern climates (Hewitt 2000). However, not all taxa moved south and found suitable habitat in unglaciated areas (Ehrlich 1961; Crespi et al. 2003). The colder climate served as an effective isolating mechanism for refugia and resulted in extinction or population bottlenecks in terrestrial and marine species. There has been continual warming since the LGM except for the Younger Dryas, a cold period that is estimated to have occurred from 12.8 to 10 ka. During maximum glaciations, elevational ranges of plants were 600–1,200 m below present and temperatures 8–14°C (Owen et al. 2003; Thompson et al. 1993).

The ecological sorting of plant species was extensive throughout the Pleistocene and throughout California. A number of lines of evidence indicate that the Sierra Nevada flora may have originated primarily from the north but also from the east. During the Miocene through the Pleistocene, the Great Basin experienced a lowering of elevations of about 1,000 m and could have survived as corridor of migration for at least some species.

Pleistocene deposits from about 1 Ma in coastal and interior Orange County support the widespread distribution of *Hesperocyparis macrocarpa* (Monterey cypress), *Pinus muricata* (Bishop pine), and *P. radiata* (Monterey pine) in a climate in that region more similar to present-day Monterey (Axelrod and Govean 1996). From 30 to 15 ka relatively mesic conditions transitioned to more xeric conditions and the spread of the deserts and desert vegetation (Axelrod 1979).

The Pleistocene megafauna were present throughout California until their extinction 15–10 ka. Ancient mtDNA and ^{14}C dating from *Mammuthus primigenius* (woolly mammoth) from 160 individuals across the Holarctic late Pleistocene analyzed via Bayesian and network methods established the distribution of the haplotypes and differentiation in the middle Pleistocene in North America from 749 to 281 ka, about 100,000 years earlier than estimated previously (Debruyne et al. 2008). *Bison latifrons* (giant bison) likely evolved from its Asian migrant via Beringia 240–220 ka but went extinct during the Wisconsin glaciation 30–21 ka. *Bison latifrons* was replaced by *Bison antiquus,* which gave rise to the modern *Bison bison* (Bell et al. 2004).

The Rancho La Brea Tar Pits document life in California from 40 to 8 ka with more than 660 species that include 59 mammals and 135 bird species. Most species are carnivores and birds, mostly birds of prey (Figure 4.1). Elsewhere in Southern California there are a number of species of birds, reptiles, fish, small mammals, and amphibians dated from 40,980 + 800 ^{14}C years before the present (ybp) consistent with other descriptions of Southern California (Brattstrom 1953, 1955; Van Devender and Mead 1978).

Figure 4.1. Photograph of a diorama from the Golden Gate International Exposition (1939–1940) by sculptor William G. Huff. Scene at Rancho La Brea with *Smilodon, Canis dirus, Camelops* sp. *Tertoris* sp., and *Bison antiquus* with background painted by Ray Strong. Photo courtesy of the University of California Museum of Paleontology (UCMP) and used with permission.

There were more bird species present during the Pleistocene than today due to climate. One hundred thirty bird species have been cataloged from the fossil beds in Rancho La Brea, McKittrick, and Carpenteria. Fossil records from other sites add another 50 Pleistocene bird species, 30 of which do not appear to have living descendants. Extinct Pleistocene taxa include the large *Bubo sinclairi* (Sinclair Owl, Strigidae), the 4.9 m wingspan *Aiolornis incredibilis* (the Incredible Teratorn, Teratornithidae), *Ciconia maltha* (Asphalt Stork, Ciconiidae), one member of the Phalacrocoracidae (pelicans), and three members of the Laridae (gulls) (Howard 1962, 1964). Representatives of Squamata from the Pleistocene Santa Ana River include Anguidae (*Elgaria* cf. *multicarinata*, southern alligator lizard), Annieliidae (*Anniella pulchra*, legless lizard), Phrynosomatidae (*Sceloporus*, fence lizard; and *Uta stansburiana*, side-blotched lizard), Colubridae (*Lampropeltis* cf. *getula*, California kingsnake; *L. zonata*, California mountain kingsnake; *Masticophis* sp., racer; *M. flagellum*, coachwhip; *M. lateralis*, striped racer; *Coluber constrictor*, common racer; *Pituophis catenifer*, gopher snake;

Thamnophis sp., garter snake; *Rhinocheilus lecontei*, long-nosed snake; and *Diadophis punctatus*, ring-necked snake) (Wake and Roder 2009). Mammals from the late Pleistocene Santa Ana River include *Mammuthus* sp., *Equus* sp., and *Bison* sp. estimated to be 19,890 ^{14}C ybp (Wake and Roder 2009).

The fossil flora of 158 species from the La Brea Tar Pits indicate a coastal maritime climate but still with primarily winter rains. Carbon data from plant material and tooth wear on carnivores suggest that the plant productivity was low and carnivores experienced extensive teeth wear due to bone consumption. The flora of the Los Angeles Basin indicates a combination of riparian, chaparral, and coastal sage scrub. In addition to the La Brea Tar Pits, other Quaternary sediments are present throughout the state, and floras from lower elevations reflect a more moderate, Mediterranean climate that became warmer to the south. Pleistocene Santa Ana River fossils indicate a riparian habitat with *Salix* sp., *Quercus* sp., *Pinus* sp., *Populus* sp., *Juglans* sp. (walnut, Fagaceae), *Platanus* sp. (sycamore, Platanaceae), and *Typha* sp. (cattails, Typhaceae) (Wake and Roder 2009).

HOLOCENE (11.5 KA–PRESENT)

Holocene climatic changes are fairly well documented in a number of locations throughout California by paleoclimatologists, who use pollen evidence in sediment cores to provide a more refined scenario of this epoch than those before it. Strongly fluctuating climates throughout the Pleistocene resulted in a mosaic of plant communities throughout the state (Axelrod 1981). Palynological data from a number of locations throughout California provide evidence for these climatic fluctuations. A 6.3 m sediment core from Little Willow Lake in Lassen Volcanic National Park gives a 13,500-year record of vegetation in the southern Cascades and the northern Sierra Nevada region. Prior to 12,500 ^{14}C ybp the area was dominated by sagebrush steppe, from 12,500 to 3,100 ^{14}C ypb by *Pinus* spp., followed by a gradual increase in *Abies magnifica*, which is present today (West 2004). More seasonality was present from 13,500 to 12,500 ^{14}C ybp. The early Holocene climate was warmer and drier than today from about 9–4 ka and is known as the Altithermal in which many species reached their northern latitudinal maximum. For nonarboreal plant species, at the juncture of the Cascades and Sierra Nevada, 24 species are at their northern limit and 14 Cascade species are at their southern limit (West 2004). Disjunct, relictual communities thought to remain from this period are in the Pit River gorge dominated by *Quercus lobata* and *Pinus sabiniana* (grey pine) and Mojave floristic elements in western Fresno County (Griffin 1966, Edwards 2004). A patchy distribution of plant communities would have resulted in disruptions to gene flow. In some instances there was lost gene flow; alternatively, in cases of increased gene flow, hybridization likely occurred among related taxa. Combined with strong selection due to rapidly changing ecological relationships, opportunities for adaptive radiation would have been high. After the Altithermal, cooling generally began, as reflected by a decrease in *Quercus* pollen in the Sierra Nevada and an increase in *Calocedrus* (Cupressaceae) and *Abies* (Edwards 2004).

At a location 41°N and 60 km southeast of Eureka between the Mad and Van Duzen Rivers, palynological and diatom data suggest establishment of modern climatic conditions. Warmer surface sea temperatures began about 4.4–3.4 ka, reflected in an increase in *Alnus* and *Sequoia* pollen, and at about 1.4 ka, a sudden increase in summer moisture resulted in an increase of *Pseudotsuga menziesii* that roughly corresponds with the Medieval Warm Period from AD 924 and 1299. After 3,100 ^{14}C ybp, the central and southern Sierra Nevada and eastern Klamath Range were cooler and moister (West 2004), followed by the drought of the Medieval Warm Period, which is also supported by *Pinus longaeva* (bristlecone pine, Pinaceae) tree rings and Mono Lake tree stumps (Stine 1994; Barron et al. 2004). In general, forests in the northern Coast Ranges, Klamath-Siskiyou region, southern Cascade Range, and northern Sierra Nevada experienced similar climatic fluctuations.

The San Francisco Bay Delta currently drains 40 percent of the state's water, the watershed including the western slope of the Sierra Nevada, parts of the Klamath and Coast Ranges and southern Cascade Mountains, and the Central Valley (Peterson et al. 1995). The pre-European modern Bay-Delta started to form about 10–8 ka during the sea level rise after the LGM. Peak summer solar energy during the Holocene was about 9 ka for the Northern Hemisphere. The late Pleistocene through early Holocene was generally dry until about 7 ka, with warming continuing from 7–3.5 ka. Sea levels globally started to slow their increase about 6 ka. The early to mid-Holocene had complicated climatic patterns, with some wet periods from 8–3.8 ka and warming around 6–5 ka. Carbon dating, palynological data, and tree ring data support 4–2 ka as wet in the San Francisco Bay region, Sierra Nevada, and White Mountains and that these areas have been cooling and drying since then (Malamud-Roam et al. 2006). Shallow lakes were present in the Mojave Desert about 3.6 ka (Enzel et al. 1989).

El Niño–Southern Oscillations affect temperature and precipitation in western North America, and in turn, California sea surface temperatures affect mainland ecosystems. Lake sediments of Swamp Lake at 1,554 m (eastern Fresno County) were used to estimate climate over the last 20 ka at 100-year intervals (Street et al. 2012). Plants, algae, and bacteria produce organic matter that has distinct signatures of C/N and $\delta^{13}C$; thus an analysis of total organic carbon, total nitrogen, C/N ratios, and $\delta^{13}C$ and $\delta^{15}N$ can provide a snapshot of organic matter production in time. Low total organic carbon cycles were postglacial, with little vegetation 19.7–18.2 and 16.5–15.8 ka. Cold and wet cycles occurred during 14.9–13.9, 13.1–11.7, and 11.0–10.7 ka. There was evidence for warm, dry plant growing conditions during the periods 17.4–16.5, 15.8–15.0, 13.9–13.2, and 11.4–11.0 ka. Three periods of increased productivity were reflected in organic matter content from 8.0 and 3.0 ka. Two hundred km south of Swamp Lake at about 50 m elevation, Tulare Lake palynological records suggest seven to eight major changes in the Holocene, with higher lake levels in the early Holocene prior to two high periods of 9.5–8 ka and 6.9–5.8 ka with some fluctuations. There was some evidence for lake level rises about 3.3 and 1.6 ka (Negrini et al. 2006).

PART II Phylogeographic Patterns
in Various Taxa

5 Conifers

The earliest fossil record of conifers is from the Carboniferous about 300 Ma. The Pinophyta, Ginkgophyta, and Cycadophyta all developed during this time, likely in Asia. Important adaptations that preceded the development of these clades are the development of vasculature to reduce dependence on water, pollen to allow the movement of sperm without water, and seeds, increasing propagule mobility. Some members of the Cupressaceae, such as *Sequoia sempervirens, Sequoiadendron giganteum,* and *Metasequoia glyptostroboides,* were present in California during the Cretaceous and throughout the Paleogene. As Quaternary climates became cooler and drier, *Sequoia sempervirens* retreated to the coast, *Sequoiadendron giganteum* became restricted to small groves in the central Sierra Nevada, and *Metasequoia glyptostroboides* was extirpated, now only occurring naturally in south central China.

There were widespread diverse conifer forests during the Pleistocene (Axelrod 1976, 1977). Mixed conifer forests that existed during the last glacial periods were about 900 m lower than today and would have had a greater degree of connectivity (Anderson et al. 2002; Hall 2007). During these cooler and wetter periods, mesic environments were more widespread, including around the Southern California mountain ranges and the western Mojave Desert, and would have provided more continuous habitat for amphibians and fish (Orme 2008). The fossil flora from the Pleistocene Bautista Formation of San Jacinto in Riverside County contains *Pseudotsuga macrocarpa, Abies concolor, Pinus ponderosa, P. lambertiana,* and *Calocedrus decurrens* (incense cedar, Cupressaceae) (Axelrod 1966).

Members of the Coniferophyta of California generally maintain high levels of genetic diversity, but there are many notable exceptions. Patterns of genetic diversity in California conifers indicate a complex array of paleoendemics, neoendemics, and widespread common taxa. California conifers are highly heterozygous and generally have less genetic diversity among than within populations, as illustrated by *Abies concolor, Pinus ponderosa,* and *Pseudotsuga menziesii.* Pleistocene relicts such as *Abies lasiocarpa, Picea engelmannii, Pinus flexilis,* and *P. longaeva* still contain high levels of within-species diversity (Hamrick et al. 1994). Pleistocene glaciations are increasingly recognized as important in shaping the genetic structure of California plant species, as has been demonstrated in *Pinus albicaulis* and *P. lambertiana* (Liston et al. 2007) and *P. monticola* (Steinhoff et al. 1990). These species demonstrate divergence into northern and southern clades with secondary contact zones in the Klamath-Siskiyou Mountains (Eckert and Hall 2006).

CUPRESSACEAE

All members of the Cupressaceae share a common ancestor that likely evolved from a mesic-loving Mesozoic ancestor but developed the ability to withstand drought sometime in the Eocene and expanded from the Oligocene forward (Pitterman et al. 2012).

Sequoia sempervirens appears in the fossil record around the Jurassic and is thought to have been common throughout western North America in the late Cretaceous. It is found in the late Miocene in Oregon but not in west-central Nevada, becoming gradually restricted to its current range in the Pliocene due to the cooling and drying trend (Edwards 2004). *Sequoiadendron giganteum* was present in Tulare Lake and Mono Lake 26–10 ka and did not reach its current range until about 4.5 ka (Woolfenden

1996). Other members of the Cupressaceae present in California from the Cretaceous and into the Quaternary, in addition to *Sequoia, Sequoiadendron, Metasequoia,* and *Taxodium* (bald cypress, Cupressaceae), are *Glyptostrobus* (water pines, Cupressaceae) and *Cunninghamia* (Chinese fir, Cupressaceae), both of which are now restricted to Asia (Edwards 2004).

Hesperocyparis macrocarpa (Monterey cypress) consists of small and isolated populations but has fairly high levels of genetic diversity, indicating a former, historic widespread distribution (Kafton 1976). Similarly, *H. forbesii* (Tecate cypress), an edaphic endemic in Southern California's Peninsular Ranges and northwest Baja California, has widely separated native populations; however, genetic diversity as measured by allozymes from 245 individuals in four populations was not associated with geographic distance (Truesdale and McClenagan 1998). A close relative to *H. forbesii, H. guadalupensis* (Guadalupe cypress), is found only on Guadalupe Island off the Pacific coast of Baja California. Paternally inherited cpDNA markers compared genetic diversity among populations of *H. forbesii* and *H. guadalupensis* and identified higher genetic diversity in *H. guadalupensis* than in its mainland congener. Contrary to expectations, *H. guadalupensis* shows no evidence of a genetic bottleneck and may reflect the environmental stability of the island habitat over a long evolutionary history. In contrast, the mainland and more northern Peninsular Ranges experienced glacial and interglacial periods during the Pleistocene (Escobar García et al. 2012).

TAXACEAE

The Taxaceae and Cupressaceae share a common ancestor that likely diverged 230–192 Ma based on the cpDNA *mat*K gene and nuclear ITS (Cheng et al. 2000). Although widespread in Asia and North America in the late Cretaceous and early Paleogene, the Taxaceae is now represented by only two taxa in California, *Taxus brevifolia* (Pacific yew) and *Torreya californica* (California nutmeg). Neither species has been the subject of a phylogeographic study; however, the population genetics of *Taxus brevifolia* indicates that the species has more among-population genetic diversity than seen in most other conifers. Lower than expected polymorphisms in

T. brevifolia for a wide-ranging woody species may be due to a combination of clonal reproduction, its patchy distribution in forest communities, understory status (which may limit pollen flow), and the germination of clustered seed in rodent caches (Scher 1996). *Torreya californica* (California nutmeg) is an ancient genus known from eastern North America in the Cretaceous and in western North America by the Paleogene (Maxwell 1992). By the Pliocene it had a circumboreal distribution but today only occurs in California in shady moist forest or woodland canyons (Howell 1992).

PINACEAE

The Pinaceae diversified in the early Cretaceous and consists of 10 extant genera. The genus *Pinus* is the best-studied member of the Pinaceae, and subclades in the genus are well established based on nuclear and chloroplast DNA data (Gernandt et al. 2009). Although phylogenetic relationships among species within subclades remain somewhat elusive, recent molecular work has shed light on a phylogeny complicated by ancient hybridizations (Liston et al. 2003; Syring et al. 2005; Eckert and Hall 2006; Gernandt et al. 2009; Willyard et al. 2009).

An analysis of cpDNA variation from four cpDNA loci (*matK, rpl20-rps18, rbcL,* and *trnV* intron) among 83 *Pinus* species indicates that it originated in Eurasia (Eckert and Hall 2006). The oldest known *Pinus* fossil, *P. belgica,* is from the Cretaceous in Europe (Alvin 1960) from 130 Ma, but it is likely a member of a stem group that precedes the subgenus *Pinus,* which arose closer to 110 Ma. *Pinus* evidently diverged from *Picea* over 100 Ma (Miller 1973; Willyard et al. 2007), and based on cpDNA calibration of silent substitutions at 2.61×10^{-10} to 4.01×10^{-10}, the *Pinus-Picea* split can be placed at 123 Ma (Gernandt et al. 2008).

An analysis of 11 nuclear and 4 chloroplast loci support division of *Pinus* into two clades, subgenus *Pinus* and subgenus *Strobus,* which diversified in the late Cretaceous to mid-Eocene. Initial subgenus *Strobus* diversification occurred about 87 Ma and corresponds with fossil evidence (Axelrod 1986; Millar 1998), with sectional and subsectional diversifications occurring about 74–34 Ma. Most members of the subgenus *Pinus* clade dispersed from Eurasia to North America in the late Cretaceous and early Paleogene and the subgenus *Strobus* in the late Cretaceous, but there were likely multiple dispersal events during 46–34 Ma (Millar 1998).

Pinus lambertiana diverged from other members of subgenus *Strobus* about 9 Ma and is one of 20 species in the clade (Gernandt et al. 2005; Syring et al. 2007). *Pinus lambertiana* occurs from Oregon south to northern Baja California. A phylogeographic survey of 84 individuals from Oregon and the Klamath and the northern Coast Ranges of California identified a common cpDNA haplotype ("N"), with plants from the Sierra Nevada and the Transverse and Peninsular Ranges having a different haplotype ("S"). There is a contact zone between the two haplotypes in northeastern California in an area of about 150 km near Mount Lassen, with an estimated divergence time of 15 Ma. Recent secondary contact suggests connections from northern and southern refugia following Holocene glaciations. Sequences of 5799 bp of cpDNA from *rbc*L, rp115, *trn*L-F, *mat*K and *trn*G provided a divergence time estimate of 0.6 Ma between *P. albicaulis* and *P. lambertiana* for the "N" haplotype (Liston et al. 2007). The southern Cascades/Klamath-Siskiyou region is proposed to be a northern glacial refugium for *P. lambertiana* (Syring et al. 2007; Eckert and Hall 2006).

Pinus albicaulis is likely a postdiversification migrant from Europe across Beringia, which was unglaciated during the Pleistocene and LGM and is a known refugium (Brubaker et al. 2005). *Pinus albicaulis* shows little differentiation between populations, even though it has a discontinuous distribution in alpine habitat (Jorgensen and Hamrick 1997). Genetic homogenization of *Pinus albicaulis* at local and regional scales may be due to Clark's nutcracker seed caching (Rogers et al. 1999).

The subsection *Ponderosae* clade within *Pinus* diversified into 17 species in the past 15 million years (Willyard et al. 2007). The *Ponderosae* is monophyletic and further divided into the Sabinianae clade and the Ponderosae *sensu stricto* clade (Gernandt et al. 2009; Willyard et al. 2007). The Sabinianae contains *P. coulteri, P. jeffreyi, P. sabiniana,* and *P. torreyana,* all in California; and in California, the Ponderosae *sensu stricto* contains *P. ponderosa* var. *ponderosa* and *P. ponderosa* var. *washoensis,* both clades supported by Bayesian posterior probabilities of 45 individuals sequenced for 1630 bp of *LEA*-like (Willyard et al. 2009). Gene trees from the nuclear gene *WD-40,* however, provided a different topology that is likely the result of recent divergence. *Pinus coulteri* sequences for *WD-40* reflected ancient hybridization with *P. jeffreyi* (Willyard et al. 2009) and are consistent with data from cpDNA data that place it sister to *P. coulteri, P. sabiniana,* and *P. torreyana* (Gernandt et al. 2009). *Neotoma* spp. middens and palynological and genetic data support two refugia for *Pinus ponderosa* during the Pleistocene, the southern Sierra Nevada and the southern Rocky Mountains (Betancourt et al. 1990; Latta and Mitton 1999).

Pinus ponderosa var. *washoensis* is endemic to a limited area of the western rim of the Great Basin, where it is distributed in small, isolated, upper elevation populations. It has been suggested that the relatively high levels of allozyme variation in this species are due to a previously widespread distribution as its geographic range expanded during the glacial periods of the Pleistocene (Niebling and Conkle 1990). *WD-40* gene trees indicate that *P. ponderosa* var. *washoensis* reflects reticulate ancestry at high elevations with *P. jeffreyi* and shares chloroplast haplotypes with *P. ponderosa* (Willyard et al. 2009).

High levels of intraspecific mtDNA variation are present in a number of *Pinus* species, a pattern not surprising given the maternal inheritance of mtDNA in these species (Strauss et al. 1993; Dong and Wagner 1993). Lower levels of intraspecific genetic variation are present in cpDNA, which is primarily paternally inherited in conifers. There is no patch structure for *Pinus* cpDNA haplotypes, but mtDNA haplotypes within patches are half-sibs, indicating that pollen is as likely to fertilize near neighbors as distant trees (Latta et al. 1998). Incomplete lineage sorting is common in the Pinaceae (Syring et al. 2007). Outcrossing, higher mean heterozygosity, large effective population sizes, incomplete mating barriers, and long generation times generally result in a lack of coalescence within *Pinus* species (Willyard et al. 2007; Rosenberg 2003). Most *Pinus* species have 90 percent of the variation contained within populations (Sorensen et al. 2001).

Small isolated populations with low levels of diversity within and among populations include some extreme examples (Syring et al. 2007); *Pinus torreyana,* known from two populations with about 9,000 individuals, has very low levels of genetic diversity. An analysis of 59 isozyme loci identified all as homozygous, suggested to be a result of reduction to less than 50 individuals during the Xerothermic period, 8.5–3.5 ka (Ledig and Conkle 1983).

Pinus longaeva (bristlecone pine) of the White Mountains represents the western extent of the species distribution. Once widespread throughout the Great Basin during the LGM, it is now limited to alpine habitat. Despite the fact that they grow in isolated alpine habitats, they retain high levels of genetic diversity because of life history characteristics of longevity, large population size, and high seed set; they show no evidence of genetic bottlenecks (Lee et al. 2002).

Pinus balfouriana (foxtail pine) occurs at 1,830–3,400 m in the southern Sierra Nevada and the Klamath Range, separated by 500 km. There is some morphological divergence but little allozyme divergence among the populations (Mastroguiseppe and Mastroguiseppe 1980; Oline et al. 2000). An analysis of four cpDNA, four mtDNA, and five nucDNA genes or regions indicates divergence during the Sherwin glacial maximum event during the Pleistocene (Eckert et al. 2008). The Sherwin glacial maximum about 1 Ma likely extirpated populations in the central and northern Sierra Nevada with dates of differentiation consistent with other alpine species. The genetic data are not consistent with the Holocene xerotherm 8–4 ka. *Pinus balfouriana* shows greater genetic diversity than other range-restricted California conifers.

Pseudotsuga menziesii is present in the fossil record from the early Miocene, 23 Ma. During the Miocene and Pliocene, it was present from British Columbia to Southern California. By the Pleistocene it had mostly retreated to the Pacific Northwest coast and Rocky Mountains (Brunsfeld et al. 2001). Fossil, cpDNA, and mtDNA data integrated in a rangewide sample of *P. menziesii* establish that the coastal and Rocky Mountain varieties split 2 Ma, with populations further diverging during Pleistocene glaciations (Gugger et al. 2010). Samples from 219 individuals at 87 localities for cpDNA (*rps7-trnL*), *ndhB* and intron, *rps15-psaC* and *ndhH*, *ndhI* and *ndhE*, and mtDNA regions *19S rDNA V*, and *nad7i1*, were tested for refugia hypotheses using coalescent methods and isolation with migration models. Haplotype diversity is high for both mtDNA (h = 0.74) and cpDNA (0.64), and nucleotide diversity is low for both genomes (π = 0.00280 and 0.00073), with no latitudinal trends. Divergence is consistent with the orogeny of the Cascades and Sierra Nevada, which resulted in a xeric trend in the Great Basin. These results are supportive of one single big glacial refugium or a number of smaller refugia from central California to parts of western Washington that were unglaciated and that are consistent with *P. menziesii* climatic tolerance and climatic models from the LGM. In general, evidence from a number of conifer taxa, *Abies concolor, Pinus longaeva,* and *P. flexilis,* indicates that refugia were elevational rather than latitudinal (Gugger et al. 2010). Based on fossil records, *Pseudotsuga menziesii* migration followed ice sheet migration beginning 18 ka.

The Klamath Range contains forests that are considered roughly equivalent to the Paleogene circumboreal forests (Sawyer and Thornburgh 1977). *Picea breweriana,* the Brewer spruce, is currently found only in the Klamath Range but was formerly widespread in Idaho, Nevada, Southern California, and central Oregon in the Pliocene and Miocene. It now occurs in roughly the north-south extent of the Klamath region at elevations of 560–2,300 m in small, disjunct populations. A sample of 26 isozyme loci from 10 populations reveals low inbreeding (F_{IS} = 0.033), and a F_{ST} value of 0.157 shows a lack of correspondence between genetic and geographic distance that is likely a reflection of some genetic drift (Ledig et al. 2005). H_E is an average of 0.129, lower than for other conifers (Hamrick et al. 1992), but it also decreases with latitudinal increases, which are estimated to be due to postglacial dispersal (Hamrick and Godt 1996; Ledig 2000; Ledig et al. 2005). N_m levels of 1.34 indicate that gene flow among *P. breweriana* populations is low and the population statistics may be a result of past genetic structure (Ledig et al. 2005).

SUMMARY

Conifers arose in the Carboniferous when Pangaea was still forming, and fossil evidence supports their once-widespread distribution. As tectonic activity continued throughout the Mesozoic and Cenozoic, California conifers migrated and diverged from their Eurasian sister taxa in the late Cretaceous and Quaternary. Conifer forests of western North America were affected by Quaternary glaciations, which resulted in much fragmentation, refugial populations, and secondary contact. The Sherwin glacial

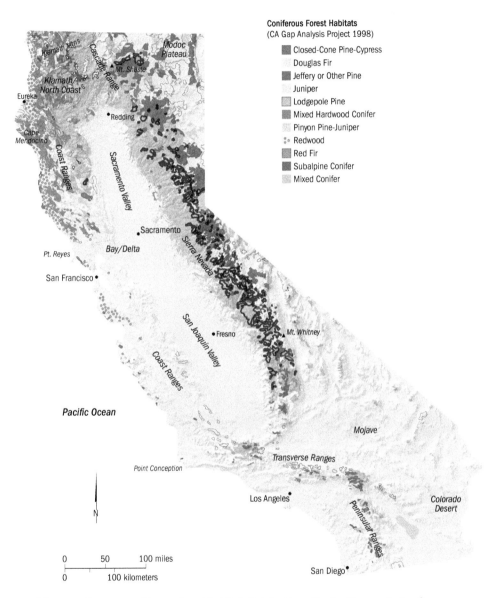

Coniferous Forest Habitats
(CA Gap Analysis Project 1998)

- Closed-Cone Pine-Cypress
- Douglas Fir
- Jeffery or Other Pine
- Juniper
- Lodgepole Pine
- Mixed Hardwood Conifer
- Pinyon Pine-Juniper
- Redwood
- Red Fir
- Subalpine Conifer
- Mixed Conifer

Map 5.1. Current coniferous forest habitat distribution. Used with permission from the California GAP analysis project 1998.

maximum of 1 Ma influenced the divergence of many high-elevation species at the ends of their ranges. Subsequent climatic fluctuations have resulted in the environmental stress and loss of populations intermediate to the northern and southern extremes of the range, in the north in the Cascades and Klamath-Siskiyou region and in the south in the southern Sierra Nevada (Map 5.1).

6 Flowering Plants

Based on molecular and fossil data, the Anthophyta arose in the late Jurassic to early Cretaceous, possibly from relatives in the Glossopteridales or Bennettitales, but did not radiate until the Cretaceous (Bell et al. 2010; Friis et al. 2011; Doyle 2012). The Rosinae (or core eudicots, except for the Gunnerales) contains the unplaced Dilleniaceae and the Pentapetalae, the latter of which includes about 70 percent of all flowering plant species (Moore et al. 2010). The Pentapetalae contains two major clades, the Superasteridae, evolving 107–98 Ma, and the Superrosids, evolving 111–103 Ma. The Caryophyllales (Superasteridae) originated about 71–63 Ma in southern Gondwana with subsequent diversification and colonization in South, Central, and North America (Moore et al. 2010). The Rosidae (Superrosids) clade is split into the Fabidae (Eurosids I) and Malvidae (Eurosid II), which together contain over 100 families, 51 of which are native to California (Moore et al. 2010).

Previous hypotheses for the geographic origins of the California flora include Asian and North American origins through the Bering and North Atlantic Land Bridges. To explain Madrean-Tethyan sclerophyllous vegetation, gene flow is hypothesized to have continued between North America and the Mediterranean from the early Oligocene to the early Miocene (Axelrod 1975). However, later studies indicate multiple origins for Madrean-Tethyan disjuncts that include Beringian migration, North Atlantic migration, and long-distance dispersal (Ackerly 2009; Wen and Ickert-Bond 2009). The influential work of Raven and Axelrod (1978) provides the framework for testing hypotheses about the evolution and diversification of the California flora, and many, but not all, of their progressive insights have held true when tested with modern molecular tools. In addition to the geographic origins of the California flora, they suggested that current plant distribution patterns are a result of climatic change that began in the Miocene.

Two major Paleogene/Neogene geofloras were long accepted in California, the Arcto-Tertiary Geoflora and the Madro-Tertiary Geoflora (see Raven and Axelrod 1978). Although these geofloras provided the foundation from which newer ideas about the origins of the western North American flora have developed, their generalizations are no longer accepted (Ackerly 2009). Instead, it is now recognized that a combination of dispersal, migration, evolutionary response to climatic, edaphic, and biotic factors, and vicariance within different lineages across space and time have formed the associations present today. Regardless, it is useful to consider the geofloras as a general explanation for the origins of some taxa, and the geoflora concept provides a general model of the migrational and evolutionary responses to changing environmental conditions. The original hypothesis for the Arcto-Tertiary Geoflora was that it consisted of primarily conifers and broad-leaved plants across North America and Eurasia from about 65–15 Ma when these continents were joined via Beringia and occurred at lower latitudes. The northward movement of both North America and Eurasia combined with mountain uplift resulted in the cooling and drying of the climate at about 15 Ma. The gradual development of mountains, ecological sorting, and climate change resulted in some of the disjunctions that currently exist in North American and Eurasian taxa (Edwards 2004). Conifers migrated to high elevations or latitudes, and the most broad-leaved tree species retreated to the eastern third of North America. Taxa that remain from this period in California include *Acer* (maple, Aceraceae) and *Cornus* (dogwood, Cornaceae); others now limited to eastern North America include *Fagus* (beech, Fagaceae), *Liquidambar* (sweetgum, Altingiaceae), and *Castanea* (chestnut, Fagaceae) (Edwards 2004).

The concept that the Madro-Tertiary Geoflora consisted of primarily sclerophyllous plants that migrated north during cooling and drying periods in the Paleogene and remained primarily intact is no longer recognized (Graham 1999). There is some evidence of late Cretaceous and early Paleocene occurrence of xerically adapted vegetation in southwestern North America, and by the mid-Eocene, it was present in more xeric areas in North America. By the end of the Oligocene, xerically adapted vegetation was widely present in the southwestern United States and northern Mexico (Axelrod 1958). As climates dried during the Miocene and into the Pliocene, the flora continued to migrate westward, and influenced by local latitudes, elevation, and proximity to the ocean, it diversified in more restricted assemblages. Most species from this period show adaptations consistent with life in xeric climates and are closely related to species from warm temperate and dry subtropic regions. However, some species may be derived from temperate species that gradually responded to increased aridity. *Ceanothus*, a genus of about 53 species, is dominant throughout a number of California ecosystems including chaparral. The stem Rhamnaceae is estimated to have arisen during the late Cretaceous or Paleocene with fossil evidence from present-day Colombia (Correa et al. 2010), with crown clade diversification supported by the fossil record by 55 Ma in the Northern Hemisphere (Burge and Manchester 2008). Two reciprocally monophyletic subgenera, *Cerastes* and *Ceanothus,* are estimated to have diverged approximately 13 Ma with further diversification within each subgenus throughout the Pliocene (Burge et al. 2011). Although diversification since the Pliocene makes resolution within the genus difficult, some clade definition is present within subgenus *Ceanothus* along the northern coast and within subgenus *Cerastes* at Point Conception (Burge et al. 2011).

Disjunct distributions between western Eurasia and western North American have been long recognized. There are a variety of explanations for these disjuncts, which include dispersal via the North Atlantic or Beringia (Raven 1972); however, these routes were sporadic with changes in climate and connectedness (Tiffney and Manchester 2001). An analysis of 66 plant clades favors a Beringian migration for most plants originating in Eurasia and migrating mostly in the last 30 Ma to western North America and not via a North Atlantic land bridge (Donoghue and Smith 2004). However, extant species distributions with disjunct populations also can be explained by independent adaptation and long-distance dispersal and, infrequently, migration (Kadereit and Baldwin 2012).

Fossil evidence provides an estimated age of 40 million years for the major *Quercus* clades (Manos et al. 1999). *Quercus* was widely present in the Northern Hemisphere by the Neogene, and in California, there were 20 species of *Quercus* in the early Miocene (24 Ma) (Keator 1998). The majestic, deciduous *Quercus lobata,* endemic to California and the western United States, occurs in remnant populations in the Central Valley and the surrounding valleys of the Coast Ranges, Sierra Nevada, and Transverse Ranges of California. The Mediterranean climate that established by the middle Pliocene in west central California and continued through the Pleistocene glaciations effected distributions of *Quercus* and the associated species of woodland habitat (Griffin 1988). A sample of 111 individuals of *Quercus lobata* from 35 localities for six cp microsatellites resulted in a G_{ST} value of 0.709 and H_T of 0.979 with local clustering and an east-west partition of 39 haplotypes. A R_{ST} value greater than the G_{ST} value supports a geographic structuring of populations within the species that was determined to be about 200 km (Grivet et al. 2006); however, the genetic connectivity of current populations supports long-distance colonization (Sork et al. 2010) (Figure 6.1). Combined with palynological data, these results suggest that the Neogene was an important period for *Quercus* woodlands as they expanded and contracted with glaciations.

The Madrean-Tethyan hypothesis states there was continuous xeric vegetation from the Eocene to Oligocene across western Eurasia and North America (Axelrod 1975). The hypothesis has been controversial and generally is no longer accepted in its entirety, but it does appear to be valid for some clades. For other taxa that may have diverged more recently, such as *Cercis,* other explanations have better support. Fossil data from North America and Europe indicate they had similar floras during Oligocene and Miocene (Donoghue et al. 2001). *Cercis* is estimated to have had a middle Miocene divergence between North America and western Eurasia that

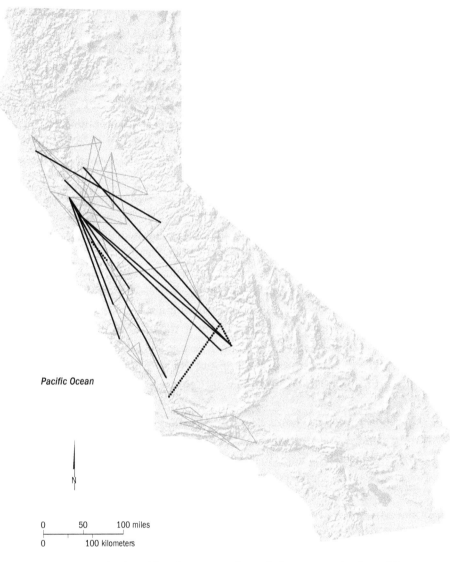

Pacific Ocean

N

0   50   100 miles

0   100 kilometers

Figure 6.1. Network of significantly connected populations among 64 sites of *Quercus lobata* shown as black lines based on PopGraph. Black lines indicate long distance dispersal greater than expected based on genetic differences and black dashed lines indicate barriers to gene flow. Used with permission from Blackwell Publishing Ltd. from Sork et al. 2010.

gave rise to *C. occidentalis* in western North America (Fritsch and Cruz 2012), a scenario consistent with *Cornus, Boykinia, Tiarella* (Saxifragaceae), *Trautvetteria* (Ranunculaceae), *Aralia* sect. *Aralia* (Araliaceae), and *Calycanthus* (Xiang et al. 1998; Milne and Abbott 2002). Best estimates split *C. occidentalis* and the eastern North America species and *C. canadensis* from the rest of the genus in western Eurasia about 15–13 Ma (Fritsch and Cruz 2012). A recent analysis of 25 examples of disjunctions between western Eurasia and western North America supports a variety of scenarios. Parallel evolution in xeric habitats in both geographic regions was found for Asteraceae-Cichorieae, Chenopodiaceae-Betoideae, *Descurainia* (Brassicaceae), Ericaceae-Arbutoideae, *Papaver* (Papaveraceae), *Platanus*, Antirrhineae-*Maurandya* group (Plantaginaceae), *Anemone* (Ranunculaceae), and *Styrax* (Styracaceae) (Kadereit and Baldwin 2012).

The Madrean-Tethyan hypothesis is considered plausible with some modifications for the *Caprifolium* clade of the genus *Lonicera* (Caprifoliaceae) (Tiffney 1985; Smith and Donoghue 2010). Nineteen species of perennial woody vines from this clade provide an estimated split between North America and Europe of 14–6 Ma, which is too recent for the Madrean-Tethyan hypothesis. However, an extinct widespread ancestor is conjectured for both North American and European clades followed by convergent evolution in the Mediterranean and in California (Smith and Donoghue 2010).

Beringia itself was dominated by temperate deciduous forests in the Miocene and coniferous forests in the Pliocene (Tiffney and Manchester 2001), with some xeric floras in isolated conditions (Stebbins and Day 1967). Molecular data and molecular clock analyses also support a combination of scenarios that include widespread distributions, some migration, and some in situ adaptation, with examples from *Liquidambar* (35 Ma, Donoghue et al. 2001), *Cercis* (32–6 Ma, Davis et al. 2002; 15 Ma, Donoghue et al. 2001), *Datisca* (Datiscaceae, 50–37 Ma, Liston 1997), and the Arbutoideae (Ericaceae, 39–21 Ma, Hileman et al. 2001). Later arrivals include *Platanus* (Platanaceae, 15 Ma, Feng et al. 2005), Antirrhineae (Plantaginaceae, 11–9 Ma, Hileman and Baum 2003), and *Kochia* (Amaranthaceae) and *Styrax* (13.8–5.0 Ma, Fritsch 1996). *Aphanisma* diverged from its sister genus *Oreobliton* in the Betoideae (Chenopodiaceae) about 15.4–9.2 Ma (Hohmann et al. 2006). *Aphanisma* is the only member of the subfamily in

North America, and sequence data and fossils support a hypothesis of long-distance dispersal with radiation into dry habitats. Alternative scenarios for disjuncts include the descent of dry-adapted species such as *Aphanisma* from mesic ancestors now extinct (Hohmann et al. 2006).

Sequence data from five cpDNA and nucDNA markers support an ancestor similar to *Campanula* (Campanulaceae) that migrated via Beringia at least six times during the Neogene and Quaternary and subsequently radiated in North America (Wendling et al. 2011). The California natives, *C. scouleri* and *C. scabrella,* are included in the mostly North American endemic "Cordilleran" clade, which occurs in mountainous environments in the Pacific Northwest consistent with a Beringian migration. Similar scenarios are proposed for *Rhododendron* (Ericaceae, Milne 2004) and *Saxifraga* (Saxifragaceae, Westergaard et al. 2010). Cooling from the Neogene through Pleistocene glaciations constricted a large circumboreal forest, and taxa from this period became gradually restricted to refugia in eastern Asia and western North America (Tiffney 1985).

Over 90 percent of *Lupinus* (Fabaceae) species occur in the New World, with only 13 species native to the Mediterranean region (Aïnouche et al. 2004). Molecular data support an Old World origin for the genus, with the New World becoming a secondary center of radiation (Aïnouche et al. 2004). Congruent *rbc*L and ITS sequences identify the Old World lupine species *L. angustfiolius* and *L. luteus* as closely related to the North American species *L. polyphyllus* and *L. nanus* and South America species *L. mutabilis*. Other relationships within the genus also indicate common ancestries between Old World and South American members of *Lupinus* (Käss and Wink 1997).

Carex (Cyperaceae) is a worldwide genus with its centers of diversity in Northern Hemisphere temperate regions. Only 30 known plant species have a bipolar distribution, 6 of which are found in the genus *Carex* (Moore and Chater 1971). Diversification via long-distance dispersal in the late Miocene through Pliocene is hypothesized for bipolar *Carex* species (Escudero et al. 2010). Molecular studies using cpDNA (*rps16*) sequences and ITS data of the subgenera *Vignen* and *Carex,* both of which occur at high latitudes in the Northern and Southern Hemispheres, suggest periodic glaciations events may have played a role in speciation (Martin-Bravo and Escudero 2012). This pattern is consistent with *Carex* as a whole

(2,000 species) as it has an inverse latitudinal gradient of species richness (Hillebrand 2004). In contrast, the distribution of *Empetrum* (Ericaceae), another bipolarly distributed species, can be explained by migration during the Pleistocene (Popp et al. 2011).

Pleistocene glaciations most certainly had an effect on the evolution of the California flora; however, the degree to which it was important is still under debate and likely depends on the climatic sensitivity of a taxon and its distribution prior to the Pleistocene. Regardless, Pleistocene glaciations are increasingly recognized as important in shaping the genetic structure of California plant species and in addition to *Quercus lobata* (Grivet et al. 2006) are demonstrated in the conifers *Pinus albicaulis, P. lambertiana* (Liston et al. 2007), and *P. monticola* (Steinhoff et al. 1990). These species demonstrate divergence into northern and southern clades during the Miocene with secondary contact zones in the Klamath-Siskiyou Ranges during the Holocene (Eckert and Hall 2006).

PATTERNS OF ENDEMISM

The extant flora of California contains a number of both paleoendemic and neoendemic species. The former consists of ancient taxa with formerly widespread distributions, for example, *Sequoiadendron* and *Sequoia*, which although once present throughout western North America during the Eocene are now restricted to the central Sierra Nevada and coastal Central to Northern California, respectively (Griffin and Critchfield 1972). Neoendemics are those species that have evolved in California and include members of the genera *Collinsia* (Chinese-houses, Plantaginaceae), *Eriogonum* (Polygonaceae), and *Lupinus* (Fabaceae) (Stebbins 1942, 1980). Edaphic endemics are species that occur only on a particular soil type, and most commonly, edaphic endemics in California occur on serpentine, but other species are limited to granitic, volcanic, or carbonate based soils. Although serpentine comprises 1.5 percent of the land area of California, virtually half its endemic plant taxa occur only on serpentine (Kruckeberg 1984). Serpentine endemism in California includes 246 specialists from 23 genera in 17 families and is associated with decrease in diversification rates (Anacker et al. 2011).

An increasing number of studies support the repeated evolution of edaphic endemism in California's angiosperm clades. *Collinsia* contains a number of edaphic endemics that appear to have arisen separately (Baldwin et al. 2011) (Figures 6.2 and 6.3). *Collinsia antonina* is endemic to silicious-shale talus in the outer Coast Ranges, *C. corymbosa* is endemic to the coastal dunes of Northern California, and the *C.* "metamorphica" complex of the central high-elevation Sierra Nevada is endemic to quartzite-rich schist and likely emerged 1.3–0.3 Ma. *Collinsia greenei* is a serpentine endemic that diverged in the late Miocene or Pliocene, consistent with the age of serpentine exposures (Baldwin et al. 2011).

Serpentine tolerance has evolved separately at least seven times beginning 7 Ma in the genus *Calochortus* (Liliaceae) and the two morphologies "fairy lantern" and "cat's ear" at least six times each. The stunning, very restricted serpentine endemic, *C. tiburonensis,* is sister to *C. umbellatus*, a more common resident of the northern Coast Ranges (Patterson and Givnish 2002, 2004)

Allium (Alliaceae) is a genus of about 750 species worldwide, 100 of which occur in North America, with centers of distribution in Texas and California (McNeal and Jacobsen 2002; Baldwin et al. 2012). An examination of 223 primarily western North American taxa used to uncover the origins of *Allium* species in western North America and the evolution of serpentine endemism included 39 of 48 taxa native to California (Nguyen et al. 2008). North American taxa are monophyletic, with 90 percent bootstrap support and a Bayesian posterior probability of 94 percent. The *Lophioprason* clade contains two subclades, one of which contains all Californian taxa and the other mostly not occurring east of the Rocky Mountains. Ancestral character state reconstruction indicates that the ancestor to *Lophioprason* was from California, with some members later migrating into the Northwest (Nguyen et al. 2008). The analysis further supports multiple origins of edaphic endemism on serpentine, gained six different times.

California has 31 endemic alpine species (Rundel 2011). Alpine habitats in California are generally considered to be those at 3,500 m or higher, with a growing season that is limited between the melt of winter snowpack and the first snowfall depending on the latitude and with intense summer drought (Sharsmith 1940). A virtually contiguous alpine habitat

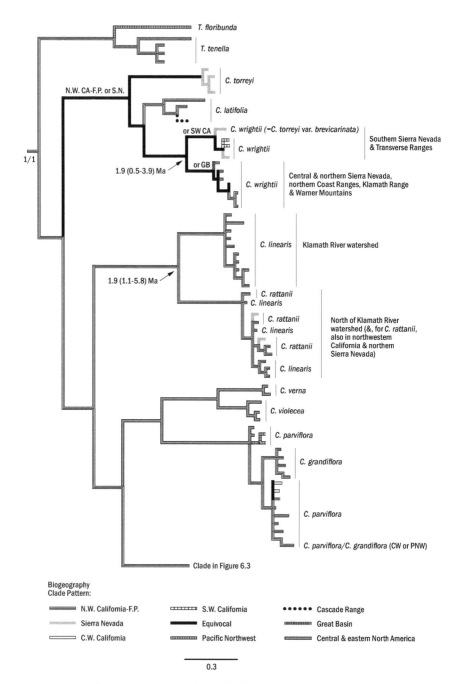

Figures 6.2 and 6.3. Mr. Bayes analysis of *Collinsia* and *Tonella* showing 0.5 majority-rule consensus tree. Geography determined using maximum clade credibility in TreeAnnotator in BEAST. Maximum divergence times also determined via BEAST. Asterisk indicates <90% Maximum Parsimony and two asterisks indicates clade not resolved using consensus tree (Baldwin et al. 2011). Used with permission from the *American Journal of Botany* and B. Baldwin.

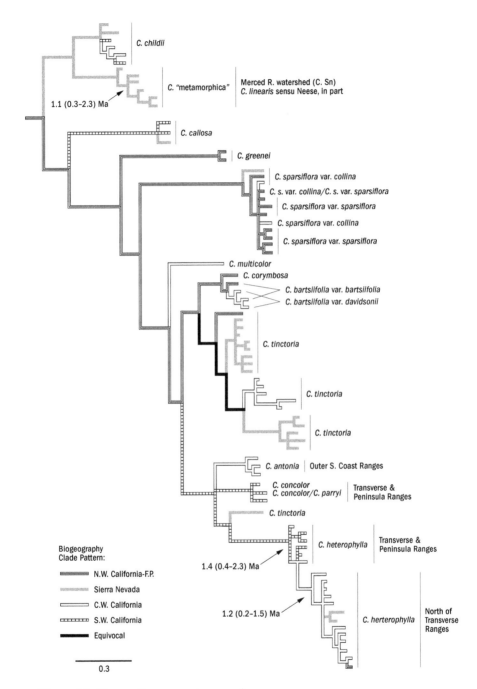

Figure 6.3. (See caption on previous page)

exists from Tuolumne to Inyo Counties in the Sierra Nevada, in addition to habitats on Mount Shasta in Northern California, on peaks in south-western California (e.g., San Gorgonio Mountain), and in the Great Basin on Mount Patterson and the White/Inyo Mountains (Rundel 2011). Additional alpine habitats at slightly lower elevations are found on Mount Eddy and Thompson Peak in the Klamath/Siskiyou Mountains and in the Panamint Mountains east of the White/Inyo Mountains in Southern California. Species that occur in these habitats can be divided into those that are endemic to subalpine and alpine habitats and those that have broader distributions.

THE KLAMATH-SISKIYOU REGION AND THE NORTHERN SIERRA NEVADA

The Klamath-Siskiyou region contains more than 281 endemic plant species and 3,500 plant species (Sawyer 2006). Pleistocene glaciations were significant forces in affecting intraspecific, interspecific, and intergeneric evolutionary patterns in plants and animals, particularly those with low dispersal ability. However, the Klamath-Siskiyou region experienced minimal glaciation and served as a refugium for many plant taxa and resulted in the loss of diversity in many clades (Smith and Sawyer 1988; Soltis et al. 1997; Sawyer 2006).

Some variation within angiosperm groups can be explained by the late Neogene fragmentation of cooler adapted floras with fluctuations of cooling and warming resulting in comingling with the more xerically adapted floras (Raven and Axelrod 1978). Oscillations in the Pleistocene glaciation cycles were a significant driving force in floristic diversification associated with range shifts, with particular disjunct population distributions found between some elements of the Klamath-Siskiyou region and the Sierra Nevada.

An examination of 74 taxa of the North American genus *Boechera* (Boechereae: Brassicaceae) using 725 bp of ITS data and phylogenetic analyses elucidates a complex evolutionary history influenced by polyploidization, hybridization, secondary contact, and Pleistocene glaciation cycles (Kiefer and Koch 2012). The earliest defined lineage of *Boechera* is *B. stricta* in southern areas of North America, including the

Great Basin and Sierra Nevada, which later differentiated but retained a genetic signature through glacial cycles and colonizations. Another genotype within *B. sparsiflora* derived from this southern genotype and was ancestral to Klamath/Siskiyou populations that survived in glacial refugia and gave rise to *B. breweri* and *B. koehleri*.

Sidalcea (checkerbloom, Malvaceae) consists of a mix of 25 annual and perennial species in western North America. An analysis of ITS (ITS1 258–296 bp, ITS2 206–226 bp) and ETS (540 bp) found that, in general, there is not extensive divergence among species, which suggests a recent radiation. *Sidalcea hickmanii, S. stipularis*, and *S. malachroides* are identified as basal perennials, and the annual habit arose four or more times (Andreasen and Baldwin 2003). Major clades supported within the genus with some geographic concurrence include the *asprella* clade in the higher-elevation Sierra Nevada to northwest Oregon, and subclade to the *glaucescens* clade; the *glaucescens* with additional taxa distributed from the Sierra Nevada to the Klamath Range; and the *oregana* clade in Oregon and mostly northwestern California (Andreasen and Baldwin 2003).

The Plantaginaceae is hypothesized to have originated in the Old World and then to have colonized the New World (Albach et al. 2005). *Collinsia* consists of annuals distributed from sea level to about 4,000 m in a diverse array of habitats. Most taxa occur in the California Floristic Province, and the genus contains 16–19 species with 6–7 varieties. *Collinsia* samples from 179 populations with sequence data from *CYCLOIDEA-1*, ITS 1 and 2, ETS, and *mat*K/*trn*K intron were analyzed with two species of the annual sister-genus *Tonella* (Baldwin et al. 2011) (see Figures 6.2 and 6.3). Examinations of the evolution of floral traits, phylogenetic and biogeographic hypotheses, hybridization patterns, and BEAST analysis of combined data sets indicate deep divergences in *Collinsia* perhaps as early as 11.71–10.35 Ma between northwestern California and the Sierra Nevada. The divergence of east-west clades occurred as early as 6.2 Ma between *C. verna* and *C. violacea* and between *C. grandiflora* and *C. parviflora*. Divergence within *Collinsia linearis* between the Klamath River and the more northerly populations was estimated at between 5.8 and 1.1 Ma. The Merced River drainage population that previously included *C. linearis* is more related to other taxa and represents an ancient lineage (*C.* "metamorphica") within *Collinsia*. Separation of clades of *C. linearis* in the northern part of the Klamath River drainage corresponds with splits in *Polystichum munitum* (Steinhoff et al. 1983) and *Pinus monticola* (Soltis et al. 1997). The *C.* "metamorphica" clade of the Sierra Nevada comprises a number of edaphic endemics estimated to have diversified within the last 2.3 million years.

Colonization of the high Sierra Nevada is suggested to have occurred from north to south, based on a greater number of endemic species in the southern Sierra Nevada (Raven and Axelrod 1978). North to south colonization of high montane taxa more generally may explain why Mount Lassen shares greater floristic affinity with Mount Shasta (Gillett et al. 1995). Similarly, the Klamath Mountains are the southern limit of a number of boreal alpine species (Howell 1944). Species with their southern limit of distribution in the northern or central high Sierra Nevada are *Carex whitneyi, Podistera nevadensis* (Apiaceae), *Claytonia megarhiza* (Montiaceae), *Thalictrum alpinum* (Ranunculaceae), *Galium grayanum* (Rubiaceae), and *Salix nivalis* (Salicaceae) (Rundel 2011).

Geographically, the alpine flora has diverse origins. The largest group of species (34.3 percent) are those that are widespread in the mountainous western United States, 20.5 percent occur in the Intermountain Great Basin, 15.8 percent in the Sierra Nevada and into the Cascade Range, and 13.6 percent widely distributed in boreal or arctic-alpine regions (Rundel 2011). Ninety-four percent of the Sweetwater Mountains alpine flora is shared with the Sierra Nevada. The White Mountains (90 percent) have a stronger affinity to the Sierra Nevada alpine flora than to the central Rocky Mountains (58 percent) (Scott 1995). Generally, these patterns support a northern origin for the alpine floras of the Cascade Range and the Sierra Nevada. Once arriving in insular alpine habitats, species became gradually isolated from their progenitor populations. In addition to allopatric divergence, mechanisms including apomixis and various forms of polyploidy undoubtedly played a role in the survival and divergence of these taxa. Montane sky-island diversification corresponds with Pleistocene glaciation (DeChaine and Martin 2005).

The alpine flora of the Sierra Nevada consists of 409 plant species of which 84 percent are herbaceous perennials in the Asteraceae, Brassicaceae, Rosaceae, Fabaceae, Polygonaceae, and Onagraceae, followed by graminoids in the Poaceae, Juncaceae, and Cyperaceae (Rundel 2011). Annual species in the alpine environment of the Sierra Nevada are represented primarily by the Boraginaceae and Polemoniaceae. Nine of the Sierra Nevada alpine species are restricted to elevations of 3,500 m, and 36 species are endemic to the Sierra Nevada at $\geq$ 2,700 m (Rundel 2011).

COAST RANGES

The northern Coast Ranges contain a wide variety of substrates that were formed during the Mesozoic as part of the Franciscan Accretionary Complex, first as sedimentation, later infused with ophiolites, and folded during plate movements throughout the Cenozoic (Dickinson 2008). Although the Coast Ranges did not experience Pleistocene glaciations, California plant taxa that descended from early arrivals experienced intense climatic change on a highly heterogeneous landscape that resulted in myriad interesting morphologies and adaptations. Short generation times and high fecundity of annual species often result in a fairly rapid response to changing environmental parameters. Long-lived species with delayed seed production should show genetic signatures of expansion and contraction. A number of species migrated south along the Pacific Northwest coast, as with the example of *Collinsia rattanii* populations in the outer Coast Ranges of California that are most closely related to Pacific Northwest populations (Baldwin et al. 2011). High levels of serpentinite soils also increase genetic differentiation. Increasingly dry summer periods in California beginning about 15 Ma (see Baldwin and Sanderson 1998) may have resulted in intense selection via the restriction of some species populations to the fine-grained, high moisture capacity serpentine soils that often contain seeps (Raven and Axelrod 1978).

Bird dispersal likely played a role in the dispersal of a number of a disjuncts as illustrated by *Howellia aquatilis* (Campanulaceae) in wetland habitat in the northern Coast Ranges. The species is known from six populations in California and a number of populations in Washington and Montana. AFLP data identified little genetic differentiation among the California and Montana populations, providing a Φ_{ST} value of 0.162 regardless of whether Montana populations were included (Schierenbeck and Phipps 2010). There is no correspondence between genetic and geographic distance across the Eel River among populations, also supporting bird dispersal.

The genus *Calochortus* contains 67 species of geophytes that range from British Columbia south to Guatemala and west to the Dakotas (Fiedler 2012). An analysis of *rbc*L and *ndh*F sequence data reveals that differentiation of the genus began about 7.3 Ma and arose in the Coast Ranges (Patterson and Givnish 2002). Further analysis of 72 taxa with *trn*T-*trn*F, *psb*A-*trn*H, and *rpl16* identifies strong geographic affinities within clades. In California these clades are the Pacific Northwest (including the Cascades and the Klamath-Siskiyou region), the San Francisco Bay Area, San Diego, the Coast Ranges–Sierra Nevada (with species surrounding the Central Valley), and southwestern California. Within each clade and almost without exception, sister species are geographically close. It is likely that the phylogeography of *Calochortus* was significantly influenced by low dispersal ability due to heavy unwinged seeds (with the exception of *C. macrocarpus*) and diverse geologic landscapes (Patterson and Givnish 2004).

Dirca occidentalis (leatherwood, Thymelaeaceae) is a disjunct endemic that occurs in a range of habitats from grassland, chaparral, and riparian woodlands to broad-leaved evergreen woodlands in California and eastern North America (Ackerly et al. 2002). Four genetically distinct populations that occur in the Coast Ranges of the San Francisco Bay Area are hypothesized to have resulted from climatic events within the last 20 ka (Graves and Schrader 2008). Colonization of *Dirca* from eastern North America is consistent with paleobotanical data that establish occurrence of ancestral Thymelaecaceae in the Eocene Rocky Mountains (Graham 1993). Today, the Thymelaecaceae has a mostly tropical distribution, however, so *D. occidentalis* may have possibly survived from more widespread mesic vegetation.

Thirty-two Coast Range populations of *Calystegia collina* (morning glory, Convolvulaceae), a serpentine endemic, were examined in part of its range in Napa, Lake, and Sonoma Counties. As with all edaphic endemics, the level of genetic variation contained within each outcrop or island population is important (Wolf et al. 2000) because of the conservation implications. If there is little variation contained in a small population, it may not be compelling to conserve the population. In contrast, populations with high levels of variation in both small and large populations are often worth conserving. In *C. collina* very little difference in genetic variation according to outcrop size was found, which was hypothesized to be the result of high levels of vegetative growth. Gene flow is low among these populations and is primarily mediated by pollinators, but some hybridization has been observed (Preston and Dempster 2012).

Notholithocarpus densiflorus (tan oak, Fagaceae) occurs as a shrub or tree in many forest types in coastal California, southern Oregon, the Sierra Nevada, the Klamath Range, and the Cascades (Nettel et al. 2009). A sample of 447 trees from 19 populations using a combination of population genetic (14 cpDNA and 11 nucDNA microsatellites) and phylogenetic markers (*trn*H-*trn*K, *trn*K1-*trn*K2, *and trn*C-*rpo*B from cpDNA) identifies a low overall population cpH$_T$ (0.40) and intrapopulation cpH$_S$ (0.03). Intrapopulational diversity explains the molecular variance at 90.8 percent, and a low F_{ST} value of 0.09 and no significant difference between R_{ST} and F_{ST} suggest that gene flow and genetic drift in some populations are important factors in explaining the distribution of genetic variance in this species. An excess of heterozygotes is consistent with a recent bottleneck followed by range expansion. The highest levels of genetic diversity (four cpDNA haplotypes) in southern Oregon, the Klamath Range, and the northern Sierra Nevada are likely a reflection of a complex terrain combined with heavy fruits which limit seed dispersal. There is one haplotype that occurs in the Coast Ranges from Humboldt to Santa Barbara Counties (Nettel et al. 2009). The nuclear genome reflects coastal groupings south of San Francisco and an interior northern Sierra Nevada, Klamath, and southern Oregon group that indicate that genetic mixing is more effective through pollen dispersal than through seed and likely reflects a Pleistocene divergence between the coastal and interior groups (Nettel et al. 2009). Holocene pollen records in coastal Northern and Central California indicate that the Klamath and Sierra Nevada were drier and colder during the early Holocene and became more mesic around 6.5 ka and then again around 5.8 ka, which resulted in a greater presence of the Fagaceae (Mohr et al. 2000; Daniels et al. 2005).

Tree species are long-lived and often have high effective population sizes and high rates of reproduction. *Cornus nuttallii* (Pacific dogwood) is endemic to North America, with a distribution from the Southern

California mountains to southwestern British Columbia. It is insect pollinated, and its seeds are dispersed by mammals and birds. A sample of 595 individuals from 20 locations from throughout the range of the species used five microsatellite loci and cpDNA markers to assess population-level divergence (Keir et al. 2011). A F_{ST} value of 0.111 and R_{ST} of 0.155 indicate that there is no isolation by distance and there is generally low diversity rangewide for both the nuclear and cpDNA genomes. A decline in diversity with latitude suggests there was long-distance dispersal from a southern refugium following glaciation. Two chloroplast haplotypes, with Nei's unbiased genetic distance (D) of 0.153 and no private alleles, also support a long bottleneck during the LGM that resulted in a north/south division in southern Oregon (Keir et al. 2011). Another example of a taxon with a large geographic range without well-differentiated populations is *Lilaeopsis occidentalis* (Mason's Lilaeopsis, Apiaceae); although the species is morphologically variable it does not have strong genetic divisions along the Pacific coast from Southern California to British Columbia (Fiedler et al. 2011).

CENTRAL VALLEY

The pre-European Great Central Valley was a mosaic of vernal pools, rivers and associated riparian vegetation, oak woodlands, estuaries, and grasslands. Very few undisturbed sites remain in the California Central Valley due to development and agricultural activities, and those habitats that remain are highly disturbed. Mountain runoff and inundation in the Central Valley increased during interglacial periods, and the species occurring there would have been subject to widespread, intermittent flooding that increased aquatic habitats such as vernal pools (Stebbins 1976, Raven and Axelrod 1978). It is estimated that only 10 percent of the pre-European vernal pool habitat remains in the Central Valley. As a result of this significant habitat loss, a number of taxa endemic to vernal pools are endangered and little is known about the historical phylogeography of the species that occurred or occur within them. *Lasthenia* (Asteraceae) is a genus of 21 taxa primarily endemic to California across a wide array of habitats, including vernal pools. A Bayesian analysis of molecular data and a Principal Components Analysis of biogeographic data identified four independent occurrences of adaptation to the semi-aquatic habitat, one of which resulted in the divergence of nine taxa endemic to vernal pools, including eight in California and a descendant species in central Chile (Emery et al. 2012).

Conservation and restoration of vernal pool taxa require the assessment of gene flow within and among populations to determine their degree of genetic isolation (Sloop et al. 2011). *Neostapfia colusana* (Colusa grass, Poaceae) is a federally threatened species endemic to vernal pools. Five microsatellites for 240 individuals from eight pools reveal high within-population genetic diversity, with high overall levels of heterozygosity (H_O = 0.68, H_E = 0.71). A F_{ST} value of 0.268 (p<0.0001) between northern and southern populations indicates that gene flow is very limited among the regions and is likely due to the loss of geographically intermediate populations. Some populations have lower than expected heterozygosity (Olcott Lake, Solano County, and Yolo County grasslands) that is likely due to geographic isolation and increased inbreeding. Similarly, *Tuctoria greenei* (Greene's tuctoria, Poaceae) is a federally endangered vernal pool endemic species evaluated using five microsatellites for 317 individuals from thirteen vernal pools. High heterozygosity (H_O = 0.77, H_E = 0.79) is also present within populations, and the species shows regional subdivision between northern and southern populations, although gene flow was not as restricted as in *N. colusana* (F_{ST} = 0.11, P = 0.0001) (Gordon et al. 2012).

The Limnanthaceae is a small clade endemic to western North America, with the exception of *Floerkea proserpinacoides* (false mermaid), which occurs in pockets throughout North America (Kishore et al. 2004). The genus *Limnanthes* consists of 18 taxa that occur primarily in wetland habitats in the Central Valley, 12 of which are listed threatened, rare, or endangered. *Limnanthes floccosa* ssp. *californica* (Butte County meadowfoam, Limnanthaceae) is a federally and state listed endangered species endemic to vernal pools along the eastern edge of the Sacramento Valley and currently known from only 14 geographically isolated populations. The species has fairly low genetic diversity with H_O = 0.10 and H_E = 0.19 and high among population genetic structure as measured by pairwise F_{ST} values of 0.12–0.79 (Sloop et al. 2011). Of particular concern for this species is its high rate of inbreeding and barriers to gene flow. Pollinator declines and reduced aquatic dispersal due to habitat loss will likely further erode genetic diversity in this species.

TRANSVERSE AND PENINSULAR RANGES/SOUTHERN SIERRA NEVADA

The Transverse and Peninsular Ranges share more than a third of their alpine flora with the Sierra Nevada (Rundel 2011); however, there are a variety of subdivisions among taxa within and among these ranges and surrounding landforms. *Collinsia heterophylla* has a clade in the Sierra Nevada and Coast Ranges that is distinct from populations of the Transverse and Peninsular Ranges (Baldwin et al. 2011). Some animal groups have a split between the Sierra Pelona and San Gabriel Mountains (Chatzimanolis and Caterino 2007), but this is not the case with *Collinsia heterophylla*. Bayesian analyses of molecular data indicate that a Transverse–southern Coast Range split occurred in *C. heterophylla* as early as 1.5–0.2 Ma and between *C. antonina* (southern Coast Ranges) and *C. concolor/C. parryi* (Transverse and Peninsular Ranges) 3.1–1.2 Ma, all of which are consistent with late Pliocene through Pleistocene uplift and climate change. *Collinsia wrightii* underwent a north-south split within the southern Sierra Nevada as early as 3.9–0.5 Ma, with populations from the Transverse Ranges and southernmost Sierra Nevada representing one lineage and populations of the central and northern Sierra Nevada, Klamath Range, high northern Coast Ranges, and Warner Mountains representing the other (Baldwin et al. 2011). Most animals are split between the central and eastern Transverse Ranges (Chatzimanolis and Caterino 2007). More detailed phylogeographic analyses of plant taxa between the Transverse Ranges and the Sierra Nevada are needed to establish the mechanisms of dispersal and distribution of plant taxa responding to vicariant, climatic, or dispersal events associated with these southern mountains.

DESERTS

Mojave Desert vegetation was dominated by *Pinus monophylla/Juniperus osteosperma* at the end of the LGM, but a drought at 14 ka resulted in its quick replacement by desert vegetation (Thompson and Anderson 2000). Desert vegetation was likely established by about 11 ka with the arrivals of current dominants, *Ambrosia dumosa* (burro bush, Asteraceae) and *Larrea tridentata* (creosote bush, Zygophyllaceae) (Axelrod 1979; Van Devender and Spaulding 1979). *Yucca brevifolia* (Joshua tree, Agavaceae) was more widespread in the Pleistocene than it is today, extending much farther south and west into the Sonoran Desert (Cole et al. 2003).

There were some introductions of taxa from the Mediterranean to the western North American deserts with resulting disjunctions (Stebbins and Day 1967). There is evidence that *Senecio mohavensis* ssp. *mohavensis* (Mojave ragwort, Asteraceae) was introduced to North America about 150 ka from the Mediterranean (Liston et al. 1989). *Plantago ovata* (plantain, Plantaginaceae) is a winter annual of Southern California deserts, southwestern North America, and the Mediterranean and is considered the only species within the *Plantago* subgenus *Albicans* section *Albicans* that occurs outside of the Old World in putatively natural populations (Meyers and Liston 2008). There has been controversy as to whether it migrated over Beringia during the Miocene or was an introduction by the early Spanish (Stebbins and Day 1967; Bassett and Baum 1969). Specimens from 585 individuals from throughout the range examined for morphology and 14 specimens for 2153 bp of cpDNA and nucDNA were analyzed via maximum parsimony and yielded an estimate of Old to New World introduction at about 650–200 ka, with ancient hybridization between Old World varieties of *P. ovata* preceding dispersal to the New World (Meyers and Liston 2008).

Encelia farinosa (brittlebush, Asteraceae) is a short-lived perennial that occurs in the Sonoran, Mojave, and Peninsular Deserts with three varieties, *E. farinosa* var. *farinosa*, *E. farinosa* var. *phenicodonta*, and *E. farinosa* var. *radians* (Fehlberg and Ranker 2009). This species group provides a good basis for phylogeographic study because it occurs on the edge of the effects of Pleistocene glaciations, and *Neotoma* (pack rat) middens provide good macrofossils. These taxa show genetic signatures that reflect migration from refugial populations. Fossil records from middens for at least 13.8 ka suggest that *Encelia* expanded from western Arizona and expanded west and north following the LGM. A sample of cpDNA sequences from the intergenic spacer (*psb*A-*trn*H) for 310 individuals from 21 locations for all three varieties and analyzed via an AMOVA reveals that 80.64 percent (P<0.001) of the variation explains differences within locations. Pairwise differences among locations are an average of 0.0013 with a range of substitution rates that estimate Pleistocene divergences occurred 0.63 and 0.21 Ma. There is no genetic support for the varieties. These results are consistent with Pleistocene glaciations and large-scale geographic structuring and NCA support past fragmentation and range expansion. Sites of glacial refugia are thought to have been the lower Colorado River Basin, the eastern Sonoran Desert, and Baja California Sur as populations there have higher levels of genetic diversity than in the expanded range of the Mojave Desert.

Yucca brevifolia is a true Mojave Desert ancient endemic within the *Clistocarpa* clade of the genus (Pellmyr et al. 2007); cpDNA data indicate that the polytomous clades within *Y. brevifolia* diverged about 6.5 Ma (Smith et al. 2008). *Yucca brevifolia* underwent range expansion and contraction during Pleistocene and Holocene climatic changes (Cole et al. 2011). Abundant fossil *Y. brevifolia* leaves from the late Pleistocene suggest a larger historical distribution than today (Van Devender 1987). Large range contractions are suspected at the last ice age; however, molecular and demographic signatures indicate large range expansion about 2 ka (Smith et al. 2011) into the Sonora Desert concomitant with expansion of the specialist moth pollinators (Cole et al. 2011; Smith et al. 2011).

The La Brea Tar Pit flora reflects a diverse array of plant communities surrounding the site, which is mostly dominated by a hot, dry climate with cooler winters. Plant communities varied from mesic to xeric, and likely depending on aspect and other local habitat features. Taxa present include *Hesperocyparis macrocarpa, Pinus muricata, P. radiata, P. sabiniana, Sequoia sempervirens, Juniperus californica*; chaparral with *Adenostoma fasciculatum* (Rosaceae), *Salvia* sp. (Lamiaceae), *Arctostaphylos* spp., *Frangula californica* (Rhamnaceae), *Xylococcus* (Ericaceae); and grasslands with *Calandrina* (Montiaceae), *Castilleja* (Orobanchaceae), and Poaceae genera *Bouteloua, Hilaria,* and *Sporobolus* (Edwards 2004). Notably, all of these plant taxa are still present in the region, whereas most of the animal taxa are extirpated or extinct.

SUMMARY

There is much excellent phylogenetic work on California plant species; however, the molecular tools used thus far are generally more useful at discerning evolutionary relationships that occurred prior to the Pleistocene. The use of population genetics tools integrated with rapidly evolving nuclear and chloroplast sequences will eventually provide a more accurate picture of plant divergence in California. For example, sequences of even rapidly evolving areas of the genome have provided little resolution for taxa such as *Arctostaphylos* with about 95 taxa that likely radiated in the Neogene. An accurate picture of diversification in those taxa that adaptively radiated in California during the Neogene will require the creative employment of a variety of genetic markers.

Some species were survivors of disjunctions among previously widely distributed taxa, and some arrived via dispersal. There is good evidence that many California plant clades are the result of Miocene migrations across Beringia. Although likely originating from the Old World, several California genera have evolutionary divergence centered in western North America and illustrate the phylogeography of flowering plants of California. Diversification on complex substrates during fluctuating climates has resulted in high levels of endemism throughout the state but particularly in the Klamath-Siskiyou and Coast Ranges. Evolutionary processes within California's many endemic species undoubtedly included allopolyploidy, autopolyploidy, and apomixis (Stebbins 1959). Secondary contact resulting in hybridization due to colonization from multiple refugia has led to interesting genetic patterns of reticulation in many taxa. Much phylogeographic work remains to be conducted on the flora of California, particularly those species that occur in the Klamath-Siskiyou region and those with distributions in the Transverse Ranges and southern Sierra Nevada. Endemics with the most restricted distributions are present in the central Coast Ranges, the Sierra Nevada, and the San Bernardino Range, with the youngest neoendemics identified from the desert and Great Basin (Kraft et al. 2010) (Map 6.1). Interestingly, the San Bernardino Mountains provide the only evidence of glaciation in Southern California (Sharp et al. 1959).

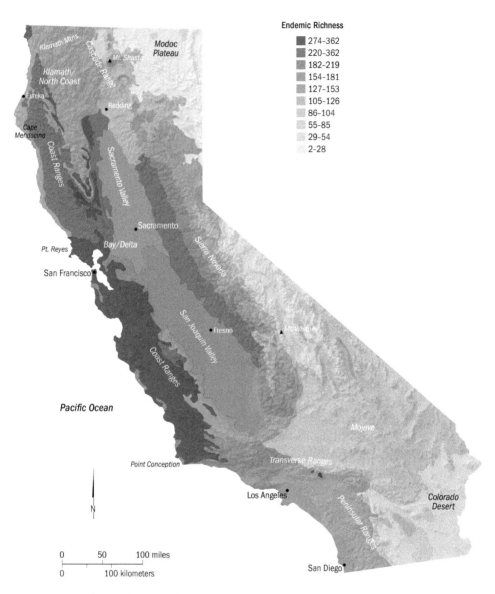

Map 6.1. Plant Endemic Richness, as total number of endemic minimum rank taxa per regional band. Used with permission from Blackwell Publishing Ltd., from Kraft et al. 2010.

Because all flowering plants share a common ancestor, it may be that plant communities dominated by Anthophyta also originated from a single location and major clades diversified out on a heterogenous landscape throughout the late Mesozoic and into the Paleogene. Alternatively, once flowers evolved, community associations with members of the Anthophyta may have arisen in more than one region. Regardless, many overlapping temporal migrations and changing latitudes and climates have led to an ecological sorting process that will require more detailed analysis to discern. Although Raven and Axelrod (1978) were correct about the dynamic equilibrium that occurred between the Madro-Tertiary and Arcto-Tertiary floras, questions remain about their initial rise and diversification as broad generalizations about the floras are no longer considered valid.

7 Insects

Based on extrapolations from global species estimates, there are approximately 100,000 insect species in California, of which about 12 percent are considered to be endemic (Kimsey 1996). California has 31 of 32 orders of insects that occur north of Mexico and about 30 percent of all insects known to the United States and Canada. Despite the established species richness, many new species of insects are still being described from California.

An examination of 770 species, most of them insects, reveals that there are more disjuncts and thus historical dispersals between eastern Asia and western North America than between eastern Asia and eastern North America. Western North America was more frequently the recipient of three major Beringian dispersals (35 Ma, 8 Ma, 1 Ma) than eastern North America for Arctic species (Sanmartín et al. 2001). Many insect species arrived in North America prior to the early Paleogene, but the contemporary genetic structure of California insects was shaped by tectonic activity throughout the Plio-Pleistocene and Pleistocene glaciations. In addition, the spread of aridity in western North America compared to the more mesic eastern North America may have resulted in more extinctions in the former.

KLAMATH-SISKIYOU REGION AND
NORTHERN SIERRA NEVADA

High botanical diversity and ancient landscapes predict that the Klamath-Siskiyou region should have a high diversity of insects. However, this area has remained relatively unexplored by entomologists until recently. Accordingly, three new species of *Grylloblatta* (*G. oregonensis, G. siskiyouensis,* and *G. marmoreus*; Grylloblattidae) have been described from the Klamath Mountains (Schoville 2012).

Acrotrichis xanthocera is a tiny feather-wing beetle in the Ptiliidae, found throughout North American forests and thought to be a low disperser. Ptiliidae species inhabit soil and litter in conifer forests and represent one of the most diverse groups of organisms in the Klamath-Siskiyou region (Moldenke 1999). An analysis of 750 bp of the mtDNA COI from 117 specimens reveals that *A. xanthocera* has low nucleotide diversity (π = 0.0494) and high haplotype diversity (h = 0.839) (Caesar et al. 2006). A generally complex history of *A. xanthocera* indicates that population structure is more associated with habitat heterogeneity than with low vagility and in fact indicates that they may occasionally be dispersed widely via wind dispersal.

Butterflies in the *Parnassius phoebus* complex (small Apollo, Papilionidae) of alpine specialists dispersed from central Asia and consists of five allopatric species each containing a varying number of subspecies or varieties (Michel et al. 2008). The complex includes *P. smintheus* in the Rocky Mountains, Cascade Range, and Trinity and Siskiyou Mountains and *P. behrii* found only in the Sierra Nevada. Butterflies in general differ from many other insects, in that even though they can disperse long distances, it is generally a rare event. Previous data from members of the *P. phoebus* complex indicate they are short-range dispersers (Schoville and Roderick 2009). A sample of 77 individuals of *P. behrii* and 32 individuals of *P. smintheus* sampled for COI, WG, and *EF1α* and analyzed with Bayesian coalescent analyses found that genetic divergences are within mountain ranges not between mountain ranges. However, there is support for recent connectivity between the Cascades/Sierra Nevada and the Rocky Mountains for members of the *P. phoebus* complex as also demonstrated with *Oeneis chryxus* (brown arctic, Nymphalidae) and *Ochotona princeps* (American pika, Ochotonidae) (Schoville and Roderick 2009; Nice and Shapiro 2001; Hafner and Sullivan 1995). A holarctic study of the *Parnassius phoebus* complex supports expansion from recent refugia. Sequence data from 824 bp of mtDNA COI from 203 samples from 72 locations found the T_{MRCA} to be 125 ka for the North American clade and 80 ka for the Eurasian-Alaskan clade. Pairwise divergence times among all locations range from 50 to 10 ka (Todisco et al. 2012). *Parnassius smintheus* and *P. behrii* are confirmed to be monophyletic species, but their genetic structure has been primarily shaped by glacial cycles during the Holocene (Todisco et al. 2012)

The alpine-inhabiting *Oeneis chryxus stanislaus* (Chryxus Arctic, Nymphalidae) from the central Sierra Nevada and *O. ivallda* from the northern Sierra Nevada can be differentiated based on allozyme data from 527 individuals from 13 populations and sequence data from 60 individuals for 480 bp of mtDNA COI (Nice and Shapiro 2001). Southern Sierra Nevada populations of *O. ivallda* share characteristics with *O. chryxus chryxus* from eastern Nevada, offering some support for the alpine island hopping hypothesis (Hovanitz 1940). Biogeographic reconstruction indicates that *Oeneis* colonized the southern Sierra Nevada near Piute Pass from the eastern Great Basin or Rocky Mountains.

The colonization of North America from Asia via Beringia by the Polyomatinae clade within the Lycaenidae was first suggested in 1945 by Vladimir Nabokov, famed author of *Lolita* and curator of Lepidoptera at Harvard University. Nabokov's hypothesis was based on genital morphology and was dismissed until 2011, when it was supported with molecular data (Vila et al. 2011). Specifically, Nabokov proposed there were five dispersals of *Polyommatus* blue butterflies from Asia into the Americas over the past 11 million years. Each dispersal event was followed by diversification, the last four of which were in North America and include the *Icaricia-Plebulina* clade (9.3 Ma), the *Lycaeides* clade (2.4 Ma), *Agriades* (1.1 Ma), and *Vaccininna* (1.0 Ma). Paleoclimatic estimates support the conditions that would have been necessary for dispersal across Beringia (Vila et al. 2011).

The *Lycaeides* butterfly species complex (Lycanidae) is composed of *L. idas* (northern blue), *L. melissa* (Melissa blue), and *L. m. samuelis*. MtDNA data from 628 individuals from 57 locations analyzed in concert with morphological variation show little correspondence between morphology and mtDNA variation, but three allopatric groups were identified (Nice et al. 2005). One clade consists primarily of *L. idas* in the Sierra Nevada and the northern Coast Ranges, another primarily *L. melissa* clade ranges from eastern and Southern California east to Wisconsin and Alaska, and the last clade is in northeastern North America. Bayesian maximum likelihood and coalescent-based analyses reveal that *Lycaeides* colonized North America in the late Pliocene with rapid range expansion within each clade at the end of the Pleistocene from glacial refugia (Gompert et al. 2006). A new homoploid species resulting from hybridization of *L. idas* and *L. melissa* was likely the consequence of a combination of genetic admixture and ecological selection in the extreme habitat of the alpine northern Sierra Nevada (Gompert et al. 2006). An analysis of 55 populations from Eurasia and North America using 450 bp of mtDNA COI and 550 bp of COII found some mixing of haplotypes in the Marble Mountains of the Klamath-Siskiyou region. It is likely that North America was colonized by different populations from Eurasia, but admixture is still very strong in areas of sympatry (Gompert et al. 2008).

The *Lycaena xanthiodes* species group (great coppers, Lycanidae) was examined using 618 bp of COII in 474 specimens from 62 locations from throughout their range with a focus on California (Oliver and Shapiro 2007). The study taxa included *L. editha*, present in the Klamath-Siskiyou Ranges, northeastern California, the northern Great Basin, and the northern Rocky Mountains, whereas *L. xanthoides* occurs from northern Mexico to south central Oregon. *Lycaena dione*, present in the northern Great Plains, was also examined due to its geographic overlap with *L. editha*. Fairly recent divergence consistent with species status, rapid speciation, low sequence divergence, and potential incomplete reproductive isolation was found between *L. xanthoides* and *L. editha* in southern Oregon and Northern California. Sequence differences of 0.4–1.5 percent support a divergence time of 0.2–0.7 Ma for the three taxa, which is much later than other members of the Lycaenidae examined here. The Transverse Ranges divide two clades of *L. xanthoides* with 1.81 percent sequence divergence, also supporting a Pleistocene rather than Pliocene divergence (Oliver and Shapiro 2007).

Hesperia comma (holarctic skipper, Hesperiidae) is a widespread, vagile species with a holarctic, distribution in a wide variety of habitats. One hundred fifty-one individuals from 49 locations were sampled from British Columbia, Eurasia, the Great Basin, and California in the Klamath-Siskiyou Ranges, Cascade Range, Sierra Nevada, and Transverse Ranges (Forister et al. 2004). Sequence data for a total of 891 bp from COII and *wingless* support strong genetic divergence between North America and Eurasia with a combined Φ_{CT} = 0.5290 (P = 0.0049). Maximum likelihood, maximum parsimony, and Bayesian analyses of COII sequences reveal three clades within western North America: Oregon and Northern California; the Great Basin, the Rocky Mountains, and British Columbia; and Southern California (Φ_{SC} = 0.2379, P<0.0001). The Oregon/Northern California and Southern California clades are separated by the southern Sierra Nevada and Transverse Ranges, and the Great Basin/Rocky Mountains/British Columbia clade is separated from the Oregon/Northern California clade by the crest of the Sierra Nevada. Genetic diversity was much higher in North America than Eurasia, indicating that colonization of Eurasia occurred from North America. Old World haplotypes in Alaska suggest a secondary dispersal to Alaska (Forister et al. 2004).

SIERRA NEVADA

There is a slowly increasing data set for the importance of glaciation processes in the evolution of alpine and subalpine environments. Alpine organisms often experience population contraction during glacial periods and dramatic population expansion during interglacial periods (DeChaine and Martin 2004). Sediment analyses from pluvial lakes on the eastern side of the Sierra Nevada substantiate fluctuations between wet and dry periods during the late Pleistocene and early Holocene (Benson et al. 2002). Neoendemics in general are expected to exhibit a variety of patterns due to isolation, colonization of extreme environments, and strong selection that can result in rapid speciation (Schoville et al. 2010). Species with low vagility will have deeper genetic breaks whether in alpine regions or along the coast.

Nebria (Carabidae) are alpine ground beetles and microhabitat specialists dependent on snow seep zones. *Nebria ingens* (219 individuals at 17 localities), *N. spatulata* (421 individuals from 33 localities), and *N. ovipennis* (196 individuals from 17 localities) from the Sierra Nevada were analyzed for 670 bp of mtDNA COI with some additional nuclear gene sequences for *N. ingens* (Schoville et al. 2012). *Nebria ingens* has the greatest elevational restriction, *N. spatulata* occurs at broader elevations, and *N. ovipennis* exists with both species. For all three species, 67–82 percent of the genetic variation is distributed among geographic groups with a strong north-south structure. All three species show genetic division at different levels across the San Joaquin River drainage and evidence of population expansion near the end of the LGM to the present.

Grylloblatta is an endemic California genus of low vagility alpine ice-crawlers that evolved from northeastern Asia and requires consistently near-freezing temperatures (Jarvis and Whiting 2006). The genus has been present in California since the Pliocene 4.53–3.9 Ma and experienced dramatic changes in climate during Pliocene and Pleistocene glaciations events. The common ancestor of the genus occurred across western North America and diversified into the northern Cascades–Rocky Mountains and California–southern Oregon clades. The California clade is thought to have dispersed from the Cascade Range, but genetic evidence indicates there were likely a number of dispersal events. Samples from 132 individuals from 30 locations, sequences of COII, *h3*, 18S, 28S (5542 bp total) and Bayesian and maximum likelihood analyses further divulge major lineages in northeastern California, the northern Sierra Nevada, and the southern Sierra Nevada (Schoville and Roderick 2010). The northern Sierra Nevada lineage diverged 1.94 Ma from the common ancestor with subclades in the Carson/Sonora Pass, Tioga Crest, and Lake Tahoe areas. The southern Sierra Nevada clade diverged from the monophyletic Sierra Nevada clade 2.52 Ma with subclades in the southwestern Sierra Nevada and central Sierra Nevada. The monophyletic northeastern California clade is composed of two taxa, *G. chandleri*, which diversified 240 ka, and *G. gurneyi*, which diversified 140 ka. White Mountain populations became isolated from central Sierra Nevada populations with a median estimate of the T_{MRCA} of about 189 ka (Schoville and Roderick 2010). Diversification patterns in *Grylloblatta* species reflect the specific habitat requirements and low dispersal ability of the taxa.

The Sierra Nevada had limited glacial activity in the LGM 31–15 ka and into the Holocene until about 3.25 ka (Bowerman and Clark 2011).

Regardless, elevation and cooler climates would still have had dramatic effects on alpine taxa, as demonstrated by alpine insect taxa. Central Sierra stratigraphic analyses from near Tioga Pass establish that the Chironomidae was represented primarily by *Heterotrissocladius* (nonbiting midge) between 14.8 and 13.7 ka which were postglacial and very cold, with temperatures increasing 4.7°C during this period (Porinchu et al. 2003). Over this period, the Chironomidae increased in richness consistent with rapid changes in habitat heterogeneity. Extreme fluctuations in climate have the potential to create strong selective sweeps and rapid radiation in alpine species; for example, a "little ice age" in the Sierra Nevada in AD 1250–1900 that increased snowfall about 3–26 cm was preceded by a drought in AD 950–1250 (Kleppe et al. 2011).

Colias behrii (Sierra green sulfur, Pieridae) is an alpine and subalpine neoendemic butterfly from the Sierra Nevada that occurs at elevations > 2,700 m. Sequence data from mtDNA (960 bp COI, 239 individuals) and nucDNA (subset of individuals for 450 bp of *WG* and 1100 bp of *EF1α* for all individuals), coalescent modeling, maximum likelihood, and Bayesian analyses dated the T_{MRCA} of *C. behrii* to 489 ka (Schoville et al. 2010). This age estimate is equivalent to the maximum divergence time of *C. behrii* from *C. meadii* from the Rocky Mountains as they share mitochondrial haplotypes. In general, genetic variation in *C. behrii* is very low but distinguishes northern, southern, and southwestern Sierra Nevadan groupings. There is a genetic discontinuity in this region near the headwaters of the San Joaquin River that is shared with *Nebria* and *Hydromantes platycephalus* (Mt. Lyell salamander, Plethodontidae) (Rovito 2010; Schoville et al. 2012). Inconsistent patterns among mtDNA and nucDNA data sets reflect incomplete lineage sorting as the mtDNA is shared with Rocky Mountain populations and the nucDNA reflects a common ancestor with Cascade Range populations. *Colias behrii* populations further analyzed using 14 microsatellite data, 1066 bp of *EF1α*, and Bayesian methods identify a bottleneck 531–281 ka during global cooling. Not surprisingly, demographic effects have a critical influence on genetic diversity in small populations of these alpine species that have limited gene flow (Schoville et al. 2012).

Greya politella (Prodoxidae) is a seed parasite and pollinator of members of the Saxifragaceae in western North America. Sequences of 648 bp of COI and AFLPs from 500 individuals from 57 localities throughout the range analyzed with Bayesian and maximum likelihood methods support four regional clades: northern Pacific Northwest, southern Oregon, southern Sierra Nevada, and the rest of California (Rich et al. 2008). A southern subclade in the Klamath Range and western Coast Ranges shows some evidence of secondary contact with the Klamath/Coast Ranges clade and the southern Oregon clade that is conjectured to have occurred about 18 ka. High levels of diversity are found in Sierra Nevada *G. politella* populations with genetic structure in the southern Sierra Nevada consistent with other taxa that have experienced significant geographic barriers along the Kern, Kaweah, and San Joaquin Rivers (Feldman and Spicer 2006).

Battus philenor hirsuta (Papilionidae) is a monophagous pipevine swallowtail, the northernmost representative of a Neotropical genus and a disjunct in California restricted to *Aristolochia californica* (Dutchman's pipe, Aristolochiaceae), itself a disjunct and endemic to the Central Valley and the foothills of the Coast Ranges and the Sierra Nevada. *Battus philenor* sequesters the toxic aristolochic acids from the plant, which renders it unpalatable to predators. Although there are a number of mimics in eastern North America, *B. p. hirsuta* has no known mimics. A sample of 422 bp of mtDNA from 74 individuals from throughout the North American range and NCA establishes that the California and Mexico population of pipevine swallowtail recently expanded into western North America (Fordyce and Nice 2003). Mexico contains the greatest number of *Battus philenor* taxa, and the butterflies likely colonized these and the California species after surviving Pleistocene glaciations.

COAST RANGES, INCLUDING THE TRANSVERSE RANGES/CHANNEL ISLANDS

The Transverse Ranges and Channel Island insect taxa reflect considerable variation in isolation and provide many examples of rapid radiation and speciation. Although there is evidence that the insect taxa *Hadrotes crassus* (rove beetle, Staphylinidae), *Hypocaccus lucidulus* (clown beetle, Histeridae), *Nyctoporis carinata* (darkling beetle, Tenebrionidae), and *Thinopinus pictus* (pictured rove beetle, Staphylinidae) disperse back and forth to the mainland (Chatzimanolis et al. 2010), the islands are home to over 100 endemic insect species (Rubinoff and Powell 2004). The first islands emerged and underwent major changes in orientation during the Miocene, other islands arose in the Pliocene or Pleistocene, and all were resubmerged about 500 ka. During the LGM, the northern islands were one landmass about 6 km from the mainland (Hall 2002; Jacobs et al. 2004).

The Channel Islands phylogeography of the darkling beetle, *Coelus pacificus* (Tenebrionidae), establishes this species as a multi-island endemic. All members of the genus *Coelus* live on sand dunes under plants; closely related taxa are *C. gracilis* from relictual sand dunes in the California Coast Ranges and *C. globosus* and *C. ciliatus*, which occur along the Pacific coast in sand dunes. Data from 235 specimens and 825 bp of mtDNA COI support *C. pacificus* as a monophyletic species consisting of two primary clades, one from the northern Channel Islands (San Miguel, Santa Rosa, Santa Cruz, and San Nicholas) and another from Santa Catalina and San Clemente, although each island had a unique haplotype (Chatzimanolis et al. 2010). High levels of haplotypic diversity on San Nicholas Island indicate either high rates of evolution or colonization from an adjacent island.

Argyrotaenia isolatissima (torix, Tortricidae) is endemic to Santa Barbara Island but has close relatives on other Channel Islands. Santa Barbara Island is the smallest of the Channel Islands and is a highly disturbed ecosystem dominated by invasive plant species. The only known host plant for *A. isolatissima* is the endemic *Hemizonia clementina* (island tarweed, Asteraceae). An analysis of 799 bp of mtDNA COI from *A. isolatissima* and its sister taxa, *A. franciscana* and *A. citrana*, both more widely distributed species with greater host plant options, suggest that the paraphyletic *A. franciscana* and *A. citrana* are closely related to the monophyletic *A. isolatissima* (Rubinoff and Powell 2004). Because Santa Barbara Island was fully inundated up to 10 ka, this species appears to have colonized early and rapidly evolved on 260 ha. Since the late nineteenth century *A. isolatissima* has maintained on increasingly small habitat fragments, now 25 ha or less.

In general, parthenogenic species are thought to colonize marginal habitats at higher latitudes and elevations and have a wider distribution; however, because of their limited ability for dispersal, they are ideal for testing phylogeographic hypotheses. *Geodercodes latipennis* is a small flightless broad-nosed weevil (Curculionidae) that reproduces from sexual and suspected parthenogenetic populations. Individuals of *G. latipennis* were collected to assess previously identified phylogeographic breaks in Southern California: a north-south break within the central Transverse Ranges, an east-west break in the Transverse Ranges between the Sierra Pelona and San Gabriel Mountains, and a break between the Tehachapi and Breckenridge Mountains (Calsbeek et al. 2003; Chatzimanolis and Caterino 2007; Phillipsen and Metcalf 2009; Feldman and Spicer 2006). The asexual lineage was tested for all biogeographic breaks; but due to distributional limitations, the sexual species was sampled only to test the north-south Transverse Ranges divergence. A comparison of sister asexual and sexual sister clades found that one asexual lineage spans 89.9 km and its sexual sister lineage spans 1.2 km. An asexual group from the northern Santa Lucia Mountains spans 16.6 km, whereas its sister lineage spans 38.3 km. Most interestingly, there is no support for phylogeographic structure in any of the sexual or asexual clades, spanning several geographic barriers (Polihronakis et al. 2010). Different patterns of range probably reflect multiple origins of parthenogenesis, with the more widely spread lineages being the oldest.

Nyctoporis carinata, the flightless darkling beetle (Tenebrionidae), has an east-west genetic break in the Transverse Ranges and isolation of southern Sierra Nevada populations consistent with a number of other species with a range of divergence times between these areas from 9.9 to 2 Ma (Segraves and Pellmyr 2001; Feldman and Spicer 2006; Polihronakis et al. 2010). Sequences of mtDNA COI and nucDNA *GFT* from over 100 individuals of *N. carinata* identify clades in the western Transverse Ranges, eastern Transverse Ranges, San Jacinto Mountains, Santa Lucia Mountains, southwestern Sierra Nevada, and Tehachapi Mountains. A scenario for a southern Sierra Nevada refugium with westward and southern migration into the Transverse Ranges is proposed for the genetic structure of *N. carinata*, although the mitochondrial and nuclear markers provide conflicting data for the Tehachapi Mountains populations and suggest secondary contact (Polihronakis and Caterino 2010).

Sepedophilus castaneus is a small, flightless, slowly dispersing beetle (Staphylinidae) that ranges from Southern California to Washington. In an effort to define finer phylogeographic patterns, sequences of 826 bp of mtDNA COI from 102 specimens from 25 localities in the Transverse Ranges and analyzed using Brooks Parsimony Analysis establish significant geographic and genetic structure, with a few minor exceptions (Chatzimanolis and Caterino 2007). Four major clades are delineated by the western Transverse Ranges, the northern and southern Santa Lucia, the easternmost Transverse Ranges, and the central Sierra Nevada. A deep divide exists between the Sierra Pelona and the San Gabriel Mountains, and within each side of the break there is some gene flow. Diversification for the phylogeography of *S. castaneus* is consistent with other species in the Transverse Ranges with three biogeographic areas, a western region (northwestern Transverse Ranges and Santa Ynez Mountains), a central region (central Transverse Ranges and Sierra Pelona), and an eastern region (San Gabriel, San Bernardino, and San Jacinto Mountains); however, no north-south break is detected for *S. castaneus*. The Sierra Pelona/San Gabriel divergence likely occurred 5–2.7 Ma based on a number of studies (Rodríguez-Robles et al. 1999; Feldman and Spicer 2006; Spinks and Shaffer 2005) and corresponds to a Pliocene embayment that now corresponds to the Santa Clara River drainage (Hall 2002).

Stenopelmatus "mahagani" (Stenopelmatidae), the Mahogany Jerusalem cricket, is a flightless, large, slow-moving endemic from coastal Southern California. The diversification of this species complex in California is estimated to have started around the Pleistocene, but there is not much known about the diversity of this group. Because they are underground and nocturnally active, their diversity is likely underestimated and there are as many as 50 putative species in California. This species has low vagility and is highly susceptible to habitat alterations; thus it is important to restore and preserve habitat for the maintenance of genetic diversity (Vandergast et al. 2007). A maximum pairwise divergence rate of 3.2 percent among 260 bp mtDNA COI haplotypes from 33 locations places the beginning of their diversification at about 1.6 Ma. Each population contains low levels of nucleotide diversity (π = 0–0.00985), but genetic structure is high with a Φ_{ST} of 0.756 (P≤0.0001) with genetic differentiation significantly correlated with geographic distance and habitat fragmentation. Bayesian analysis and rooting suggest the tentatively named *S.* "mahagani" originated in the Chino and Puente Hills, colonized the southern Santa Ana Mountains, and subsequently colonized the higher mountains and coastal areas. The weather in this area was cooler, and because this species is mesic adapted it expanded throughout this region in the Pleistocene. Model estimates indicate that urban fragmentation has increased the Φ_{ST} value by 0.087, on average (Vandergast et al. 2007).

DESERTS

Comparative phylogeographic study among *Yucca brevifolia*, two obligate pollinators (*Tegeticula antithetica* and *T. synthetica;* yucca moths, Prodoxidae) and two species of parasitizing yucca moths (*Prodoxus weethumpi* and *P. sordidus*, Prodoxidae), provides evidence for the effects of climate change on species distributions and genetic structure (Smith et al. 2011). GPS records were collected for > 5,000 *Y. brevifolia* individuals and sequence data from five noncoding regions of cpDNA (*trn*T-L spacer, *trn*L and *trn*L intron, *trn*L-F spacer and *clp*P). For the insects, DNA sequences from 1400 bp of mtDNA COI, 400 bp of *ND5*, and 490 bp of *EF1α* were collected. Coalescent-based DNA sequence analyses and distribution modeling of *Y. brevifolia* trees since the LGM indicate that demographic changes in all species were prior to climatic change during the Holocene (Smith et al. 2011). For the insect species some divergence is apparent, with average F_{ST} values of 0.12 and AMOVA values of 80–97 percent for variation within populations. The *Y. brevifolia* variation is quite different, with AMOVA values showing less than 13 percent of the variation within populations and high global F_{ST} values of 0.87 that strongly correlate with geographic distance (R^2 = 0.378; p<0.001). The differences in F_{ST} and AMOVA values between *Y. brevifolia* and their associated insects reflect differences in dispersal ability. Demographic and genetic analyses indicate that all of the species had concerted population growth in the late Pleistocene beginning about 0.2 Ma, were little affected by the LGM, and reached their current distribution about 50 ka.

Currently only three species of *Reticulitermes* (subterranean termites, Rhinotermitidae) are known from California, *R. hesperus, R. flavipes*, and *R. tibialis*. Evidence from 94 samples for 428 bp of mtDNA and 16S rRNA supports conjecture that the eastern distribution of *R. hesperus* is limited by the Sierra Nevada, the southwestern deserts, and the Cascade Range, with *R. tibialis* found only in and around the Mojave Desert (Tripodi et al. 2006). Although this sample is not broad, interesting anomalies and variation indicate there may be more diversity of this genus in California.

SUMMARY

California insects surveyed for their phylogeographic structure display a variety of patterns depending on their vagility, habitat requirements, and reproduction. Although insects have been present since the Paleozoic, they experienced significant extinction at the end of the Permian. Modern insect lineages expanded and migrated in the middle Mesozoic with the expansion of freshwater ecosystems and via associations with angiosperms during the Cretaceous. Later colonizations of western North America by insects were likely the result of repeated migrations across Beringia and from eastern North America. Molecular evidence from taxa in the Klamath-Siskiyou region, the southern Cascades, and the northern Sierra Nevada support migration from Asia. Endemism in the Sierra Nevada is surprisingly low at 0.9 percent and occurs primarily at higher elevations (Kimsey 1996); however, mountainous species, particularly alpine taxa, reflect the effects of Pleistocene climatic change on their genetic structure. Butterflies in particular have the capacity for long-range dispersal, but because they are vulnerable to environmental change they provide a good model for the study of phylogeographic change. Low-dispersing species have emerged as particularly useful in providing genetic signatures of past vicariant events.

8　Fishes

Fossil records of western North American contain primarily archaic fishes. Fossil fish are known from the late Triassic in the Hosselkus limestone formation in Shasta County (*Acrodus*, Hybodontidae) and in the Panache Hills of Fresno County from the late Cretaceous (*Enchodus*, Enchodontidae). Incomplete skeletons from the late Cretaceous are known from Alameda County and the Chico Formation in Butte County. Modern fishes do not appear in western North America until the late Paleocene and early Eocene, with diversification increasing in the Oligocene. Catostomidae (suckers) are present in Nevada and British Columbia by the Miocene, and ancestral fish fauna in the Great Basin about 17 Ma are probably the source of extant members of the Cataostomidae in California (Sigler and Sigler 1987). Since the beginning of the Paleogene, over 1,200 Chondrichthyes and 497 Osteichthyes fossil specimens have been described from California (ucmpdb.berkeley.edu).

By the Pleistocene eight modern families of freshwater fishes are present in California: Salmonidae, Ictaluridae (freshwater catfish), Cyprinidae, Gasterosteidae, Centrarchidae, Cottoidae (sculpins), Aphredoderidae (pirate perches), and Cyprinodontidae (Sigler and Sigler 1987). Currently there are 141 fish taxa native to California, 100 of which occur in waterways that drain into the Pacific Ocean and 60 percent of which are endemic to the state (Moyle 1976). Of the major fish families in California, by far the most widely studied genetically are the Salmonidae, Cyprinodontidae, and Cyprinidae.

The Pacific coast of California has experienced dynamic changes before and since the formation of terrestrial California, and the genetic structure of the marine fauna has been shaped by these events. The contemporary CCS begins around southern British Columbia moving south, ends at southern Baja California, and is part of the North Pacific Gyre, a large circular current in the northern Pacific (Map 8.1). The CCS continues to play an important role in shaping phylogeographic groupings. At present, the CCS is driven in the winter by the Aleutian Low, which pushes storms into Northern California, and in the summer the CCS is driven by the North Pacific High. Upwelling is driven by the northeasterly winds, which have a cooling effect, and in the summer create a negative temperature gradient that results in the fog common along the coast (Mann and Lazier 2006). Upwelling provides the plankton-rich basis for the coastal fisheries and avifauna and marine mammal populations. North of Cape Mendocino, downwelling occurs in the winter, and to the south, upwelling occurs during the winter (Checkley and Barth 2009).

The California Bight marks the beginning of the California Transition Zone. It is an important biogeographic boundary that starts at Point Conception (Santa Barbara County) and continues south to the area of San Diego. Phylogeographic studies of the Oregonian and Californian faunas place a barrier to gene flow in the middle of the Bight (Burton 1998; Dawson 2001). However, as a biogeographic break, the transition is more gradual than previously estimated (Horn et al. 2006). Although the CCS flows south, in the California Bight there are a lot of shoals and islands that moderate a northward flow nearshore. The Bight favors species with limited dispersal ability and low fecundity (Dawson 2001).

The northern Channel Islands experienced Miocene-Pliocene uplift, and by the Plio-Pleistocene, the Los Angeles area had transitioned from a marine to a terrestrial environment (Ingersoll and Rumelhart 1999). Climatic fluctuations resulted in extinctions of estuary biota from the

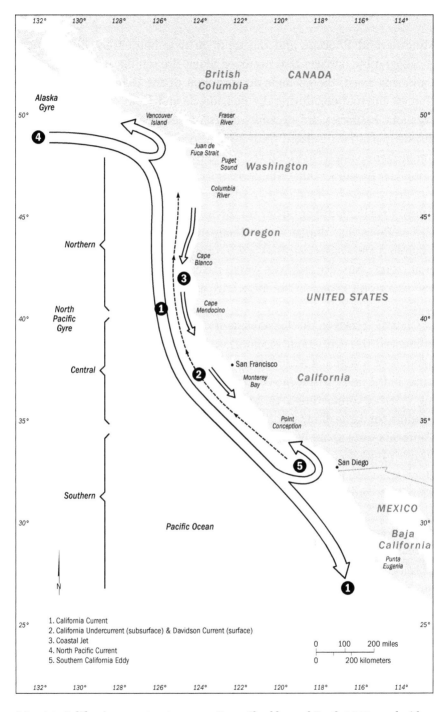

Map 8.1. California current system map. From Checkley and Barth 2009, used with permission from Elsevier, Ltd. Current begins in the North Pacific.

Miocene and Pliocene and the input of new biota from other areas. Biogeographic barriers that occurred along the middle Pliocene Central California coast, the geologic development of the California Bight, and climatic fluctuations during the Pleistocene and Holocene affected estuary biota. Estuaries and streams were colonized following glaciations and during glaciations isolated from others, resulting in the development of unique genotypes from Monterey Bay to Bodega Head for members of the Gobiidae (Barstow 2002). Similarly, isolation of resident fish species such as *Gila intermedia* (Gila chub) and *Cyprinodon macularius* (desert pupfishes, Cyprinodontidae) from the Bonneville and Lahontan systems, as well as those that migrated into them such as members of the Cottidae (sculpins) and *Oncorhynchus clarkii*, resulted in genetic isolation (Smith et al. 2002). Smaller estuaries would have been more frequently isolated, and as a result genetic differences have emerged for a number of species as discussed below.

Climatic cycles in the Pleistocene resulted in changes in sea level and influenced the extinction, recolonization, and new colonization of local habitats at a regional scale for marine, estuarine, and terrestrial organisms. In the late Pleistocene 0.7 Ma, Los Angeles was submerged at varying extents, connecting a variety of estuaries and perhaps providing a conduit for gene flow. Air and sea surface temperatures of Southern California were about 6–10°C cooler during Pleistocene glaciation events (Davis 1999). Refugia during the Pleistocene worked for some taxa like *Oncorhynchus*, but extinctions prior to and during the Pleistocene were also likely. During melting periods, increased water would allow riverine systems to be connected to the ocean and colonized by anadromous fishes like *Gasterosteus aculeatus*, which have residence in postglacial lakes and appear to have experienced many colonization events (Thompson et al. 1997). Fossil evidence reflects an increase in sea surface temperatures during the Pleistocene-Holocene that resulted in movement of the marine fauna toward the poles (Addicott 1969).

The Gulf of California has been in its present form since about 1 Ma. Before the Pleistocene, Baja California was a series of islands with a number of connections between the Gulf of California and the Pacific via seaways present from about 7.5 to 1.0 Ma (Murphy and Aguirre-Léon 2002). A Pliocene seaway across Baja California was a barrier to terrestrial taxa and provided an important link between the Gulf of California and the Pacific.

PRIMARY FRESHWATER FISHES

Primary freshwater fishes are represented by five clades in California, the Salmonidae, Cypriniidae, Catostomidae, Centrarchidae, and Embiotocidae, the largest of which is the Cypriniidae with 15 species (Schoenherr 1992) (Map 8.2). The Catostomidae are the second most commonly found fish in California, with fossil evidence from the Pleistocene (Savage 1951). They are closely related to and similar in their distribution mechanisms to the Cyprinidae and are well adapted to a number of habitats. Since it was drier in the Miocene, fauna adapted to warmer freshwaters were extirpated from Southern California. The San Joaquin River has only a few primary native fish, which include members of the Cyprinidae and Centrarchidae that had to persist through the remaining waters of the late Miocene and early Pliocene until it cooled again. During the Pleistocene, prior to European settlement in California, hundreds of diverse freshwater streams led to adaptations to local environments and resulted in the high levels of fish endemism found in California (Moyle 2002). The evolution of the recently evolved icthyofauna in the higher-order streams of California and their subsequent anthropogenic disturbances provide important lessons on how humans can recklessly reverse natural evolutionary processes.

Samples from 30 populations of *Oncorhynchus mykiss* (rainbow trout) from a number of California drainages, Nevada, and Idaho using 1055 bp of the mtDNA D-loop and 1566 bp of nucDNA in six single copy regions identify 10 different groups. Among-group variability is 35 percent and 65 percent for nuclear and mitochondrial variability, respectively (Bagley and Gall 1998). Based on a sequence divergence of 0.1–0.3 percent for single copy nuclear DNA per million years, *O. mykiss* and *O. clarkii* (cutthroat trout) diverged 6.1–2.3 Ma. Within *O. mykiss*, clade divergence ranged from 2.0 to 0.6 Ma consistent with Pleistocene divergence. MtDNA supports the origin of *Oncorhynchus mykiss aguabonita* (golden trout) from the Sacramento–San Joaquin Delta. Two separate colonization events established *O. m. aquabonita* populations in the Volcano Creek and Kern

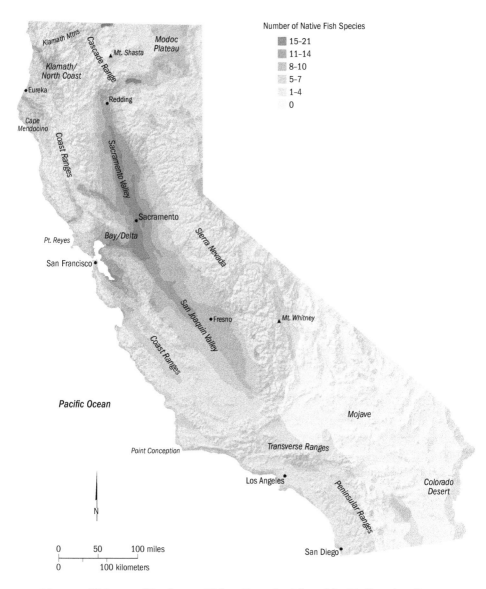

Number of Native Fish Species

- 15–21
- 11–14
- 8–10
- 5–7
- 1–4
- 0

Map 8.2. Richness of Freshwater Fishes. From the Atlas of the Biodiversity of California, used with permission from the California Department of Fish and Wildlife.

River of Kern County but show evidence of gene flow through secondary contact. Natal philopatry is a strong isolating mechanism in these species; however, postglacial contact and human introductions are indicated through some haplotypes. The Upper McCloud River (Siskiyou County) populations also are divergent from other populations (Map 8.3).

The evaluation of 1,559 *O. mykiss* individuals from 17 localities along tributaries to the San Francisco Bay Delta and 5 *O. mykiss* hatchery strains using 18 microsatellite loci reveals significant population structure among all population localities. Notably, Central Valley populations are closely related whether sampled below or above dams, which indicates that hatchery strains have not yet replaced native strains. However, above-dam populations are more similar to San Francisco Bay populations. Significant gene flow is reflected among tributaries to the Sacramento River, but, in contrast, above dams along the Kings River, a tributary to the San Joaquin River, there are two distinct genetic clusters in its tributaries (Garza and Pearse 2008).

The Cyprinidae is widespread throughout the world and is the largest family of freshwater fish in California. Cyprinids migrated to North America during the Miocene probably as the result of three or four separate introduction events. In the context of the mountain barriers in California, how the minnow family became widespread is probably best explained by headwater stream capture. Streams that are above grade erode backward into their beds; thus at the summit between two drainages, headwaters erode toward each other. If one stream cuts into the rock faster than the other, it is possible for the stream on one side to intercept the stream on the other side of the ridge, also intercepting the resident fishes. Another mechanism for Cyprinidae diversification occurred in drainages that were covered with water and eventually became isolated. The well-studied *Gila* spp. are differentiated throughout their range, indicating that they are a highly adaptable, plastic species. Many closely related species of *Gila* within the Colorado River system are divergent based on adaptations to various degrees of turbulence and water speed (Schoenherr 1992).

Lavinia (roach, Cyprinidae) is a widely distributed genus in a variety of stream habitats. All subspecies are listed as species of "special concern" (Aguilar and Jones 2009). Two very different clades for *Lavinia symmetricus* (California roach) exist in the Gualala and Pit Rivers, separated from

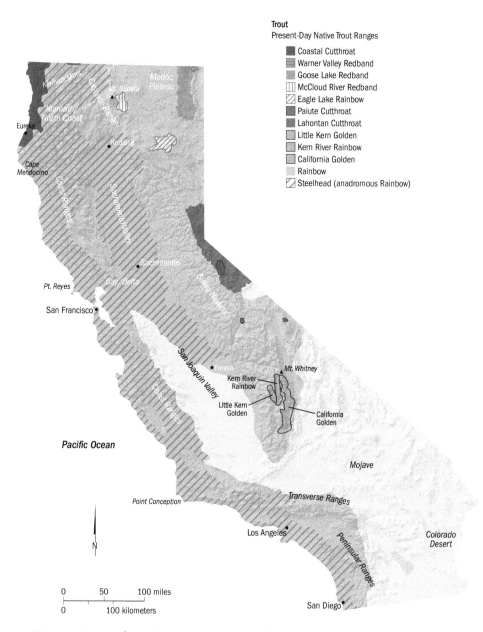

Map 8.3. Present day native trout ranges. From the *Atlas of the Biodiversity of California,* used with permission from the California Department of Fish and Wildlife.

other clades by 6–3 Ma with 12.1 percent corrected sequence divergence for mtDNA sequences. Prior to the Pliocene, the Pit River was connected to the Klamath River and not associated with Sacramento River as it is today (Dupré et al. 1991). Isolation of the Pit River drainage also may be evident in phylogeographic patterns in other taxa such as the *Catostomus microps* (Modoc sucker) and *Cottus pitensis* (Pit River sculpin) (Moyle 2002). Different clades of *Lavinia* are also found in the Navarro River, Tomales Bay, the Red Hills, and the Russian River–Clear Lake Basin. Gualala River individuals are separated from other mtDNA haplotypes by 4.9 Ma. Monterey Bay and Sacramento–San Joaquin Delta lineages are the same for both *L. symmetricus* and *L. exilicauda*, which probably resulted from hybridization, as the San Francisco and Monterey Bays were connected until as recently as 1.5 Ma (Dupré 1990). Unique haplotypes are found for both species in the upper Tuolumne River and southern San Joaquin populations.

Estimates for diversification of Great Basin and Death Valley Cyprinidae are based on mtDNA substitution rates of 1–2 percent per million years (Bermingham and Lessios 1993). Consistent with paleontological records, mtDNA substitution rates indicate that the ancestor to *Mylopharodon conocephalus* (hardhead) and *Ptychocheilus grandis* (Sacramento pike minnow) was introduced to North America from 4.7 to 2.4 Ma. Speciation between *Lavinia symmetricus* and the *L. exilicauda/Mylopharodon* clades was between 2.9 and 1.5 Ma and the *L. exilicauda/Mylopharodon* split between 0.5 and 0.25 Ma. Estimates for divergence vary with sampling error but are consistent with glaciation events (Redenbach and Taylor 1999).

SECONDARY FRESHWATER FISHES

The Salmonidae evolved for living in the cold, nutrient-poor waters of recently glaciated areas, and the Salmoninae was present in British Columbia by the Eocene, having evolved from Mediterranean ancestors (Stearley and Smith 1993). Glacial advances and retreats have resulted in transient and isolated habitats. Salmonids are opportunistic feeders, capable of moving through salt water and adaptable in behavior and life history patterns (Moyle 1976). The life history of *Oncorhynchus* spp. includes a coastal stream, maturation to adulthood in the ocean, and a long return to spawn in its natal stream. *Oncorhynchus* spp. have high natal philopatry that contributes to their high levels of genetic structuring and local adaptation in populations (Taylor 1991).

The genus *Oncorhynchus* is represented by about 18 taxa in California (Map 8.4). Most of the species remain in freshwater throughout their lives, but some species in the northeast Pacific are anadromous. Nuclear and mtDNA sequences indicate that North American salmonids diverged from Asian salmonids in the Pliocene and that *O. clarkii* and *O. mykiss* diverged in the late Miocene or early Pliocene (McKay et al. 1996; Brunelli et al. 2013). *Oncorhynchus* spp. originating in the Pliocene expanded from the sea into the mountain riverine systems during the Pliocene and Pleistocene. A drier climate during the Miocene may have restricted stream flow, which would have contributed to genetic isolation. High natal philopatry reinforces genetic isolation but is not exclusive, which allows for the colonization of new habitat (Milner et al. 2000). *Oncorhynchus mykiss* is endemic to the Pacific Ocean and coastal waterways of the eastern Pacific Coast and northeastern Asia. Steelhead is the anadromous form; the nonanadromous freshwater form is rainbow trout. Steelhead spend one to seven years in freshwater, migrate to the ocean for one to three years, and then spawn in freshwater. Populations are temporally separated in spring, summer, fall, and winter runs and differ in degree of anadromy, temperature, alkalinity tolerance, and morphological traits (Abadía-Cardoso et al. 2011).

The genus *Cyprinodon* includes about 20 known species that occur in the deserts or semideserts of North America. *Cyprinodon* species are unique in their ability to tolerate extremes in temperature, salinity, and dissolved oxygen (Schoenherr 1992). These species live in remnant populations in relic aquatic habitats of the Death Valley drainage system. The Colorado River has been speculated to be the source of ancestors of the Death Valley *Cyprinodon*, becoming separated from Death Valley about 1.0–0.1 Ma. During Quaternary pluvial periods, many *Cyprinodon* populations have been part of more complex fish communities in more mesic habitats formerly present in this region (Miller 1950). During more arid times, habitats became fragmented into marshes and springs, populations became isolated, and now nearly all species are allopatric (Miller 1981).

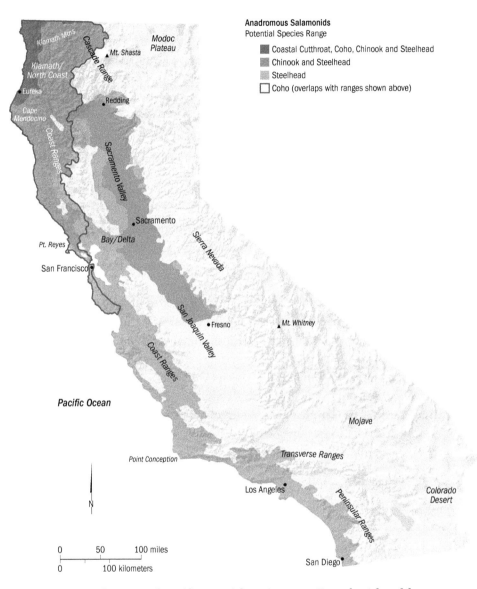

Map 8.4. Anadromous Salmonids potential species range. From the *Atlas of the Biodiversity of California*, used with permission from the California Department of Fish and Wildlife.

The Death Valley populations of *C. salinus* and *C. diabolis* have been isolated for at least 20 ka, and *C. nevadensis* has been isolated for 5–4 ka (Soltz and Naiman 1978). Five morphologically distinctive pupfish species, *Cyprinodon macularius* from the Colorado Basin, *C. radiosus* from Owens Valley, and *C. salinus, C. diabolis, C. nevadensis nevadensis, C. n. amargosae*, and *C. n. mionectes* from the Death Valley region, surveyed for genetic divergence, reveal very little allozyme differentiation (Turner 1974). In six remnant populations of *Cyprinodon macularius* from the Salton Sea in California, the Colorado River Delta in Baja California, and the Sonoyta Basin in Mexico, overall differentiation is low, with an average genetic similarity of 0.970 (Turner 1983). Thirty-two allozyme loci in the *C. nevadensis* complex exhibit a low level of divergence among these morphologically divergent species separated by large stretches of desert (Echelle and Echelle 1993a, 1993b).

An analysis of 334 bp of the mtDNA D-loop from 278 Death Valley *Cyprinodon* individuals from three species indicates that ancestral pupfish populations were likely quite large (Duvernell and Turner 1998). Present among-population variation indicates Death Valley pupfish are more closely related to the more southern species, *C. fontinalis*. *Cyprinodon radiosus* found in the Owens Valley is more closely related to *C. macularius* from the Colorado River region.

Members of the Cottidae are able to tolerate salt water, but they spend most of their time in freshwater. They are the largest group of euryhaline fish in California. This group has origins from the north and entered California by dispersal down the coast and from the northern Great Basin via the Pit River system and during eras of higher water (Schoenherr 1992).

MARINE FISHES

Along the California coast habitats are diverse, and there is an array of different species groupings; however, there are some broadly distributed species. Pelagic fish include *Trachurus symmetricus* (jack mackerel, Carangidae), *Engraulis mordax* (northern anchovy, Engraulidae), *Sardinops sagax* (Pacific sardine, Clupeidae), *Scomber japonicus* (mackerel, Scombridae), Xiphiidae and Istiophoridae (billfish), *Thunnus alalunga* (albacore tuna,

Scombridae), and *Merluccius productus* (hake, Merlucciidae) (Checkley and Barth 2009). A number of global phylogeographic studies on broadly distributed pelagic fish indicate that Indo-Pacific populations show little genetic structure (Díaz-Jaimes et al. 2010), in contrast to estuarine species, which often show strong genetic structure.

Members of the Sebastidae (rockfish) radiated during upwelling peaks of the late Miocene and subsequently, or initially depending on the taxon, in the Pleistocene (Jacobs et al. 2004). There are 66 species of *Sebastes* that occur along the Pacific coast from Alaska to Costa Rica, and these predatory fish reside in a variety of habitats, including rocky shores and kelp beds (Kendall and Gray 2000). They give live birth to thousands of larvae with mobility continuing until the juvenile stage, and the adults are sedentary (Love et al. 2002). Samples of 245 *Sebastes mystinus* (blue rockfish) individuals collected along the eastern Pacific coast at eight locations from Washington to California and sequenced for 498 bp of mtDNA CR expose phylogenetic divergence in the vicinity of Cape Mendocino (Cope 2004). Significant genetic structuring is present between the Washington-Oregon populations and the California populations; however, the Farallon Islands population is differentiated from the California populations and is more closely associated with the Washington-Oregon populations. Genetic structuring does coincide with the Cape Mendocino break for the California populations, but there is no structuring associated with Point Conception. The Cape Mendocino location is important because it is the area that has the maximum amount of upwelling in the CCS (Cope 2004). Cyt *b* sequences from 27 species of *Sebastes* from Monterey Bay suggest a divergence of 5 Ma when upwelling was at its highest in this region, but others support a radiation in the late Miocene about 9 Ma. Secondary radiation within the subgenus *Sebastomus* is estimated to have occurred in the Pleistocene (Rocha-Olivares et al. 1999). Given the degree of habitat specificity, it is not surprising that these nearshore and estuarine fish show more population substructuring than would be expected for offshore fishes (Cope 2004).

MtDNA CR sequences in *Embiotoca jacksoni* (black surfperch, Embiotocidae), a rocky shore inhabitant without dispersing larvae, show 1–2 percent sequence differences from populations in the Los Angeles region and more southern coastal locations when compared to the Channel

Islands and northern coastal populations (Bernardi 2000). The northern clade is further divided into a clade from Santa Catalina, San Clemente, and San Nicholas Islands and one from the northern Channel Islands north along the coast to San Francisco. Sequence divergence data suggest that these populations have been diverging for at least one million years; the subclades, hundreds of thousands of years. An exclusive Santa Catalina clade with deep divisions among haplotypes suggests it has been present in this area for many Pleistocene glacial cycles. These same Santa Catalina haplotypes are present on San Clemente, Santa Barbara, and San Nicholas Islands, however, which suggests recent colonization from a Pleistocene Channel Islands refugium. The phylogeographic break along the south coast just north of Los Angeles with the northern and Channel Island clade suggests there was also a Pleistocene refugium along the coast south of Los Angeles. These clades are separated contemporarily by the nonsuitable habitat of sandy beaches. Current patterns within the California Bight likely play a role in the separation of these clades (Bernardi 2000)

Phylogeographic data from *Eucyclogobius newberryi*, *Embiotoca jacksoni*, and *Tigriopus* (Copepoda) suggest strong divergence in the California Bight region but with difference in timing of geographic differentiation (Jacobs et al. 2004). *Eucyclogobius newberryi* lives along the California coast in the brackish, cool water of estuaries closed in the summer. The larva stage is limited to the estuary, but dispersal can occur infrequently via winter flooding (Barlow 2002). Strong genetic isolation for at least one million years is identified by a 4 percent sequence divergence in the mtDNA CR and cyt *b* data (Dawson et al. 2002). A strong phylogeographic boundary within the California Bight is explained by their residence in closed estuaries. During the late Quaternary, there was year-round flow from large rivers that drained the surrounding Los Angeles terrain, which when combined with occasional low sea levels, a steep continental shelf, rocky habitats, and strong wave action limited gene flow in the *E. newberryi* (Jacobs et al. 2004).

Two sister sympatric and competitive species of *Embiotoca newberryi*, *E. jacksoni* and *E. lateralis* (striped surfperch), have no planktonic dispersal stage. A low species-to-genus ratio and deep phylogenetic divergence of 11.8 percent measured by cyt *b* sequences suggest they radiated during the Miocene (Bernardi 2000). *Embiotoca jacksoni* occurs from Northern California to Baja California, and *E. lateralis* occurs from Alaska to Santa

Barbara and again from Punta Banda, Baja California, to Punta Cabras, Baja California. *Embiotoca jacksoni* has a strong genetic break in the region in which non-native *Morone saxatilis* (striped bass, Moronidae) is absent and also shows a phylogenetic break. *Embiotoca lateralis* is a benthic invertebrate substrate feeder but competes with *E. jacksoni* for crustaceans on algae. Based on mtDNA CR data analyzed via the neighborjoining method with Kimura's two-parameter distance, strong genetic structuring is present for both species into three major clades, the northern region, the Channel Islands, and the southern region (Bernardi 2005). Divergence distances between the northern and southern clades indicate a separation of about 328 ka for *E. jacksoni* and 150 ka for *E. lateralis*. There is some gene flow indicated between the northern and Channel Islands populations, with a F_{ST} value of 0.39 for *E. jacksoni* and 0.40 for *E. lateralis*. The Illinoian glaciation (310–128 ka) lowered sea levels and coincides with the *E. jacksoni* break south of Santa Barbara; their current presence there is likely due to recolonization and is preventing *E. lateralis* from recolonizing the area (Bernardi 2005).

Lampetra tridentatus (Pacific lamprey, Petromyzontidae) is anadromous with an ammocoete stage as long as five years in which they filter feed on sediments in freshwater (Potter 1980); it occurs throughout the North Pacific, including North America from the Aleutian Islands to Rio Santo Domingo, Baja California (Goodman et al. 2008). The predatory adult stage is in the Pacific Ocean, after which they return to freshwater waterways, spawn, and die (Beamish 1980). Samples from 1,246 ammocoete and adult individuals along 2,600 km of the Pacific coastline examined using restriction site analysis of mtDNA indicate high levels of gene flow among populations. AMOVA results reveal 99 percent of the variation is contained among locations (Goodman et al. 2008). Although the specific mechanism for high levels of gene flow has not been identified, it is conjectured to be due to passive oceanic dispersal.

Coryphaena hippurus (Dolphinfish, Coryphaenidae) is a pelagic cosmopolitan species that is highly migratory. Despite their high levels of migration, three phylogroups can be identified from an AMOVA of 493 individuals from 12 locations (Díaz-Jaimes et al. 2010). MtDNA sequences from 750 bp of *ND1* reveal that although there are very low interocean differences and most haplotypes are dispersed among all oceans, there is some east-west differentiation in the Pacific Ocean, with a Φ_{ST} of 0.002 (P = 0.664). The little genetic structure within this species can be explained by high rates of dispersal and large population sizes.

A nearshore resident that utilizes estuaries and the open coast for all life cycle stages, *Paralichthys californicus* (California halibut, Paralichthyidae), occurs from the Pacific coast of Washington south to Baja California. Evidence from 375 individuals from 14 locations sequenced for 681 bp of cyt *b* identifies a low overall nucleotide diversity (π = 0.0026 ± 0.0017) and a very low Φ_{ST} of 0.0030 (P = 0.22). There is no evidence of a phylogeographic break at or near the California Bight, and data support a single genetic population (Craig et al. 2011).

A cosmopolitan, bentho-pelagic resident, *Galeorhinus galeus*, the tope (Triakidae), becomes sexually mature at 8–10 years old with 6–54 pups per birth and a life span of 40 years (Olsen 1954). A global sample of 116 individuals sequenced for about 1000 bp of the mtDNA CR reveals a high Φ_{ST} of 0.84 (P<0.000001), which indicates there is little migration in this species (Chabot and Allen 2009). Interestingly, North American populations are more closely related to populations in the United Kingdom than South American populations, a pattern consistent with a thermal barrier established in the Pliocene.

The length of the larval stage in pelagic species has been long associated with genetic structure (Palumbi 1994). *Hypsopsetta guttulata* (diamond turbot, Pleuronectidae) occurs from Cape Mendocino to Baja California and again in the Gulf of California. Molecular data from 11 populations show that a long pelagic larval stage is more correlated with gene flow than the adult stage, which is limited to soft-bottom habitats and has little movement (Schinske et al. 2010). Genetic structure was identified among the Pacific coast sites, particularly north and south of Point Conception (F_{ST} =0.262), and between the Pacific coast and the Gulf of California.

Eastern Pacific species are ancestral in the genus *Syngnathus* (Syngnathidae), which is most speciose in Southern California. *Syngnathus leptorhynchus* (bay pipefish) is a nearshore resident that ranges from Prince William Sound, Alaska, to Bahia Santa Maria, Baja California. Six microsatellite loci, 579 bp of 16S rDNA, and 480 bp of the mtDNA CR reveal that southern populations in California separated from northern populations (Oregon, Washington, and Alaska) about 0.130 Ma (Wilson 2006). MtDNA haplotypes provide Φ_{ST} values of 0.488–0.627 (P<0.001) between California and the other populations. The specific evaluation of Point Conception as an isolating mechanism shows that although sampling was not dense enough to establish Point Conception as the isolating mechanism, there is genetic structure across this boundary.

There is no evidence of *Sardinops* (sardine, Clupeidae) fossils in California from 3 to 0.1 Ma, but they were present by 9–3 ka as evidenced by Native American middens and sediment cores (Casteel et al. 1975) and are now extant in temperate upwelling zones. Sequence data from a 500 bp segment of the mtDNA CR from *S. ocellatus caeruleus* show large-scale geographic subdivisions with divergence about 0.5 Ma between South Africa, Australia, Japan, Chile, and California. All five regions have high haplotypic diversity; however, the population structure of California populations of *S. ocellatus caeruleus* is low, with the closest populations to California off the coast of Chile (Bowen and Grant 1997).

SUMMARY

The California Transition zone is a long area of coast from near Point Conception to San Diego with specific phylogeographic breaks near Los Angeles and Monterey Bay (Dawson 2001). Point Conception provides a significant barrier to Oregonian and California faunas. Twenty-one species of Gastropoda, barnacles, and algae have their range limits at Point Conception (Doyle 1985). A later survey and summarization of 41 coastal marine animals for phylogeographic patterns found 24 species with phylogeographic structure below the species level (Dawson 2001). Of the 30 percent of taxa that are endemic to California, about half have a close taxonomic relationship with Oregon taxa (Briggs 1974).

Phylogeographic breaks are present along the coast for marine taxa, particularly those that are poor dispersers. Drier, more seasonal climate about 14–11 Ma led to development of estuaries that were seasonally closed and led to low nucleotide and high haplotype diversity that reflects a rapid expansion of range (Avise 2000). Life history factors are important in phylogeographic patterns for species with low dispersers having strong genetic signatures.

The habitats in which California's freshwater fish species occur include large rivers, estuaries, and desert pools and represent millions of years of evolution in a geologically complex environment. Prior to the European settlement of California, hundreds of distinct, undisturbed streams provided the basis for diverse metapopulations of fish species. The high levels of endemism of the icthyofauna of California combined with the high rate of species extinction means not only a loss of local biodiversity, but global biodiversity. Overfishing, introduced species, and water diversion have lowered water tables, have changed flow rates and turbidity, and mean doom for many of these species.

9 Amphibians

Amphibians were among the first vertebrates to colonize land in the Devonian, approximately 375 Ma, and diversified in the Carboniferous and Permian, subsequently declining with the evolution and increase of reptiles. The three major, extant clades are Anura (frogs), Caudata (salamanders), and Gymnophiona (Caecilians), which diversified prior to the breakup of Pangaea (San Mauro et al. 2005). The first Anuran fossil (*Prosalirus bitis*) known from North America is from early Jurassic deposits in Arizona, and by the Paleocene, Anuran fossils are common in western North America (Holman 2003). The earliest fossil Caudata in North America are from 25 Ma (van Frank 1955), but molecular clock data place them in North America by the Eocene (Wake 2006).

KLAMATH-SISKIYOU REGION

Members of the Salamandridae are sensitive to changes in temperature and moisture and thus have left strong genetic signatures in a number of cases. *Notophthalmus* (eastern North American newts) differentiated from *Taricha* by the middle Miocene based on fossil evidence and mtDNA distances (Estes

1981). Their common ancestor likely came over Beringia about 30–25 Ma and became separated by mountain uplift and rain shadows. Thirty-six individuals of *Taricha torosa* (California newt) sampled from 22 locations and sequenced for 375 bp of cyt *b* establish that within *Taricha*, species divergence occurred about 15 Ma, subspecies diverged about 9 Ma, and haplotype clusters within subspecies diverged 5–2 Ma. Southern California populations are basal and remain from remnant populations that migrated and diverged north (Tan and Wake 1995). An additional analysis of 778 bp of cyt *b* in *T. torosa* subspecies confirms a split of 13–7 Ma between *T. t. torosa* and subclades. *Taricha t. torosa* subclades are estimated to have diversified 3.4–2.6 Ma, with further divergence 2.3–1.9 Ma into coastal California, San Francisco and Monterey, and Southern California. The southern Sierra Nevada was colonized 1.7–1.4 Ma from Southern California *T. t. torosa* populations. Maximum likelihood estimates of allozyme data from 45 loci clustered similarly and support these divisions (Kuchta and Wake 2005). *Taricha t. sierrae* in the Sierra Nevada and southern Cascades have more population structure, likely driven by climatic fluctuations. Although there are a few anomalous populations, overall nucleotide diversity for *T. t. torosa* is 0.049 and for *T. t. sierrae*, 0.022 (Kuchta and Wake 2005).

The center of diversification for the Plethodontidae is believed to be in the southeastern United States. However, an important phylogeographic break has been identified for a number of taxa from the Klamath-Siskiyou region south to the Coast Ranges in the Clear Lake area (Lake County) (Rissler and Smith 2010). *Plethodon stormi* and *P. elongatus* (woodland salamanders, Plethodontidae) are the only endemic members of the genus in the Klamath-Siskiyou region of Northern California and southern Oregon, with *P. stormi* found only in the Siskiyou Mountains. *Plethodon elongatus* is found throughout southwestern Oregon and northwestern California and has three major clades, which are similar in divergence depth to the *P. stormi/P. elongatus* division. Sequences from 385 bp of cyt *b*, 679 bp of *ND4*, and 670 bp of the entire ATPase 6 gene for 98 individuals from 71 locations found average percent divergences range 7.67 to 9.49 between the two species (Mahoney 2004). Genetic divergences among the taxa are similar to other Pacific Northwest amphibians in the area such as *Ensatina*, *Taricha*, and *Rana* (Moritz et al. 1992; Tan and Wake 1995; Macey et al. 2001).

The Klamath River serves as a barrier for some populations of *P. elongatus* but not all. *Plethodon elongatus* variation is clinal for both mtDNA and morphology from coastal Northern California inland along the Klamath River, with coastal *P. elongatus* populations having an affinity with Oregon populations. The inland populations of *P. elongatus* are of the same mtDNA haplotype, with the exception of a Trinity River Basin haplotype that is more related to some coastal populations and some inland Klamath River populations. There is a zone of contact between *P. elongatus* and *P. stormi* along the upper Klamath River; however, despite some gene flow between the species, they retain morphological and genetic differences, suggesting recent contact (Mahoney 2004).

Lassen National Park (Shasta County) is the genetic break for Oregon *Ensatina eschscholtzii oregonensis* and the Sierra Nevada *E. e. platensis* (Jackman and Wake 1994). Samples from 385 individuals from 224 locations analyzed using Bayesian methods and compared to previous allozyme work establish the degree of divergence and secondary contact in these taxa (Stebbins 1949, Kuchta et al. 2009). The *E. e. oregonensis* clade from Northern California has subclades in the San Francisco Bay area and surrounding Coast Ranges, which are estimated to have split 8.9–5.1 Ma and precedes the development of the Coast Ranges. The coastal clade was present during later development of the Coast Ranges, 2–0.600 Ma. Despite their divergence, some hybridization does occur between these taxa when in sympatry.

Dicamptodon tenebrosus (coastal giant salamander, Dicamptodontidae) occurs from the coast of British Columbia south to the northwestern coast of California, in and near streams. Consistent with other taxa, coalescent modeling identified a southern refugium in the Klamath-Siskiyou Mountains, with postglacial expansion north in the early to middle Pleistocene (Steele and Storfer 2006).

SIERRA NEVADA

Hydromantes platycephalus, is a mid- to high-elevation species in the Sierra Nevada with a putatively distinct sister species, *H. brunus*, which occurs at low elevations along the Merced River. Peripatric speciation should result in paraphyly among closely related clades within a species at many loci (Harrison 1991), and the distribution of these species is appropriate for assessing whether their divergence resulted from peripatric or allopatric speciation. The hypothesis for peripatric speciation was tested and supported for these taxa using data from two mtDNA fragments and five nuclear loci from 89 individuals from 53 locations. A STEM coalescent analysis establishes that *H. brunus* is nested within *H. platycephalus* and arose within the latter species' northern range from 0.7 to 0.3 Ma, probably from a small number of individuals. In the Sierra Nevada, glaciations began about 2.5 Ma, and from 0.66 to 0.15 Ma major glaciations events lasted as long as 100,000 years. Estimates for the divergence between the northern and southern lineages of *H. platycephalus* are between 4 and 3 Ma (Rovito 2010).

The *Rana aurora/draytonii*, or red-legged frog complex (Ranidae), was examined using 400 bp of mtDNA cyt *b* sequences from 108 frogs from six taxa (Shaffer et al. 2004). The evaluation of species boundaries within this complex confirms that although closely related, these taxa are separate, with some overlap in Northern California. *Rana aurora* ssp. *draytonii* has been extirpated from its former range in the San Joaquin Valley and has only six remaining populations in the Sierra Nevada. Two major clades are recognized for *R. aurora*, *R. a. aurora* in California from Mendocino County north to British Columbia and *R. a. draytonii* from southern Mendocino County to Baja California. There is some overlap between subspecies in Elk Creek in Mendocino County in a 5 km area, but elevation to species status is supported based on deep divergence and reciprocal monophyly (Shaffer et al. 2004).

Rana boylii (foothill yellow-legged frog) (Lind et al. 2011) was formerly distributed from northern Baja California to southern Oregon from sea level to about 1,830 m in riparian habitat in the foothills of the Sierra Nevada and Coast Ranges (Stebbins 2003). It has undergone a major decline in the past century, especially in Southern California where it has been extirpated from much of its range. Data from 1026 bp of cyt *b* and 499 bp of ND2 were sequenced from 77 individuals from 34 locations and analyzed using Bayesian methods from seven major hydrological drainages. A F_{ST} value of 0.3996 (P = 0.000) indicates a significant amount of variation among the regions. Within the Sacramento–San Joaquin Basin, an AMOVA identifies 60.56 percent of the variation among river basins. Most populations fall within a major clade, but some populations at the periphery of the range in Oregon and in Southern California reflect deep divergence. In particular, populations south of the San Joaquin River in the Tulare Lake Basin are divergent from those in the north and are consistent with a number of other taxa. Three clades within the major clade indicate a break in Mendocino County, also consistent with other taxa.

Ambystoma macrodactylum sigillatum (southern long-toed salamander) occurs in high alpine habitat from southern Oregon to the Lake Tahoe region of the Sierra Nevada. To evaluate the importance of landscape on genetic structure, 1,142 individuals from 29 breeding sites were sampled in two watersheds for 18 microsatellite loci. The average F_{ST} across all populations is high at 0.27 and virtually the same for each watershed. No admixture is detected between the watersheds and supports a hypothesis for rare dispersal in Washington-Oregon that ultimately will affect the phylogeography of this and similar clades with comparable life history characteristics, particularly in extreme environments (Savage et al. 2010).

COAST RANGES/TRANSVERSE RANGES

Ambystoma californiense, the California tiger salamander (Ambystomatidae), is sister to other tiger salamanders and endemic to California's Central Valley from Solano and Sacramento Counties south to Tulare and San Luis Obispo Counties. This federally listed species is a fairly wide disperser after the larval stage, and the historical range is grassland and open-canopy oak woodlands. Samples from 696 individuals from 84 populations, sequenced for 852 bp of mtDNA *THR-DL1* and analyzed via maximum likelihood, maximum parsimony, and neighbor-joining methods, establish six major clades: southern San Joaquin Valley, Central Valley, central Coast Ranges, Bay Area, Santa Barbara County, and Sonoma County (Shaffer et al. 2004). The Santa Barbara and Sonoma clades have been isolated for 0.74 Ma and 0.92 Ma respectively. The orogeny of the Sierra Nevada provided the vicariant events separating *Ambystoma californiense* from other *Ambystoma* species with the late uplift of 5 Ma, which is used as the calibration point to separate *A. californiense* from other taxa. The San Joaquin

Valley–Pacific Ocean connection closed in the early Pliocene with Coast Range uplift and opened again in the late Pliocene with some estuarine environments around 5–2.5 Ma. These geologic changes isolated respective populations latitudinally in the Coast Ranges and further isolated the Santa Barbara clade around 2.5–2 Ma. Vernal pools, which historically formed a ring around the Central Valley and generally more so in the northern end of the valley, provided habitats for *A. californiense* that became isolated from each other about 1 Ma (Shaffer et al. 2004).

Batrachoseps (Plethodontidae), the slender salamander, has been present in western North America since the Eocene; species within the genus are well differentiated, and there is rarely any hybridization between species (Wake 2006). *Batrachoseps attenuatus* (California slender salamander) sequences from 178 individuals from 123 locations for mtDNA cyt *b*, 16S, *cox1*, *ND4*, and a nuclear gene (*RAG-1*) reveal a high correlation between geographic and genetic distance and recent demographic expansion as measured by Fu's F_S tests (Martínez-Solano et al. 2007). The minimum age for all species of *B. attenuatus* is 7.6 Ma, with two major biogeographic barriers identified in the Sacramento–San Joaquin Delta and the San Andreas fault region that correspond with Miocene/Pliocene and Pleistocene waves of divergence. Five major mtDNA clades with divergences of 5.5–9.5 percent have a high diversity of mtDNA haplotypes within each clade with each haplotype ranging no more than 35 km. NucDNA variation, however, does not correspond to the mtDNA haplotypes. These results suggest that *Batrachoseps* had historical distribution across the Central Valley and dispersed from a clade north of the Sacramento–San Joaquin Delta and into the Sierra Nevada, whereas southern populations were separated by the Monterey Embayment. Interestingly, there is a separation of clades that occurs west of the San Andreas fault in the Salinian Block and east on the Franciscan complex that likely corresponds to geologic events in this area. The Point Reyes Peninsula, on which this split occurs, has moved northwest 70–90 km over the past 6–8 million years and 180 km over a 60-million-year period. The peninsula has been above sea level for the past five million years (Burnham 2005).

Rana muscosa (Ranidae) populations are estimated to have declined more than 99 percent in their historical range, and the species is listed as federally endangered (Schoville et al. 2011). In the San Gabriel, San Bernardino, and San Jacinto Mountains, the species is limited to nine populations, with an additional population in the southern Sierra Nevada. MtDNA microsatellites establish that evolutionary lineages have been long separated, with divergences that correspond with glacial cycles. The Sierra Nevada population split from the main lineage 1.42 Ma, the San Gabriel population about 0.289 Ma, and the San Jacinto and San Bernardino populations about 47 ka (Schoville et al. 2011). Overall levels of genetic diversity in this species are very low compared to other species of *Rana*.

The Transverse Ranges are thought to be the center of origin for present-day *Pseudacris cadaverina* (stream-dwelling frogs, Hylidae), with northern, central, and southern groupings. A study of 221 individuals from 46 populations using 1100 bp sequences from cyt *b* and tRNA-GLU provides evidence that genetic diversity is distributed in accordance with geographic features in the eastern Transverse Ranges related to watersheds, mountain ranges, and aridity shaped by Pleistocene events (Phillipsen and Metcalf 2009) (Map 9.1). The three primary haplotypes are consistent with the genetic structure beginning in the Pleistocene, with currently limited gene flow in adjacent populations. The availability of appropriate mesic habitat would also have limited their occurrence in specific areas, but during wetter times some individuals did disperse, and divergence times of the major clades are about 0.6–0.259 Ma.

The ring species closure model for *Ensatina eschscholtzii* (Stebbins 1949) was examined using 385 individuals from throughout the range (Kuchta et al. 2009). A Bayesian analysis of cyt *b* sequences upholds the classical model of secondary contact and "closure" of the ring of seven subspecies, which predicted that peripatric speciation can occur around a central barrier such as the Central Valley. At the southern closure of the ring, *Ensatina eschscholtzii* and *E. klauberi* are in sympatry but have limited to no hybridization, whereas other subspecies that occur in sympatry in the Coast Ranges and the Sierra Nevada interbreed. Although with a wide confidence interval, the lower estimate of 8.0 Ma for the origin of the complex is in agreement with previous estimates and predates much of the formation of the Coast Ranges (Parks 2000). Three basal clades are now recognized with internal subclade structure: the

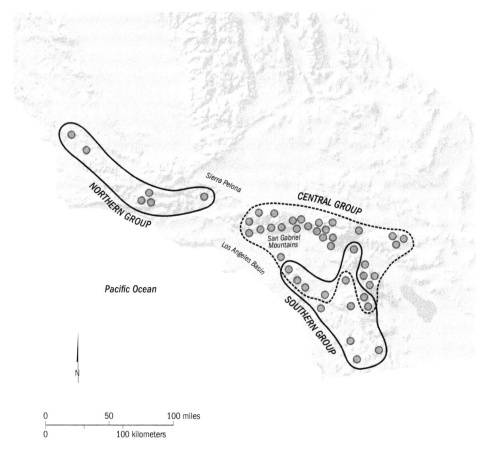

Map 9.1. Collection locations for *Pseudacris cadaverina*. Outlines show haplotype association groupings. Hypothesized phylogeographic breaks are indicated by named landscape features. Used with permission from Elsevier from Phillipsen and Metcalf 2009.

coastal clade, which primarily includes the San Francisco Bay Area, central coast, and Sierra Nevada foothills; the oregonensis clade, which is predominantly found along the northern coast of California; and clade "A," which occupies central Northern California to Washington. Inland clades are diverse and reflect an old evolutionary complex with significant geographic structure, particularly in the San Francisco Bay Area (Kuchta et al. 2009).

DESERTS

The *Anaxyrus boreas* species group (western toads, Bufonidae) was examined using 2225 bp of the 12S rDNA and the mtDNA CR from 288 individuals from 58 locations. The phylogeny is not reflected in current taxonomy but established three major clades in the Pacific Northwest, Great Basin, and southwestern deserts (Goebel et al. 2009). In California, both the southwestern desert and Pacific Northwest clades were present, with the southwestern clade identified as the tentative ancestor to the species complex. The Klamath-Siskiyou Mountains likely served as a refugium for populations that later spread to the Northwest.

Sampling of 100 individuals of *Pseudacris regilla* (Pacific tree frog, Hylidae) from throughout their range but primarily from the Baja Peninsula and analysis for 609 bp of cyt *b* identify three clades: Washington and Oregon; central California through eastern Oregon, Idaho, and Montana; and a southern group from Southern California to Baja California (Recuero et al. 2006). The Baja clade is divided into southern and northern clades corresponding with a mid-peninsula seaway, but incomplete sampling limits resolution in the other clades.

The late Pleistocene Santa Ana River fossil bed from about 40,980 ^{14}C ypb (Van Devender and Mead 1978) notably includes *Ensatina eschscholtzii* and *Batrachoseps* sp. among 9 families, 17 genera, and 12 species of amphibians and reptiles. This fossil bed represents the most diverse herpetofauna from the Southwest with the exception of Rancho La Brea. Plethodontidae members besides *Ensatina* include *Aneides lugbris* (arboreal salamander). Also present are Bufonidae (*Bufo boreas* western toads), Hylidae (*Pseudacris regilla*), and Ranidae (*Rana draytonii, R. boylii,* and *R. muscosa*), which have been found at other sites.

SUMMARY

Anurans were common in western North America by the Paleocene, and Caudates were likely present by the Eocene. Both lineages are sensitive to changes in temperature and moisture, and because dispersal events are rare due to habitat availability, they leave strong genetic signatures.

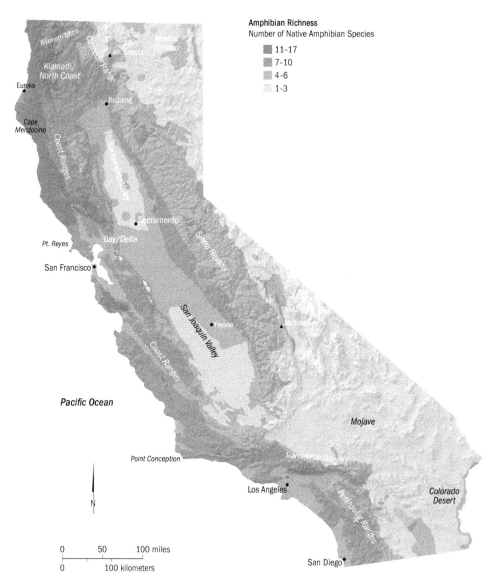

Map 9.2. Number of native amphibian species. From the *Atlas of the Biodiversity of California,* used with permission from the California Department of Fish and Wildlife.

Plethodon, Ensatina, Taricha, and *Rana* share similar Miocene divergence times in the Klamath-Siskiyou region and undoubtedly resided there as refugial populations during later glaciations in surrounding regions.

A number of taxa were present along the coast prior to the formation of the Coast Ranges, and the subsequent orogeny and associated embayments of the Coast Ranges had a strong influence on clade divergence. *Rana aurora draytonii* and *R. a. aurora* divergence in and around the San Francisco Bay is consistent with other species (Rodríguez-Robles et al. 1999, 2001; Maldonado et al. 2001). A number of northwestern species also have their southern limit in this area. *Ambystoma californiense* diverged from other *Ambystoma* species approximately 5 Ma. The Northern California divergence of *Ensatina eschscholtzii* in the San Francisco Bay Area and the Coast Ranges occurred 8.9–5.1 Ma.

Batrachoseps attenuatus clades showed evidence of divergence 7.6 Ma, as did subclades of *Taricha torosa* about 9 Ma. The uplift of the Coast Ranges, not surprisingly, influenced subclade structure in all these groups, as illustrated in *Taricha torosa* divergences of 5–2 Ma. Later Pleistocene waves of divergence were illustrated in *Batrachoseps attenuatus, Ambystoma californiense,* and *Hydromantes brunus/shastae* in the Coast Ranges, Sierra Nevada, and Northern California.

The Transverse Ranges breaks are important in many amphibian species and are also present in insects and reptiles and small mammals. The Santa Clara River Valley Embayment in the Pliocene is estimated to have existed about 3.3–1.6 Ma and likely served as an important biogeographic barrier in this region (Feldman and Spicer 2006). The Transverse Ranges population structure for *Pseudacris cadaverina* was shaped 0.6–.250 Ma.

As elsewhere in the world, the amphibians of California are under extreme threat of extinction due to habitat loss, invasive species, disease, and climate change. The extreme loss of mesic and wetland habitats throughout California but particularly in the Central Valley has isolated populations and removed opportunities for adaptive radiation and gene flow (Map 9.2).

10 Reptiles

Technically a paraphyletic clade, due to the placement of Chelonii, "Reptilia" is currently the preferred term that includes Anaspida (Chelonii or turtles, terrapins, and tortoises) and Eureptilia, which includes Aves, Crocodilia, Squamata (lizards and snakes), and Rhynchocephalia (1 species in New Zealand) (Modesto and Anderson 2004). The Squamata contains 61 families with 9,004 species. The Reptilian clade Iguania is nested within Squamata and forms the lineages Acrodonta, which originated in the Old World, and Pleurodonta, a predominantly New World clade. Pleurodonta is monophyletic. Twenty-nine nuclear protein–coding genes from 47 igauanian and 29 outgroup taxa analyzed via maximum likelihood and with BEAST identified most families within the clade as monophyletic in North America, including Crotaphytidae, Phrynosomatidae, and Iguaniidae. A crown root of 123 Ma for Iguania identified Pleurodonts from the Cretaceous Northern Hemisphere and indicated that they radiated rapidly as they expanded south (Townsend et al. 2011).

Squamates have been present in North America since the Cretaceous and were common in Paleogene/Neogene western North America (Holman 2000). Fossil evidence establishes the presence of Erycinae (rubber boas and close relatives) in North America 35 Ma (Kluge 1993),

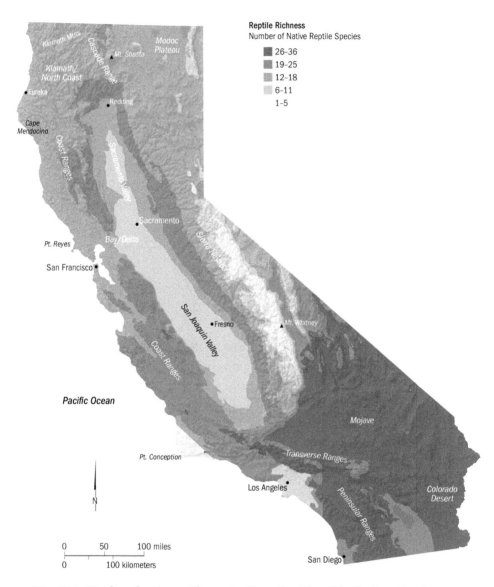

Map 10.1. Number of native reptile species. From the *Atlas of the Biodiversity of California,* used with permission from the California Department of Fish and Wildlife.

and molecular data indicate ancestral taxa to modern rubber boas evolved 29–10 Ma (Rodríguez-Robles et al. 2001). Viperidae originated in Africa and later diversified into the Crotalinae (pit vipers) during the Oligocene in Asia. Fossil Crotalinae were present in the New World by 22–19 Ma. *Crotalus* split from *Sistrurus,* with the earliest *Crotalus* fossils from the middle Miocene, and molecular data support that divergence as 12.7 Ma (Douglas et al. 2006).

The earliest members of Reptilia in California are represented by fossil marine reptiles in the Triassic formations of Shasta County (Hilton 2003). A number of reptilian fossils were described from the Cretaceous in Fresno County (ucmpdb.berkeley.edu). Fossil Chelonii have been found in the late Cretaceous marine rocks from a number of counties that were coastal during the Mesozoic. Virtually all the fossil Chelonii in California were marine dwellers in the clade Chelonioididea and members of the Dermochelyidae, Cheloniidae, and Toxochelyidae from the Chico Formation of Butte County of the early Cretaceous (Hilton 2003). The Dermochelyid turtles are most similar to the modern-day leatherback turtles and are the oldest Dermochelyids from the eastern Pacific (Parham and Stidham 1999). One possible exception to the marine nature of turtles in the California Mesozoic is the genus *Basilemys* (Nanhsiungchelyidae); however, this remains controversial (Hilton 2003).

Compared to other major clades, Squamata and Chelonii diversity in California is low, with 84 named taxa, only 5 of which are endemic; however, most taxa have low population sizes or reside in fragmented populations, and 42 percent are considered "special status" by the California Department of Fish and Wildlife (CDFG 2011). Most taxa (57/84) are found in the Mojave and Colorado Deserts (Map 10.1).

AREAS OUTSIDE THE MOJAVE AND COLORADO DESERTS

Charina bottae (rubber boa, Boidae) is widespread in western North America, occurring in grasslands, chaparral, woodlands, and forests. Samples from 38 individuals and 783 bp of mtDNA sequences identify the northern and southern clades *Charina bottae bottae and C. b. umbratica,* respectively, consistent with current classification. The northern clade is further divided into the Sierra Nevada and northwestern subclades, separated by Lassen National Park, whereas the southern clade shows little differentiation among the mtDNA haplotypes (Rodríguez-Robles et al. 2001). Mount Lassen was glaciated during the Pleistocene when the perennial snow line was estimated to have been about 2,000 m below the present one. Because Mount Lassen was and is volcanically active, there have been repeated cycles of extirpation and recolonization from both northern and southern source populations. An uncorrected percent sequence divergence between the southern and northern clades of *C. bottae* of 5.77 supports a separation time of 12.3–4.4 Ma. This geographic separation is also thought to be important in the Northern and Southern California clades of *Lampropeltis zonata* (California mountain kingsnake) and *Diadophis punctatus* (Rodríguez-Robles et al. 1999).

Thamnophis sirtalis (common garter snake, Colubridae) is a broadly distributed complex of 12 subspecies in northwestern North America. Sequences of 2217 bp from *ND2, ND4*, and cyt *b* analyzed via conservative maximum parsimony and maximum likelihood found that despite a lot of phenotypic variation there are three major clades. The Pacific Northwest coastal and intermountain clades and all California clades are phenotypically variable due to local adaptation, and the results establish that the current taxonomy is invalid (Janzen et al. 2002). Vicariant events and dispersal after glaciations were important in establishing the current genetic structure of *T. sirtalis*. In the California clade, a little more structure is present in the San Francisco Bay Area, but in all clades, decay indices are small, which reflects ongoing gene flow and recent population expansion.

Analysis of mtDNA from 50 populations of *Sceloporus occidentalis* (western fence lizard, Phrynosomatidae) identified four major clades, Marin County north to Washington (*S. o. biseriatus* and *S. o. taylori*), San Joaquin and southern Sierra Nevada (*S. o. longipes*), Southern California/ Great Basin (*S. o. bocourtii*), and Santa Cruz Island (*S. o. becki*). A lack of divergence between *S. o. longipes* and *S. o. bocourtii* indicates recent dispersal (Archie 2010). Samples of 181 individuals of *S. occidentalis* analyzed for sequence variation in 969 bp of *ND1* in the Merced and Tuolumne River drainages at a range of elevations substantiate two clades in the Yosemite National Park strongly differentiated by river drainages (62 percent between river basins) but not by elevation (7.3 percent between high and low elevations) (Leaché et al. 2010). These results suggest that the high-elevation subspecies *S. o. taylori* evolved twice in respective river basins via expansion up the Merced and Tuolumne River basins after the retreat of the last glaciation 20–15 Ma. A broad sample of *S. occidentalis* individuals from 150 populations in the Transverse Ranges, Coast Ranges, and the Los Angeles Basin sequenced for 1550 bp of mtDNA identified five major clades (Archie 2010). These clades are the western Transverse Ranges south of the Cuyama River; the Los Angeles Basin and San Diego; the southern Coast Ranges north of the Cuyama River drainage; Pine and Topatopa Mountains; and the Ridge Basin and Sierra Pelona. There are also a number of minor clades associated with geologic activity that began in the Miocene.

Lampropeltis zonata (California mountain kingsnake, Colubridae) has two clades, one in Southern California and another that extends from northern Baja California to the central coast, southern Sierra Nevada, San Francisco, and most of the Sierra Nevada. The currently classified seven subspecies were not supported by maximum parsimony and maximum likelihood analyses of 787 bp mtDNA (Rodríguez-Robles et al. 1999). The ancestor of *L. zonata* evolved 14.6–5.2 Ma during the cooler middle to late Miocene, and the northern and southern clades diverged 5.7–2.0 Ma when divided by inland seas.

The western pond turtle, *Actinemys marmorata* (Emydidae), is found along the Pacific coast from Washington to northern Baja California west of the Cascade Range and the Sierra Nevada. There are currently two named subspecies, one in Northern California and another in Southern California, but mtDNA analyses confirm the taxonomy is not consistent with their phylogeny (Spinks and Shaffer 2005; Spinks et al. 2010). Sequences from 3376 bp of mtDNA (*ND4*, fragment of *GAPDH*, *R35*) and nucDNA (*RELN*, *c-myc*, *TGFB2*, *HNF1α*) document four major clades: a coastal clade from San Luis Obispo to Oregon and Washington; a San Joaquin Valley clade east of the Coast Ranges, west of the Sierra Nevada, and from the San Francisco Bay Delta to the Tehachapi Mountains; a clade in the Transverse Ranges in Santa Barbara and Ventura Counties and the San Luis Obispo coast; and some members in Santa Barbara and south of the Transverse Ranges to northern Baja including the Mojave

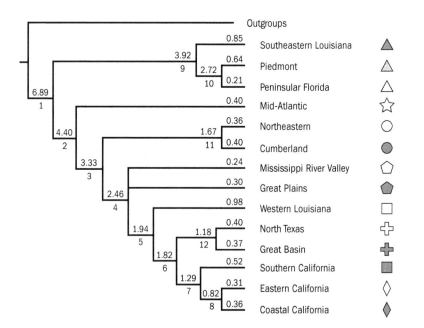

Figure 10.1. Diadophis punctatus population lineages in North America. Numbers above the branches are origin dates (mean) for each lineage. Used with permission from Elsevier from Fontanella et al. 2008.

Desert. The northern and southern clades are estimated to have split 8–2 Ma, with some interesting evolutionary anomalies between them. When the Central Valley Embayment subsided in the Pliocene and early Pleistocene, gene flow occurred from Northern California and the San Joaquin Valley as individuals migrated into the Coast Ranges around the northern San Joaquin Valley into the foothills of the southern Sierra Nevada and into the inner eastern Coast Ranges. Divergence times based on cyt *b* indicate that *Actinemys marmorata* may have started diverging 9–8 Ma, but divergence was more apparent by 4 Ma in the Monterey area that connected the Central Valley to the Pacific Ocean until the late Pliocene (Shaffer et al. 2004).

Diadophis punctatus is a complex of 14 subspecies across the continental United States sampled for 1117 bp of cyt *b* and 659 bp of *ND4* from 286 individuals and analyzed with coalescent and noncoalescent methods (Fontanella et al. 2008) (Figure 10.1). The California clade is well supported with three subclades, Southern California (π = 0.0170), Sierra Nevada/Cascade Range (π = 0.0078), and coastal California (π = 0.0044). Populations of *D. punctatus* reside in mesic habitats, they do not disperse well across arid lands, and their divergence times are concordant with other western North American reptiles among these regions.

DESERTS

Xantusia vigilis is a species complex of five species with very low vagility. An examination of all five species of *Xantusia* (desert night lizard, Xantusiidae) and seven subspecies of *X. vigilis* from 122 individuals and 615 bp of *ND4* and 307 bp of cyt *b* analyzed via Bayesian, maximum likelihood, and NCA methods substantiates high levels of genetic divergence among species (21.3 percent) and suggests that the genus was present before Wisconsin glaciations 110–10 ka (Sinclair et al. 2004). A sample of 500 *X. vigilis* individuals from 156 locations and cyt *b* and nucDNA sequences support current taxonomy in this complex and the San Jacinto and Yucca Valley clades as ancestral (Leavitt et al. 2007). The San Gorgonio Pass in the Peninsular Ranges is an important transition zone for a number of taxa as it was covered by the Gulf of California 6.5–6.3 Ma

(Dorsey et al. 2007). However, some people contend that vicariance between Mexico and Baja occurred 5.5–4 Ma, peaking at 3 Ma (Riddle et al. 2000). *Xantusia* started diversifying in southern Baja about 18.5 Ma, and Baja California began to split from Mexico about 13.2 Ma. Further diversification within the complex occurred when the Channel Islands split from the mainland in the middle Miocene with *X. riversiana* splitting from *X. vigilis*. *Xantusia vigilis* continued to diversify about 7.7 Ma with population expansion, and further divergence occurred with Pleistocene uplifts 1.5 Ma. These findings are consistent with the fossil record, which supports a late Miocene development for the biota of the desert Southwest.

Sauromalus spp. (chuckwallas, Iguanidae) are large herbivores native to southwestern North America that extend into the Mojave and Sonoran Deserts of California and thirty-one islands in the Gulf of California. They occur only among rock outcrops and have limited dispersal. Genetic, fossil, and geologic data indicate that *Sauromalus* spp. were widespread through the southwestern United States during the late Pleistocene but not before (Lamb et al. 1992). The genus is hypothesized to have originated in the Sonoran region of Mexico during the Pliocene (Hollingsworth 1998). MtDNA restriction site analysis of 51 *S. obesus* individuals from throughout the Colorado River Basin analyzed via UPGMA and PAUP identifies a strong local structure of clonal assemblages and indicates that subsequent to their arrival in the region the Colorado River has served as a significant barrier to dispersal. An evaluation of 902 bp of cyt *b* from three subspecies of *S. obesus*, two subspecies of *S. ater*, *S. hispidus*, *S. varius* and *S. australis,* using maximum-likelihood analysis found strong support for *S. ater* and *S. australis* but virtually no support for the subspecific structure within *S. obesus* (Petren and Case 1997). Further genetic and morphological analyses provided the basis for the inclusion of *S. obesus* in *S. ater* (Hollingsworth 1998). The gigantic size of *S. hispidus* and *S. varius* was found to have occurred prior to their divergence and in response to their insular environment in the Gulf of California.

The *Sceloporus magister* species complex (spiny lizards) contains six widespread and geographically variable species with four major clades. The genus is estimated to have emerged 25 Ma in the early Miocene of California, and based on fossil evidence they were widespread during the late Pleistocene (Robinson and Van Devender 1973; Cole and Van Devender 1976). A combination of relaxed phylogenetic analysis and divergence times of mtDNA and nucDNA for both coding and noncoding loci establishes that divergence within the S. *magister* group is estimated to have a posterior mean age of 6.23 Ma (Leaché and Mulcahy 2007). The transfer of Baja California from the North America Plate to the Pacific Plate began in the middle to late Miocene and was the precursor to a number of divergence events that further separated clades in the S. *magister* complex (Carreño and Helenes 2002). The sister taxon of S. *magister*, S. *orcutti*, has a posterior mean age of about 4.11 Ma, with a lot of admixture and gene flow between different desert regions along the Colorado River. The results are not in agreement with current subspecific classification.

Crotalus atrox (western diamondback rattlesnake, Viperidae) is widespread in arid and semiarid habitats from southeastern California across the desert Southwest east to the southern Plains and south through the Chihuahuan Desert and the Tamaulipan Plain of Mexico. The Gulf of California is important for separation of ancestral C. *atrox* mainland and Baja peninsula clades for this wide disperser. The desert Southwest is thought to be a formerly continuous expanse of desert ("Mojavia"; Axelrod 1958) from the Miocene through the early Pliocene, and as a result, a number of taxa have hypothesized dispersal from east to west. Fossil records establish that C. *atrox* spanned this area from 3.7 to 3.2 Ma; however, fossils are not present west of the Continental Divide until the late Pleistocene, and a split between eastern and western clades is present 1.36 Ma. Proposed Pleistocene refugial locations for the western clades include the southern Sonoran Desert and the lower Colorado River valley (Castoe et al. 2007). A minor suture zone, called the California Desert–Pacific Slope Suture Zone, between coastal vegetation and the Sonoran Desert is suggested to be a zone of secondary contact between lineages formerly separated by a larger Gulf of California during the Pliocene (Remington 1968; Devitt 2006).

An evaluation of *Crotalus* evolution in western North America using mtDNA ATPase 8 and 6 genes shows a distribution concordant with a Baja California/Mexico split for the widespread C. *cerastes* (sidewinder) and C. *mitchellii* (speckled rattlesnake) in the late Miocene to early Pliocene, with later Pleistocene effects supported by deep sequence divergence within each species (5.8–6.4 percent). *Crotalus mitchellii* and *C. tigris* (tiger rattlesnake) also diverged during the split of Baja California from Mexico. Shallow divergence exists between *C. tigris* (Sonoran Desert) and *C. ruber* (red diamond rattlesnake of southwestern California and all of Baja California), which have narrower distributions (Douglas et al. 2006).

Hypsiglena torquata (night snake, Dipsadinae) consists of a complex of 17 subspecies throughout the western United States and Mexico. Sequences of 800 bp of mtDNA *ND4* and two tRNA regions were analyzed from 178 individuals from 132 locations from throughout the geographic range with all subspecies and closely related taxa. Bayesian analyses reveal that the taxonomic treatments coincide with ring speciation around the Gulf of California (Mulcahy 2007).

The Bouse Embayment was a significant drainage and barrier to gene flow that eventually receded and formed the Colorado and Gila Rivers in the late Miocene. One hundred thirty-one individuals of *Charina trivirgata* (rosy boa, Boidae) were sampled from the southwestern North America species range. Sequences from 1105 bp of mtDNA and nucDNA analyzed via Bayesian analysis and NCA delineate three major clades (Wood et al. 2008). Two clades (A and B) are separated by the Colorado River, and the third clade (C) consists of disjuncts in Baja California, south central Arizona, and northern Mexico. Clades A and B developed in the late Miocene to early Pliocene as the desert continued to dry from 30 to 15 Ma (Van Devender and Spaulding 1979). Clade C split from the others about 7.4 Ma with the formation of the Gulf of California, and clade A split from clade B 4.8 Ma due to the Bouse Embayment/Colorado River. The phylogeography of *C. trivirgata* is not consistent with the current taxonomy.

Fifty-six individuals from 22 locations of *Gopherus agassizii* (gopher tortoises, Testudinidae) were sampled throughout their range in the southwestern United States and northwestern Mexico. Sister taxa of *G. agassizii* also sampled, and based on fossil evidence divergent since the Miocene, were *G. berlandieri* (north central Mexico), *G. flavomarginatus* (southern Texas and northeastern Mexico), and *G. polyphemus* (southeastern United States) (Bramble 1982). Restriction site polymorphisms of mtDNA examined via UPGMA and Wagner network analyses were generally concurrent with the fossil data and supported genetic separation between eastern and western North American *Gopherus* since the Miocene. Populations of *G. agassizii* west and north versus east and south of the Colorado River are also strongly divergent and coincide with a Pliocene divergence due to the Bouse Embayment (Lamb et al. 1989). A recent review of morphological, genetic, and ecological data names populations east and south of the Colorado River as a new species, *G. morafkai* (Murphy et al. 2011).

Phrynosoma mcallii (flat-tailed horned lizard, Phrynosomatidae) is endemic to the Salton Trough, a remnant of the Bouse Embayment. Samples of 781 bp of *ND4* and two tRNAs from 82 individuals were analyzed via NCA and compared to related taxa (Mulcahy et al. 2006). The Colorado River is confirmed to have provided a significant barrier to gene flow in this species during the Pleistocene as the river developed, with higher genetic diversity retained in the southeastern extent of the range.

Phrynosoma coronatum (coast horned lizard) occurs from the northern Central Valley to Baja California, although many Central Valley populations have been extirpated. An analysis of *P. coronatum* from combined climatic and sequence data from over 4500 bp from five gene regions from the mtDNA and nucDNA identifies five phylogeographic important lineages with deep coalescence. Three of these clades occur in California: northern Baja California/southwestern California from Ensenada, Mexico, to San Diego County; San Diego County north to the Los Angeles Basin, San Gabriel Mountains, and Mojave Desert; and Northern California from the Los Angeles Basin north along the Coast Ranges in the Central Valley (Leaché et al. 2009).

Trimorphodon biscutatus (western lyresnake, Colubridae) is widespread across the southwestern deserts of the United States. Divergence times estimated via a relaxed Bayesian molecular clock of mtDNA sequences in 91 individuals identify the division of the Nearctic and Neotropical biotas. Five clades in California include the Baja California clade and the Sonoran/Chihuahuan Desert clade that are sister to the western Mexico clade (Devitt 2006). The late Oligocene through middle Miocene marked the development of major biogeographic regions in the western United States and likely was the time *Trimorphodon* diverged from other genera. Most Neotropical taxa are at their northern limit in

Mexico, and consistent with this pattern, the Baja California clade diverged from the Sonoran/Chihuahuan/western Mexico clade about 6.5 Ma, and the Sonoran/Chihuahuan Desert clade diverged from western Mexico about 5.4 Ma (Devitt 2006).

Urosaurus (brush and tree lizards, Phrynosomatidae) are sister to *Sceloporus* and are widespread along the Pacific coast of Mexico, Baja, and North American deserts. In California, there are three species: *U. lahtelai* found primarily in Baja California, *U. graciosus* in the Mojave and Sonoran Deserts, and *U. ornatus* in California, primarily near the Arizona border. Twenty samples sequenced for 6639 bp from both mtDNA and nucDNA support a western Mexico origination for *Urosaurus*, with clade divergence developing when Baja California separated from mainland Mexico. A combination of vicariant and dispersal events resulted in further diversification (Feldman et al. 2011).

SUMMARY

The data suggest a Miocene/Pliocene split for some species of reptilians in California and a Pliocene/Pleistocene split for other species. Coalescent and noncoalescent measures reveal deep genetic divisions in the *Charina bottae* complex, *Contia tenuis* (sharp-tailed snake), *Diadophis punctatus*, *Elgaria multicarinata*, and *Lampropeltis zonata*, that correspond with vicariant events; however, each species shows different responses in terms of dispersal (Feldman and Spicer 2006). Important events in Squamata divergence in California include the formation of the Transverse Ranges, the Central Valley–Pacific connection at Monterey Bay (late Pliocene/ mid-Pleistocene), the San Francisco Bay Delta (mid-Pleistocene), southern Sierra Nevada glaciations (Pleistocene), and the Bouse Embayment. All species studied indicate a long-term stability in the southern range, which also has the highest level of endemism. The Pleistocene development of the Colorado River provided a significant barrier to gene flow, demonstrated in *Gopherus agassizii*, *Phrynosoma mcallii*, and *Uma scoparia* (Mojave fringe-toed lizard) but not in *Dipsosaurus dorsalis* (desert iguana) and *Sauromalus ater* (chuckwalla); regardless, the latter two species retained higher diversity in the southeastern extent of the range.

11 Birds

Of the 233 families and over 10,000 species of birds and 914 breeding birds that occur in North America, California has 641 species, 2 of which have been extirpated and 15 of which are introduced. Thirty-three species or subspecies are listed by the state or federal government as threatened or endangered.

Conservative estimates of fossil and DNA sequence data indicate birds radiated in the late Cretaceous (Clarke et al. 2005; Slack et al. 2006). Independent lineages represented in western North America 95–85 Ma include *Hesperornis*, *Ichthyornis*, and members of the Enantiornithes (extinct primitive birds), but none survived the Cretaceous-Paleogene extinction (Clarke 2004; Longrich et al. 2011). Both *Ichthyornis* and *Hesperornis* have been found in the Chico Formation of Butte County. Shorebird and seabird lineages likely diverged about 74 Ma (Slack et al. 2006). During the Paleocene and Eocene, there was an "explosion" of bird diversity, with about 16 orders developing in temperate North America and Greenland and notably including Trochilidae (hummingbirds), Cathartidae (New World vultures), Vireonidae (vireos), Parulidae (New World warblers), Emberizidae (emberizids), Icteridae (icterids), Cardinalidae (cardinals and allies), Odontophoridae (New World Quails), and Tyrannidae (tyrant flycatchers) (Feduccia 1995, 2003). By the Miocene, most modern bird genera were present, but genetic distance and molecular clock data estimate Tyrannidae ancestors arose during the late Miocene or early Pliocene 9 Ma (Zink and Johnson 1984).

California currently contains avifauna that originated from six biogeographic realms (Mayr 1946):

1. Palearctic (Old World) taxa including both early colonizers such as Strigidae (true owls), Corvidae (crows), and later colonizers such as Tytonidae (barn owls);

2. Nearctic taxa that may have originated in North America such as Phasianidae (grouse), Troglodytidae (wrens), and Parulidae (New World warblers), although the later likely originated in Central America;

3. Holarctic taxa including some shorebirds;

4. Neotropical taxa including Trochilidae (hummingbirds) that originated in South America;

5. Pantropical taxa that likely originated in the Southern Hemisphere such as Tyrannidae (tyrant flycatchers); and

6. cosmopolitan taxa, those with widespread distributions, including Picidae (woodpeckers), Apodidae (swifts), and Accipitridae (hawks).

Pleistocene glaciations in the Nearctic biogeographic zone likely resulted in higher bird extinctions compared to the Neotropics (Lijtmaer et al. 2011). Twenty-two of 47 sister species sampled revealed a sequence divergence of 2.8 percent, which indicates they were present by the Pliocene; and 30 percent of Nearctic species sampled had divergence rates of 1 percent, indicating a Pleistocene speciation of about 0.7 Ma (Lijtmaer et al. 2011). Coalescence times among closely related taxa are 0.1–2.0 million years, indicating diversification probably happened in refugia over one or more glaciation cycles. Rapid expansion after glaciation would be reflected in little haplotype divergence among populations.

There are both resident and migratory bird species in the CCS that show differences in the genetic structuring of populations. Residents include Alcidae (auks, puffins, murres) and Pelecaniformes (pelicans and allies), and migratory bird species include taxa from the Procellariiformes (tube-nosed swimmers) (Jacobs et al. 2004). Members of the Alcidae are associated with nutrient-rich, cold water conditions. These and other nonairborne fliers that wing-swim evolved where prey was plentiful, in nutrient-rich upwelling cold water (Day and Nigro 2000; Jacobs et al. 2004). Southern California has evidence of 11 fossil taxa of alcids, which indicates a rapid radiation 13–7 Ma (Warheit 1992). Migratory birds travel extensive distances but still show genetic structure based on breeding grounds and wintering areas. Intuitively, migratory species should exhibit a more uniform genetic distribution throughout the range, as documented in *Artemisiospiza belli* (Sage Sparrow, Emberizidae) (Johnson and Marten 1992).

PASSERIFORMES

Passerine birds appear to have diversified in the early Paleogene, but their genetic structure was affected in varying degrees by Pleistocene glaciations dependent on their degree of philopatry and migrational patterns. Members of the Passeriformes vary in migratory patterns and include strictly nonmigratory to fully migratory species.

Pleistocene glaciations resulted in the isolation and differentiation of some taxa and for others, recent expansion. Phylogeographic breaks of 28–30°N with mtDNA distances of 3.5 percent have been identified for *Auriparus flaviceps* (Verdin, Remizidae), *Campylorhynchus brunnei-capillus* (Cactus Wren, Troglodytidae), *Toxostoma lecontei* (LeConte's Thrasher, Mimidae) consistent with a seaway across the Baja peninsula about 1 Ma (Zink et al. 2001). *Polioptila californica* (California Gnatcatcher, Polioptilidae) does not have such a break, suggesting that it resided in a southern refugium with recent northward expansion to 34°N.

Interspecific haplotype divergences in *Parus* (chickadees, Paridae) of 3–7 percent suggest speciation in the Pleistocene occurred about 5.2–3.7 Ma. Mitochondrial DNA divergence rates of 2 percent occur between all geographically separated congeners and support postglacial expansions (Gill et al. 1993). More recent differentiation occurred in the late Pleistocene, and secondary contact after the last glaciations probably accounts for the different forms observed in zones of sympatry (Cicero and Johnson 1995).

Poecile gambeli (Mountain Chickadee, Paridae) is a highly philopatric, common, permanent resident with little migration in altitude during the winter (Spellman et al. 2007). Samples from 320 individuals from 31 locations sequenced for 694 bp of mtDNA *ND2* and analyzed with maximum parsimony and maximum likelihood identified two strongly supported clades in the Sierra Nevada/Cascades and Great Basin/Rocky Mountains that are estimated to have diverged 0.61–0.53 Ma. Within the Sierra Nevada/Cascade clade, the subclade that forms the northern populations has been isolated from the Transverse/Peninsular Ranges for 60,000 years. Recent population growth, range expansion, and some gene flow limit further interpretation of the subclades.

Aphelocoma californica californica (Western Scrub-jay, Corvidae) ranges year-round in much of western North America. A comparison of *A. c. californica* with other subspecies in North America, *A. c. woodhouseii* (central Mexico), *A. c. sumichrasti* (southern Mexico) and *A. insularis* (Island Scrub-jay) using 389 bp of mtDNA COI and analyzed with maximum likelihood and Bayesian methods identified two major clades of *A. californica*, one from Northern California and one from Southern California (Delaney et al. 2008). The *A. californica* clade split from its sister species *A. insularis* about 325 ka; however, the divergence time between *A. c. californica* and *A. c. woodhouseii* is estimated to be about 500 ka. Although there is evidence of hybridization between the two subspecies, it has been recommended that they should be elevated to species (Delaney et al. 2008). Hybridization between these taxa is consistent with a contact zone "hotspot" in which many species from southern Oregon and the Sierra Nevada hybridize (Hewitt 2000). Birds, in general, follow the east/west split of many other species, but because they more easily disperse, the hotspot is not as strong. Habitat and climatic differences correspond to the *A. c. woodhouseii/californica* split but also the split between the *Baeolophus inornatus* (Oak Titmouse, Paridae) and the *B. ridgwayi* (Juniper Titmouse) and the divergence of *Pica nuttalli* (Yellow-billed Magpie, Corvidae) and *P. hudsonia* (Black-billed Magpie) (Delaney et al. 2008).

Cardellina pusilla (Wilson's warbler, Parulidae) is a widespread migratory songbird in western and northern North America with a large breeding range and three geographic subspecies. In general, migratory songbirds have very little natal philopatry (<10 percent), suggesting that gene flow can occur long distance, but *C. pusilla* shows some evidence of historical isolation. Samples from 338 individuals were taken from a combination of breeding and overwintering sites and sequenced for 343 bp of COI and 316 bp of cyt *b*. Separation between western and eastern haplotypes suggests divergence during the LGM, with subsequent expansion northward followed by admixture (Kimura et al. 2002). Similarly, *Catharus ustulatus* (Swainson's Thrush, Turdidae) is a long-distance migratory species that breeds in the boreal forests of North America and the riparian woodlands of the Pacific coast and winters from southern Mexico to northern Argentina. MtDNA data used to test a hypothesis for expansion following the LGM and the geographic isolation of haplotypes provide the basis for some infraspecific divergence (Avise and Walker 1998; Lovette 2005). Current migratory routes differ for coastal and inland populations, with a net sequence divergence of 0.69 percent, and suggest that western populations migrated down the coast and the inland boreal forest populations migrated across North America during the LGM and then underwent range expansion (Ruegg and Smith 2002; Ruegg et al. 2006).

Samples from 127 *Agelaius phoeniceus* (Red-winged Blackbird, Icteridae) individuals reflect shallow diversification and recent expansion following a glacial retreat 18 ka, probably from a single population (Ball et al. 1988; Avise 2008b).

Mitochondrial DNA and AFLP analyses from 264 *Junco phaeonotus* (Yellow-eyed Junco, Emberizidae) individuals also support rapid diversification following rapid postglacial range expansions and recolonization to the north. *Junco phaeonotus* shows evidence of northward expansion from Mesoamerica following the LGM, with the eventual speciation of *J. myemalis* (Dark-eyed Junco), which subsequently diverged into five different geographic entities in the United States and Canada (Milá et al. 2007).

A nonmigratory resident of conifer forests from British Columbia to Southern California, primarily above 850 m, *Picoides albolarvatus* (White-headed Woodpecker, Picidae), consists of two subspecies, *P. a. albolarvatus* in the northern region of the range and *P. a. gravirostris*, which occurs in alpine habitat from the San Gabriel Mountains south to San Diego County. Seventy-eight individuals sampled from 24 locations and cyt *b* data indicate that there is a phylogeographic division within the species in

the Transverse Ranges. A Φ_{ST} value of 0.33 (P<0.001) among all populations and 0.27 (<0.001) among subspecies establish differences among the
northern and southern subspecies. NCA establishes that there is restricted
gene flow and allopatric fragmentation among populations. Of particular
note are the Warner Mountain populations, which, consistent with geologic data, place the split of those populations during the Pliocene
(Alexander and Burns 2006).

A widely distributed coastal resident from Oregon to Baja California in
chaparral and shrub habitats, *Chamaea fasciata* (Wrentit, Sylviidae) individuals travel less than 400 m from their natal range and spend their
adult life on 0.40–1.0 ha areas (Baker et al. 1995). Based on 1.6–2.0 percent divergence per million years for mtDNA for birds, *C. fasciata* evolved
from ancestors 8.1–6.5 Ma (Fleischer et al. 1998). The species is suggested
to be a southern refugee from the Pleistocene that experienced a recent
range expansion throughout the Holocene as it gradually migrated north
and northwest along the coast and northeast through the Sierra Nevada
(Burns and Barhoum 2006). Although five subspecies are recognized,
their geographic structure is not reflected in their classification. Currently,
these taxa have a ring distribution that includes the Klamath Mountains,
the Coast Ranges, the Sierra Nevada, and the Peninsular Ranges excluding the Central Valley. A sample of 1143 bp of cyt *b*, 71 bp of tRNA-lysine,
68 bp of COII, and 23 bp of COIII from 61 individuals in 20 populations
revealed that the San Gabriel and San Bernardino Mountains within the
Transverse Ranges may have contained two refuge populations (Burns
and Barhoum 2006).

A combined analysis tested statistical congruence for phylogeographic
patterns in *Chamaea fasciata*, *Toxostoma redivivum* (California Thrasher),
and *Picoides albolarvatus gravirostris* in western North America. All three
species showed intraspecific sequence divergences of less than one percent,
and molecular clock data support respective intraspecific differentiation
within the past one million years. The haplotype networks for *T. redivivum*
and *P. albolarvatus* are similar in geographic distribution and separate
Southern and Northern California haplotypes just north of the Transverse
Ranges, although *P. albolarvatus* occurs at higher elevations. Similar to *C.
fasciata* in life history traits and ecological and geographic distributions,
Toxostoma redivivum displays mtDNA haplotype distributions that also

indicate ancestry from southern refugia in the Transverse and Coast Ranges (Sgariglia and Burns 2003). However, the *T. redivivum* shows a north/south split around the Transverse Ranges, unlike *C. fasciata*, which shows a split within the Transverse Ranges (Burns et al. 2007) (Figure 11.1). NCA suggests that *C. fasciata* underwent range expansion, *T. redivivum* experienced allopatric separation, and *P. albolarvatus* underwent a number of long-distance colonization events that resulted in restricted gene flow.

Corvus corax (Common Raven, Corvidae) has an extensive range throughout the Northern Hemisphere. Sequences from 72 individuals for mtDNA COI and cyt *b* and microsatellite variation establish that the western clade is divergent from the Holarctic clade. Separation of the California and Chihuahuan Desert clades is well supported, and a cyt *b* sequence divergence of 4 percent suggests they split 2 Ma (Omland et al. 2000)

A common resident associated with the *Pinus ponderosa* forests of western North America, *Sitta pygmaea* (Pygmy Nuthatch, Sittidae) samples from 202 individuals from 20 populations for 1041 bp of the *ND2* analyzed with maximum-likelihood suggest that they maintain large populations for long periods of time. Nonparametric bootstrapping using estimates of N_e identifies the largest estimates of nucleotide diversity in western and Southern California (Spellman and Klicka 2006). Coalescent simulations identify one primary refugium for *S. pygmaea* during the Pleistocene, with subsequent spread since the last glacial maximum into its current range.

ANSERIFORMES

All members of the Anseriformes share adaptations to an aquatic habitat, and there are all levels of migration within the clade. The earliest confirmed member of the clade is the genus *Vegavis* present in late Cretaceous Antarctica (Clark et al. 2005). Molecular analyses confirm the Anseriformes as monophyletic, as is the Anatidae (Livezey and Zusi 2007).

Recent speciation between *Anas rubripes* (American black duck, Anatidae) and *Anas platyrhynchos* (Mallard) is estimated to have occurred about 400 ka, placing it within the last Pleistocene glacial episode (Avise

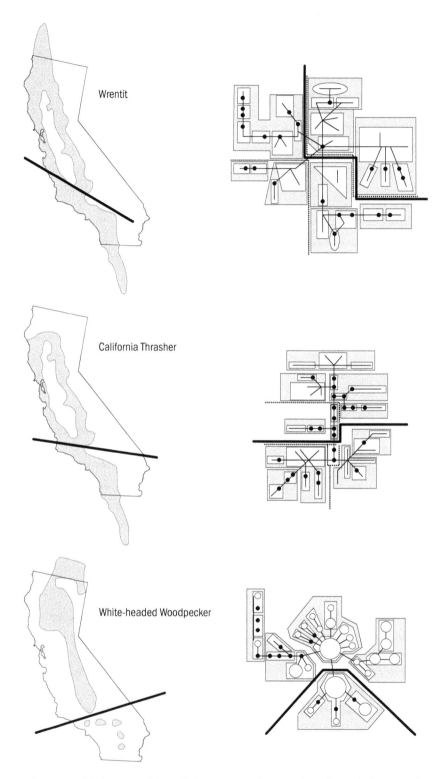

Figure 11.1. Phylogeographic splits between northern and southern clades around the Transverse Ranges consistent with other taxa. Used with permission from Kevin Burns with data from Burns et al. 2007.

et al. 1990). *Chen caerulescens* (Snow Goose, Anatidae) is a migratory species with two phases, blue and white, that share breeding grounds but do not share wintering grounds. MtDNA distance analyses indicate snow geese arose before or during the early Pleistocene, but the two phases are genetically indiscriminate (Avise et al. 1992).

CHARADRIIFORMES

The Charadriiformes is broad and contains widely distributed families (Robbins 1983). *Larus californicus* (California Gull, Laridae) is thought to have split from ancestral populations during rapid speciation caused by Pleistocene events, suggesting that evolution of the gulls occurred well before the Pleistocene.

GALLIFORMES

The examination of distinct populations in *Callipepla californica* (California Quail, Odontophoridae) establishes that there are low levels of local differentiation and suggests that extended periods of geographic isolation do not characterize this species (Zink et al. 1987). The distribution of *C. californica* is restricted to California (including Santa Catalina Island), Baja California, and southern Oregon. However, congenerics surround their distribution and hybridization is possible if contact occurs (Robbins 1983; Zink et al. 1987). Genetic evidence does not support hybridization between congeners in Mexican populations, and because Baja California populations are genetically similar to Northern California populations, it is proposed that radiation occurred from north to south into Baja California. *Callipepla californica* are thought to have first appeared on Catalina Island 12 ka.

Dendragapus fuliginosus (Sooty Grouse, Phasianidae) is found along the Pacific Coast Ranges and Sierra Nevada in highland areas, and *D. obscurus* (Dusky Grouse) is found in the Great Basin and Rocky Mountains. These taxa were originally described as separate species but were recombined in the twentieth century based on molecular data that indicate that the Sierra Nevada and Great Basin clades split about 30 ka (Barrowclough et al. 2004). A G_{ST} value of 0.66 separates the haplotypes, and almost zero gene flow between the two taxa supports their separation.

STRIGIFORMES

Strix occidentalis (Spotted Owl, Strigidae) lives in mature forests of the western United States and consists of three subspecies. Samples from 93 individuals from 13 populations from throughout the range sequenced for 1105 bp of mtDNA from COI and COII reveal primary divisions. Current subspecies classifications are consistent with genetic divergence into three groups; the Cascades and northern Coast Ranges, the Sierra Nevada, and the southwestern United States and northern and central Mexico (Barrowclough et al. 1999).

An analysis of *Strix nebulosa nebulosa* (Great Gray Owl) using 30 microsatellites and 1938 bp of mtDNA found strong evidence for recent population bottlenecks and low N_e (Hull et al. 2010). *Strix n. nebulosa* is dependent on mature montane forests and occurs almost exclusively in the Cascade Mountains north of the Klamath Basin, with a disjunct population in the southern Sierra Nevada. In California, *S. n. nebulosa* is a state endangered raptor known statewide to have no more than 200 individuals (Rich 2000). Based on F_{ST} and Φ_{ST} values of 0.09–0.17 between the Sierra Nevada and Pacific Northwest populations and substitution rates, the Sierra Nevada/Pacific Northwest split is estimated to have arisen 26.7 ka in the vicinity of the Klamath Basin. The populations remained isolated until postrefugial migration (Hull et al. 2010; Haig et al. 2004). Overall, these data suggest a number of recent bottlenecks and founder events with no evidence of hybridization between the Sierra Nevada population and others.

FALCONIFORMES

The Falconiformes is a paraphyletic group that is most closely related to the Psittacopasserae (Passeriformes and Psittaciformes [parrots]) (Suh et al. 2011). The Falconidae remains within the Falconiformes, but the

Accipitriformes is now separated and consists of four clades, three of which occur in North America, the Acciptridae, Cathartidae (New World vultures), and Pandionidae (Osprey). The Accipitridae are cosmopolitan in distribution and include eight clades, among them Buteoninae (Buteos, eagles), Elaninae (kites), Aegypiinae (Old World vultures), and Accipitrinae (true hawks).

Buteo lineatus elegans (Red-shouldered Hawk, Accipitridae) is a non-migratory Pacific Coast species that occurs from Oregon south to Baja California. It is one of five subspecies, with the four others occurring in eastern North American forests. Samples from 53 *B. l. elegans* compared to 59 samples of the other subspecies and 21 microsatellite loci identify a significant excess of heterozygotes in *B. l. elegans*. This significant excess of heterozygotes is evidence of a recent bottleneck in western populations that is estimated to be recent and the result of the loss of over 95 percent of riparian habitat in California since European expansion. There is no detectable effect of Pleistocene or post-Pleistocene climatic changes on their population structure, which is not the case in eastern populations. The F_{ST} value measured by microsatellites between eastern and western populations is 0.18, and an additional sample of sequence data from 375 bp of mtDNA CO from 25 Pacific and 54 eastern individuals yields similar results with a $\Phi_{ST} = 0.11$ (P<0.001) (Hull et al. 2008).

Gymnogyps californianus (California Condor, Cathartidae) fossils are common in the California Pleistocene. An analysis of hundreds of specimens from Rancho La Brea assigned to the species *G. amplus* revealed no change in the past 35,000–9,000 ^{14}C ybp. The modern *G. californianus* is thought to have existed with the ancient form (Syverson and Prothero 2010).

SUMMARY

The avifauna of California initially emerged as the result of colonization from a number of different biogeographic areas and increased in diversity during the Paleogene. Upwelling during the Miocene was likely important in the radiation of the Alcidae and other shorebirds. Pleistocene glaciations and climatic changes had effects on phylogeographic patterns in many species, including Aves (Avise and Walker 1998; Hewitt 2004; Zink et al. 2004). Worldwide, it is estimated that there were 21,000 bird species at the beginning of the Pleistocene, but by the end of the Pleistocene, only 10,000 species remained. Boreal birds that had large distributions in glaciated North America diverged in the Pleistocene due to selection in different ecological habitats, and many species show evidence of refuge in Northern California (Weir and Schluter 2004). However, birds that had large ranges in sub-boreal or tropical environments diverged prior to the Pleistocene. There remain questions about whether Pleistocene glaciation cycles resulted in speciation rates higher than background rates (Zink and Klicka 2006).

Species that show evidence of retreat to southern refugia during glaciation events include *Agelaius phoeniceus* and *Junco phaeonotus*. Refugial populations of *Chamaea fasciata, Toxostoma redivivum*, and *Picoides albolarvatus gravirostris* were present in the Transverse Ranges region. Habitat and climatic differences correspond to divergence of *Aphelocoma c. woodhouseii* and *A. c. californica, Baeolophus inornatus* and *B. ridgwayi, and Pica nuttalli* and *P. hudsonia* between California and central Mexico (Delaney et al. 2008).

Migratory species in general have more uniform genetic distributions than nonmigratory species; however, their genetic structure is affected by breeding range. Species that show evidence of divergence during the LGM with subsequent admixture are found in *Cardellina pusilla* and *Catharus ustulatus*.

12 Mammals

The mammal fauna of the North American Cenozoic was influenced by several important events, both biological and geological. Periodic and extensive connections with Asia via the Bering Strait began as early as 40 Ma in the Eocene and continued episodically into the late Pleistocene, with peaks in the Miocene (21–18 Ma and 8.7 Ma) and Plio-Pleistocene (5–2 Ma) (Sutcliffe 1985; Zhanxiang 2003; Ge et al. 2012) (Figure 12.1). Three mammal ages are distinguished in the North American Plio-Pleistocene: the Blancan (4.75–1.9 Ma), the Irvingtonian (1.9–0.250 Ma), and the Rancholabrean (0.250–0.011 Ma).

At 3–2 Ma, during the Blancan, plate tectonic forces were responsible for establishing a land bridge between North and South America, and resulted in what is now known as the Great American Interchange (Novacek 1994; Benton 2005) (Figure 12.2). Land mammal families from South America that migrated to North America include Megatheriidae (extinct ground sloths), Megalonychidae (two-toed sloths), and Nothrotheriidae (ground sloths), Glyptodontidae and Pampatheriidae (extinct relatives of armadillos), Dasypodidae (extant armadillos), Myrmecophagidae and Cyclopedidae (anteaters), Erethizontidae (porcupines), Didelphidae (opossums), and Toxodontidae (extinct large-bodied ungulates), but the only taxa that

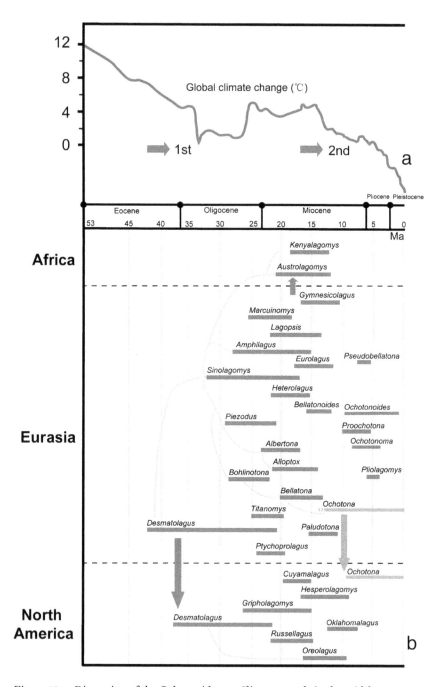

Figure 12.1. Dispersion of the Ochotonidae. a. Climate trends in the mid-late Cenozoic. b. Ochotonidae fossils; light gray = extant, dark gray = extinct. c-e. Ochotonidae fossils. f. Current distribution of *Ochotona*. Used with permission from Elsevier from Ge et al. 2012.

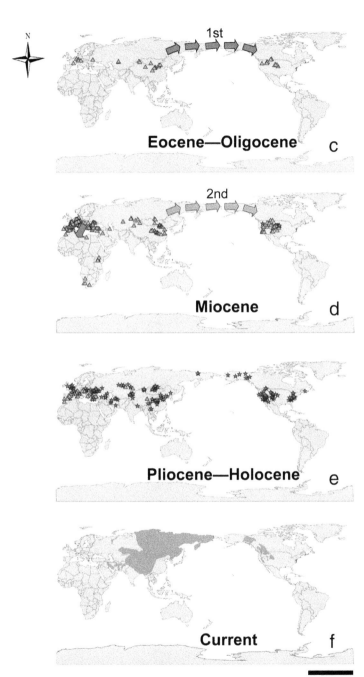

N

1st

Eocene—Oligocene c

2nd

Miocene d

Pliocene—Holocene e

Current f

4000 miles

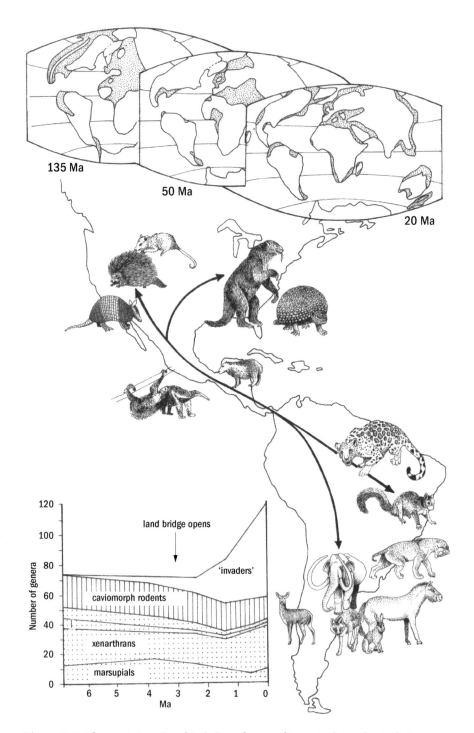

Figure 12.2. The great American biotic interchange of terrestrial vertebrates between North and South America. Used with permission from Wiley and Sons from Benton 2009.

survive in North America today are Erethizontidae, Didelphidae, and Dasypodidae.

The Irvingtonian age is named after a fossil fauna in the southeastern San Francisco Bay region from around 1.3 Ma. It includes many of the same representatives as the Blancan: extant mammal clades and large mammals that will become extinct at the end of the Rancholabrean. Although the clade radiated in North America during the Miocene and Pliocene, the fossil and ancestor to modern pronghorns, *Tetrameryx irvingtonensis,* is included in this fossil fauna.

The Rancholabrean fauna consisted of a very large and diverse group of animals represented by the clades Soricomorpha (Soricidae, Talpidae), Chiroptera (Vespertilionidae), Pilosa (Mylodontidae, Megatheriidae, Megalonychidae), Rodentia (Sciuridae, Geomyidae, Heteromyidae, Cricetidae), Lagomorpha (Leporidae), Perissodactyla (Tapiridae, Equidae), Artiodactyla (Camelidae, Tayassuidae, Antilocapridae, Cervidae, Bovidae), Carnivora (Mustelidae, Ursidae, Felidae, Canidae, Procyonidae, Pantheridae), Proboscidea (Mammutidae, Elephantidae), and Hystricomorpha—all variously from Asia and South America (Stock 1992).

Cordilleran ice sheet and alpine glacier fluctuations beginning 1 Ma resulted in the isolation of populations along the western coast from the main continental populations in the early to middle Pleistocene (Arbogast 1999). Species characteristics such as migratory ability, mating strategies, generation time, and birthrate and environmental parameters such as mating and foraging habitat are important in establishing genetic structure in mammal populations. Early protein and molecular studies provide evidence for phylogeographic discontinuity between coastal and interior lineages in California that was likely shaped by climatic change. These lineages include several genera of squirrels (*Glaucomys, Tamias, Spermophilus, Sciurus, Tamiasciurus*), *Odocoileus hemionus* (mule deer), and *Ursus americanus* (American black bear) (Arbogast 1999). By the end of the Wisconsin Quaternary glaciations (<10 ka), many large mammals became extinct throughout North America, including Mammutidae (mastodons), Elephantidae (elephants, mammoths), Megatheriidae, Megalonychidae, Nothrotheriidae, Camelidae, Equidae, Castoridae (giant beavers), *Bison latifrons* (giant bison), and *Ovibos moschatus* (musk-oxen) (Nilsson 1984). Native Americans probably contributed directly or indirectly to the extinction of the megafauna of North America, although recent findings implicate climatic factors as the cause of *O. moschatus* extinction (Campos et al. 2010).

UNGULATA/ARTIODACTYLA

Odocoileus (deer) probably migrated from Eurasia via Beringia in the late Miocene (5.7–4.2 Ma) and subsequently radiated from southern refugia in the Pleistocene (Gilbert et al. 2006). In contrast, *Cervus* (elk) probably endured north of Beringia and migrated to North America before and after the LGM (Flagstad and Roed 2003).

   Odocoileus hemionus californicus (California mule deer) and *O. h. columbianus* (Columbian black-tailed deer) exist throughout western North America from 23° to 60°N in a wide variety of habitats, and the two morphologically distinct types have a 6–7% mtDNA divergence (Carr and Hughes 1993). MtDNA CR and cyt *b* sequence data from 1,766 individuals in 70 populations throughout western North America support strong genetic differences between *O. h. californicus* and *O. h. columbianus* separated by the Cascade Range (Latch et al. 2009). *Odocoileus h. californicus* likely existed in multiple refugia during the LGM in southern areas free of ice, and *O. h. columbianus* existed in one northern refugium. In general, very high levels of genetic variability probably result in the mix of haplotypes and metapopulation dynamics (Leaché et al. 2009).

   In a detailed study of population structure of *Odocoileus hemionus* in California, 587 individuals were sampled from a wide array of habitats, from coast to alpine, in all ten bioregions (Pease et al. 2009). Fifteen microsatellites identify five different genetic groups. However, there is genetic overlap among the five clades, which indicates that their divergence is mediated by ecological factors. There is some evidence for a genetic split due to the Monterey Embayment during the Pleistocene, but not enough time has elapsed to have developed reciprocal monophyly between the clades. MaxEnt predictions for habitat suitability indicate that matrilineal relationships and habitat availability are important in genetic grouping. A coalescent analysis of 65 individuals for 624 bp of the mtDNA CR indicates subdivisions around the LGM for eastern versus central California populations of 27 ka, central versus southern populations of 13 ka, and eastern versus southern populations of 13 ka (Pease et al. 2009).

California has three subspecies of *Cervus elaphus*: *C. e. nannodes* (Tule elk) native to oak woodlands and grasslands of the Central Valley that were decimated during the mid-nineteenth century from 500,000 individuals to one remaining breeding pair discovered in 1874 and from which all extant members are descended; *C. e. roosevelti* (Roosevelt elk) from the northwestern California coast; and *C. e. nelsoni* from northeastern California and the Rocky Mountains. Mitochondrial evidence from 676 individuals and 16 microsatellites from northern California and surrounding states indicate that *C. e. nannodes* is more closely related to *C. e. nelsoni*. An AMOVA establishes 24.18 percent shared variation among subspecies, with 68.81 percent variation within populations. An exact test (P<0.0002) confirms that differences among subspecies are greater than variation among populations within subspecies. The F_{ST} value between *C. e. nannodes* and *C. e. roosevelti* is 0.30 (P = 0.05); between *C. e. nannodes* and *C. e. nelsoni*, 0.28 (P = 0.05); and between *C. e. roosevelti* and *C. e. nelsoni*, 0.14 (P = 0.05). Not surprisingly, *C. e. nannodes* has reduced genetic variation compared to the other subspecies. The hypothesis that *C. elaphus* colonized California after the LGM is consistent with their generally low levels of variation.

UNGULATA/PERISSODACTYLA

Pleistocene caballine horses, including the *Equus caballus* (domestic horse), are confirmed to be a single species divided into two major clades (Weinstock et al. 2005). One clade was widely distributed from central Europe across to North America both north and south of the ice sheets with much physical variability, but a second clade was present only in North America and likely disappeared by 5 ka. *Equus caballus* is found only in the former clade (Weinstock et al. 2005).

CHIROPTERA

There is little agreement on whether the Chiroptera is monophyletic. Regardless, there are two recognized clades, the Megachiroptera and

Microchiroptera, that likely evolved in the Eocene (Simmons et al. 2008). The oldest bat fossil in North America is *Onychonycteris finneyi* dated from about 52 Ma, but bats are not commonly found in the fossil record because of their light bones and habitat requirements. In California, bats are represented by three major clades, Phyllostomidae, Vespertilionidae, and Molossidae. *Desmodus rotundus* (vampire bat, Phyllostomidae) is known from Pleistocene California, but is now present only in tropical Central America and Mexico.

Myotis californicus (California myotis, Vespertilionidae) and *M. ciliolabrum* (western small-footed myotis) are found throughout western North America. *Myotis leibii* (eastern small-footed myotis) is found in the eastern United States and southeastern Canada. Although the three species are morphologically different, they are not genetically divergent, as reflected by sequences from 758 bp of cyt *b* (Rodriguez and Ammerman 2004). There is, however, strong bootstrap support for two monophyletic groups in a combined analysis of the three species, but *M. californicus* and *M. ciliolabrum* are not monophyletic. Within-species sequence divergences are 0.0–4.8 percent, and interspecific divergences are 2.1–2.5 percent, much less than the >10 percent divergence between other *Myotis* species. These values are consistent with early speciation and incomplete lineage sorting.

Migratory and nonmigratory populations of the *Tadarida brasiliensis mexicana* (Mexican free-tailed bat, Molossidae) in Texas and California are genetically similar and strongly support gene flow between the two groups (McCracken and Gassel 1997). However, generally very little is known about the phylogeography or population genetics of Chiropterans in California.

CARNIVORA

Mustelidae

A phylogeny of the Mustelidae using 12,000 bp from 22 gene segments using maximum parsimony, maximum likelihood, and Bayesian inference establishes that the Mustelidae originated in Eurasia and colonized North America multiple times (Koepfli et al. 2008). The root age of the

Mustelidae is 28.5 Ma, the crown age 24 Ma, with a 95 percent highest posterior density of 0.274–0.486, which suggests rate heterogeneity among clades. The first members of the Mustelidae to colonize North America in the late Miocene (11.2–5.3 Ma) via Beringia, substantiated with fossil evidence, include a number of now extinct "paleomustelids" (Baskin 1998; Koepfli et al. 2008). Miocene colonization coincides with a burst of diversification that includes extant members, followed by another burst of diversification during the Pliocene, but indications are that these diversification events occurred in the Old World and that subsequent dispersal of Mustelidae into the New World occurred in nine separate events. Extant *Lutra* and *Mustela* arrived in the late Miocene–early Pliocene, but it is unclear whether some traits were present before their arrival or after, and the number and nature of dispersal events remain elusive. *Lontra* species (New World river otters) were first found in North America 5.9–4.5 Ma and likely resulted in the diversification into modern *Lontra* and *Mustela/Neovison* (Tedford et al. 2004). Mustelidae endemic to California include *Lontra canadensis*, *Mustela nigripes*, and *Martes pennanti*.

Martes americana (American marten) colonized late-successional coniferous forests in California and Oregon likely from Pleistocene refugia in southeastern and southwestern North America (Slauson et al. 2009); the species was heavily hunted and trapped in the late nineteenth and early twentieth century (Zielinski et al. 2001). Currently, *M. a. sierrae* is distributed in the southern Cascades and the Sierra Nevada, and *M. a. humboldtensis* is distributed in the northern Coast Ranges. An analysis of 1,428 bp from mtDNA cyt *b*, tRNA-Pro, and partial mtDNA CR sequences from 26 individuals from Oregon and California identifies two distinct clades, one from the Oregon Cascades, coastal California, and Oregon and another from the California Cascades, the Sierra Nevada, and southeastern Alaska (Slauson et al. 2009). Although the phylogeography of *M. americana* would improve via additional sampling, it is evident that there is enough differentiation between the subspecies that they would benefit from being managed as separate populations.

The presumed range of *Martes pennanti* (fisher) is from Alaska south to western Oregon and Washington, parts of Idaho and Montana, the northwestern coast of California, and the Sierra Nevada in mid- to lower elevation late-successional forests with dense canopies. In California, the northern Sierra Nevada populations are presumed to have been extirpated due to habitat loss and hunting. A sample of 112 individuals from six populations from British Columbia to the southern Sierra Nevada found an overall F_{ST} of 0.42, with the greatest difference (F_{ST} = 0.60) between the Northern California coast and the southern Sierra Nevada. Two southern Sierra Nevada populations less than 100 km apart in contiguous forest have limited gene flow (N_m = 0.02); however, these populations are separated by the Kings River, an established barrier to gene flow. Genetic, paleontological, and archaeological data suggest that *M. pennanti* colonized California within the last five thousand years from eastern North America via British Columbia. Decreasing levels of heterozygosity and increasing population structure south in the Sierra Nevada support the hypothesis of a stepping-stone model of migration (Wisely et al. 2004).

Felidae

The felid-like clade first appeared about 35 Ma in the Oligocene in Eurasia. Ancestors of the extant *Felinae* arose in the Miocene (~9 Ma) in western Eurasia and migrated to North America 8.5–8.0 Ma, coinciding with additional Eurasian carnivore migrations. *Panthera onca* (jaguar) likely diverged from a common ancestor in Eurasia and migrated to North America in the Pleistocene, with fossils dating to 0.850 Ma; however, mtDNA estimates that extant *Panthera* arose 0.510–0.280 Ma (Eizirik et al. 2001). *Puma* originated in the late Miocene (8–5 Ma) from a common ancestor with *Acinonyx jubatus* (African cheetah) and *Puma yaguaroundi* (American jaguarundi) (Johnson and O'Brien 1997; Pecon-Slattery and O'Brien 1998).

Smilodon (saber-toothed cats), along with other large mammals, was present during the Pleistocene from 2.5 to 0.01 Ma. Three species are known successively in North America, *Megantereon* from 4.5–2.5 Ma and likely ancestral to *S. gracilis* from 2.5–0.5 Ma and *S. fatalis* from 1.6–0.01 Ma (Christiansen and Harris 2005). *Smilodon* likely preyed on large mammals, as evidenced by the approximately 2,000 individuals identified from the La Brea Tar Pits. Felids present in the La Brea Tar Pits are *Smilodon fatalis*, *Homotherium serum* (scimitar cat), *Panthera atrox* (American lion), *P. onca*, *Puma concolor* (cougar, mountain lion), and *Lynx rufus* (bobcat) (www.tarpits.org).

Genetic data from 315 specimens of *Puma concolor* from throughout North, Central, and South America and sequenced for three mtDNA gene regions and 10 microsatellite loci identified six phylogeographic groupings in the Americas, with only one grouping present in North America (Culver et al. 2000). The North American populations are more genetically homogeneous than those from Central and South America and appear to be derived from founders descended from Pleistocene ancestors that colonized eastern South America 0.200–0.300 Ma.

Microsatellite data from 12 loci for 431 *Puma concolor* individuals from throughout their habitat in California and northwestern Nevada analyzed via STRUCTURE identified most individuals as part of one of two major clusters either in southwestern California or on the northern coast (Ernest et al. 2003). Twenty-eight percent of the individuals from the central coast clustered with the southwestern cluster and none with the northern coast. There were no differences among F_{ST} and R_{ST} between males and females. F_{ST} values reflect strong gene flow (0.0–0.04) in Northern California between the Klamath-Siskiyou region and the Modoc Plateau, the northern Sierra Nevada and the Cascade Range, the northern and southern Sierra Nevada, the Modoc Plateau and the Sierra Nevada, and the eastern and western Sierra Nevada. The Tehachapi Mountains provide an important connection between the Sierra Nevada, the Peninsular Ranges, and the Central Coast as reflected by moderate gene flow (F_{ST} = 0.05–0.15) across the Central Valley between the central Coast Ranges and the southern Sierra Nevada, the northern Coast Ranges and the northern Sierra Nevada, and the southern Sierra Nevada and the Peninsular Ranges. The most divergent populations were between the northern Coast Ranges and the central Coast Ranges, which are separated by the San Francisco Bay Delta (F_{ST} = 0.26) (Ernest et al. 2003). Although habitat connectivity has been altered by human habitat disturbance and fragmentation, current population structure is likely strongly influenced by historical processes.

Canidae

Members of the Canidae are habitat generalists with longevity, great dispersal ability, and large unfragmented populations. They are expected to have high levels of heterozygosity, and in fact *Canis lupus* (gray wolf) and *Canis latrans* (coyote) do have high levels of heterozygosity across their range. In contrast, *Vulpes vulpes* (red fox) and *Vulpes velox* (swift fox) have reduced heterozygosity due to habitat fragmentation and reduced dispersal ability. Despite severe habitat losses, habitat fragmentation, and competition from *V. vulpes*, *Vulpes macrotis mutica* (San Joaquin kit fox) has been able to maintain genetic variation through dispersal (Schwartz et al. 2005).

Vulpes vulpes is indigenous to North America, Europe, Asia, and North Africa in a wide variety of habitats and evolved during the last glaciation from two refugia, Beringia and the conterminous United States. MtDNA sequence data from 220 *V. vulpes* specimens for cyt *b* (354 bp) and mtDNA D-Loop (342 bp) establish that Holarctic and Nearctic clades diverged about 0.4 Ma (Aubry et al. 2009). There are three clades within the Nearctic clade; two derived about 20 ka in the southwestern mountains and eastern North America, respectively, and an older, more widespread clade diverged about 45 ka. Initially, North America was colonized over Beringia, and populations migrated south during the Illinoian glaciation (0.3–0.130 Ma) and then returned north, forming the Nearctic clade. Other populations went south during the Wisconsin glaciation (0.11–0.01 Ma) (Péwé and Hopkins 1967) and gave rise to the North American part of the Holarctic clade. The montane lineage occurs in the Cascade Range and shares a disjunct population with other Pacific Northwest populations of boreal mammals. NCA and AMOVA analyses make it clear that the montane foxes were derived from the southern refugial populations about 17.3 ka. Northern refugial populations stayed in Alaska and western Canada, and the southern populations stayed in central and eastern Canada. The latter clade likely resided in a separate northern refugium that came in secondary contact with other North American *V. vulpes* populations as it would have become separated from Eurasian populations and expanded as the glaciers retreated. These populations would also have moved in elevation as the climate warmed. Prior to European settlements in California, *V. v. cascadensis* and *V. v. necator* resided in the Sierra Nevada and other western states. They now occupy low-elevation areas with human occupation and have been moved around (Aubry 1984). There is a present-day cline of morphological variation in North America that is due to secondary contact and admixture from the two lineages (Aubry et al. 2009).

During the late Wisconsin glaciation, the treeline dropped by 1,000 m in the northwestern United States (Barnosky et al. 1987). Southern refugia were more open, with *Picea* forests, and following glacial retreat, many animal species migrated north and still occur in subalpine habitats and some low-elevation habitats in Northern California (Jacobsen et al. 1987; Perrine 2006). Other boreal animals diversified similarly to *V. vulpes*, for example, *Martes americana* (Stone et al. 2002), *Tamiasciurus* spp., *Glaucomys* spp. (Arbogast et al. 2001), *Myodes rutilus* (northern red-backed vole) (Runck and Cook 2005), and *Ursus americanus* (Stone and Cook 2000). These taxa either came from the east after the Wisconsin glaciation or were in mountain refugia. During glacial retreats, a northern ice-free corridor between the Cordilleran and Laurentide ice sheets allowed the eastern populations to colonize the Northwest, but this was not so for western mountain populations (Aubry et al. 2009; Runck and Cook 2005).

Vulpes macrotis mutica formerly occupied most of the San Joaquin Valley and surrounding areas but has lost more than 50 percent of its habitat since the 1930s (Schwartz et al. 2005). *Vulpes vulpes*, non-native to the Central Valley, has ecologically replaced *V. m. mutica* in some of the latter's former range (Cypher et al. 2000). Microsatellite data from 317 individuals from eight locations surprisingly reveal low mean F_{ST} values (0.043) among populations with low mean R_{ST} values (0.041), suggesting minimal populational subdivision and some genetic drift. Heterozygosity levels are similar across all populations of *V. m. mutica*, but differences among some loci and the R_{ST} value less than the F_{ST} suggest this species may be in the early stages of genetic isolation (Schwartz et al. 2005).

Early Pleistocene North American forms of *Canis lupus* have been identified as the Eurasian *C. etruscus*. The early Pleistocene North American form of *Canis latrans* is the Eurasian *C. lepophagus* (Nowak 1979). *Canis lupus* first appears in the fossil record about 0.7 Ma (Kurtén 1968). *Canis latrans* first appears in the fossil record about 1 Ma (Kurtén and Anderson 1980). Current estimates for divergence between the two species is 1 Ma from a corrected sequence divergence between wolves and coyotes of 9.6 percent with an estimated divergence rate of about 10 percent per million years (Nowak 1979). A mean sequence divergence of 2.9 percent gives *C. lupus* a coalescence value of about 0.290 Ma and 0.420 for *C. latrans*; however, other data suggest much older coalescence times. The more recent values are suggested to be from population fluctuations during Pleistocene glaciation cycles with reductions and expansions in population size that would have reduced the effective population size (Avise et al. 1984). *Canis dirus* (Dire wolf) remains have been identified from the La Brea Tar Pits and in the San Francisco Bay Area, and *Canis latrans orcutti* (extinct coyote) has been identified from La Brea (Moratto 1984).

Canis lupus used to inhabit a wide array of habitats throughout Europe, Asia, and North America. However, the species was extirpated by hunting from much of its former range, including California (with the recent exception of one migrant from Oregon in 2012 known as OR-7). *Canis latrans* is better adapted to human-modified habitats and has expanded into former wolf habitat (Macdonald 1984). Both species subsequently have had genetic intermixing, which has affected their genetic structure (Vila et al. 1999). Prior to the western expansion of humans across North America, there were an estimated two million *Canis lupus* individuals, down now to about 20,000. The current lack of phylogeographic structure in *C. lupus* populations may be due to the many expansions and contractions their populations experienced during the various ice ages. *Canis latrans* has expanded its range in the past few hundred years, but its genetic variability actually reflects the decrease in population numbers that were experienced during the LGM (Vila et al. 1999). Rapid expansion of *C. lupus* after glaciation combined with rapid and extensive migration distances erased previous genetic structure. In California, *C. latrans* has a high rate of genetic diversity even though it was recently colonized. California *C. latrans* genotypes group with some populations from Texas and some from Alaska, and microsatellite analyses also confirm a lack of genetic structure in their populations.

Ursidae

The current genetic structure of *Ursus americanus* (black bear) as measured from 540 individuals for 13 microsatellite loci from throughout their habitat in California identifies three and perhaps four genetic clusters via

STRUCTURE analysis (Brown et al. 2009). *Ursus americanus* is present in all regions of California except the Mojave Desert and the agricultural Central Valley. The extirpated *Ursus arctos californicus* (grizzly bear) formerly occupied the Coast Ranges and Southern California, which are now occupied by *U. americanus*. Between four geographic regions (northern coast/Klamath; Cascades/northern Sierra Nevada; central Sierra Nevada/ Southern California; and southern Sierra Nevada/central coast), F_{ST} values range from 0.03 between the Cascades/northern and central Sierra Nevada/Southern California to 0.10 between the Cascades/northern Sierra Nevada and the southern Sierra Nevada/central coast. Manipulations of K values in the STRUCTURE analysis still result in a break between the central and southern Sierra Nevada, indicating historical effects on gene flow. The central Coast Ranges were likely colonized by southern Sierra Nevada populations following the extirpation of *U. a. californicus*. Although *U. americanus* populations in California are not entirely discrete, the maintenance of habitat corridors, particularly in the Coast Ranges, the Sierra Nevada, and through the Transverse Ranges and Tehachapi Mountains will be important in maintaining their genetic diversity (Brown et al. 2009).

Rodentia

The phylogeography of *Microtus longicaudus*, the long-tailed vole (Cricetidae), is consistent with Pleistocene vicariant events common to other rodents (Mönkkönen and Viro 1997). A sample of mtDNA cyt *b* from 111 individuals with a focus on sampling north of 54°N identifies California populations as part of a California, Idaho, and Montana clade (Conroy and Cook 2000). *Microtus longicaudus* is dependent on mesic habitats and shares distribution with plants and animals that also occur in these refugia in high elevations along the Pacific coast. One would expect that any clades persisting near glacial refugia would be more deeply diverged than those from rapidly expanding populations (Bernatchez and Wilson 1998); however, diversity in the Pacific Northwest seen in the *Microtus longicaudus* may be the result of serial bottlenecking that may be common in animals in populated areas that are deglaciated (Sage and Wolff 1986). Expanding populations from multiple refugia following glacial retreat would be complicated by admixture, depending on the length of time of separation (Soltis et al. 1997).

Microtus californicus occurs from southern Oregon to Baja California and is divided into 17 subspecies, which are not supported by a Bayesian analysis of mtDNA cyt *b* and nuclear intron *AP5* data from 178 individuals (Conroy and Neuwald 2008). Consistent with other species, the molecular data divide *M. californicus* into northern and southern haplotype groupings that in this case have 41 steps between the two clades. An AMOVA establishes that most of the genetic variation is present within the respective northern and southern clades (88.64 percent). Nuclear haplotypes are not entirely concordant with the mtDNA haplotypes and indicate that there may be different dispersal patterns between males and females. The northern clade occurs from the northern border of California south to Santa Barbara, Kern, and Ventura Counties; and the southern clade overlaps with the northern clade in Santa Barbara County and in the Transverse Ranges east to the Tejon Pass. The average sequence divergence of 4.46 percent for cyt *b* is within the range of interspecific divergence for other *Microtus* species (0.54–13.35 percent; Jaarola et al. 2004), indicating that the *M. californicus* should be split into two separate species.

Glaucomys sabrinus and *G. volans* (New World flying squirrels, Sciuridae) examined via mtDNA cyt *b* sequences indicate two distinct lineages of *G. sabrinus*, one composed of populations from western California, Oregon, and Washington and one larger lineage from the remainder of the species east of the Pacific states. Discontinuity within the range of *G. sabrinus*, as with other species, indicates there was likely an early to middle Pleistocene separation between the ancestral populations in boreal forests of North America (Arbogast 2007). *Glaucomys volans*, on the other hand, is represented by one lineage.

A rare species found in riparian communities and coastal marshes 39°N south to Baja California, *Sorex ornatus* (ornate shrew, Soricidae), is a small mammal that lives 12–16 months. It has a high metabolism, requires constant feeding, and does not undergo long-distance dispersal. There are nine subspecies, two historically widespread and seven that occur in small patches of wetland habitat distributed from the coast to montane areas. Relictual populations remain in Sierra de la Laguna in northern Baja California and on Santa Catalina Island (Owen and Hoffman 1983).

Sequences of cyt *b* from 251 individuals from 20 locations identify three strongly supported major haplotypes in Southern, Central, and Northern California similar to other small vertebrate species. Four amino acid changes define the clades. Average sequence divergence values of 4.11 percent between southern and central populations and 7.59 percent between northern and southern populations increase with geographic distance based on a Mantel test. The northernmost basal clade is *S. ornatus* and likely was derived from *S. vagrans* from the boreal forests in western North America before the Pleistocene. MtDNA analyses indicate that the northern clade of *S. ornatus* is actually *S. vagrans* (vagrant shrew) in lowland form after excluding the possibility of ancient or recent hybridization or incomplete lineage sorting (Maldonado et al. 2001). Basal to the southern and central clades are the Santa Catalina, Sierra de la Laguna, and Bodega–Grizzly Island clades, which are suggested to be true *S. ornatus*. Northern California populations have higher nucleotide diversity within populations than do central and southern populations. Overall nucleotide diversity for Northern California populations was 0.012, lowest in the central (π = 0.002) and southern populations (π = 0.005). Repeated cycles of cold during the Quaternary would have limited wetland habitats throughout the range of *S. ornatus* and *S. vagrans* and undoubtedly influenced the genetic structure of these populations. The extensive loss of habitat for *S. ornatus*, already restricted by high specialization, places the survival of some populations in peril (Maldonado et al. 2001).

Neotoma fuscipes (dusky-footed woodrat, Cricetidae) resides in chaparral, woodlands, and mixed conifer forests at low elevations from Oregon to Baja along the Pacific coast and in the foothills of the Sierra Nevada (Murray and Barnes 1969; Matocq 2002). There are two additional relictual populations in the Granite Mountains in the Mojave Desert and in the Central Valley east of Fresno. Samples from 116 individuals from 41 locations and mtDNA CR and cyt *b* sequences establish that these mtDNA lineages form distinct nonoverlapping clades: a northern clade from Oregon along the coast to Sacramento and the San Francisco Bay Delta; a west central clade from the southern Sacramento/San Joaquin Delta to the inner coastal Diablo Range; a southern clade that starts on the east side of the Sierra Nevada and extends east to the Mojave, west to the Santa Lucia Mountains, across the Tehachapi Mountains, and south to Baja California;

and an east central clade in the central Sierra Nevada foothills from Murphy's (Calaveras County) to Shaver Lake (Fresno County) and in the San Joaquin Valley. The four subclades have strong, 97 percent bootstrap support from the mtDNA CR. No differences in rates of evolution among the clades are indicated, and sequence differences support their divergence during the late Pliocene to early Pleistocene. The northern and west central clades diverged from the southern and east central clades about 2 Ma, which coincides with the drainage of the Central Valley through the Monterey Bay. Subsequently, the northern and west central clades diverged 1.8 Ma, and the southern and east central clades diverged about 0.7 Ma (Matocq 2002). Lower levels of genetic variation in the northern clades suggest recent expansion into this range. Estimates of nucleotide differentiation and population statistics indicate that the northern clade may have undergone a bottleneck and recent population growth, particularly in the north. Later divergence of the northern and west central clades of *N. fuscipes* likely coincided with the formation of the Sacramento–San Joaquin Delta during the Pleistocene and is consistent with a number of other taxa. During the Pleistocene, the San Joaquin and Kings Rivers were glaciated and would have prevented long-distance migration, which explains the southern east central clades. The population in the Mojave Desert may indicate a former, larger distribution (Matocq 2002) (Figure 12.3).

Cytochrome *b* data support three major clades of *Tamiasciurus*, the New World tree squirrels: a western linage, *T. douglasii* (British Columbia, Washington, Oregon, and California), a lineage of *T. mearnsi* (northern Baja California), and a southwestern lineage, *T. hudsonicus* (New Mexico and Arizona). Divergence values of 1.0–2.4 percent suggest late Pleistocene divergence of these three species. Because of close genetic distance as measured by allozymes, *T. hudsonicus* and *T. douglasii* are probably one species with existing genetic structure likely a result of Pleistocene forest fragmentation (Arbogast et al. 2001). Genetic data indicate that *T. mearnsi* is most closely related to *T. douglasii* and that they were likely connected via forests between Sierra San Pedro Mártir and the Sierra Nevada when forests were 600–1,000 m lower than present (Lindsay 1981).

Consistent with the examples from *Sorex ornatus*, *S. vagrans*, *Neotoma fuscipes*, *Microtus californicus*, and *M. longicaudus*, Quaternary climatic fluctuations affected the genetic variation of less vagile species, but those

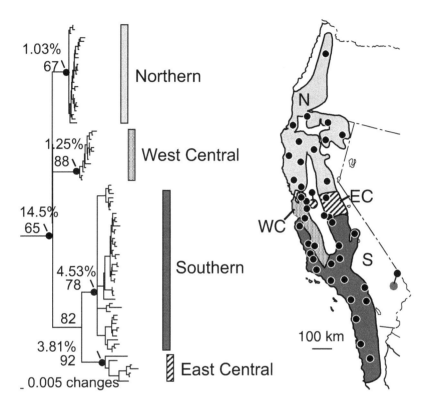

Figure 12.3. Neotoma fuscipes neighbor-joining tree based on mtDNA. Clades shown have greater than 50 percent support out of 1000 replicates. Numbers on map indicate collection locations. Used with permission from Blackwell Science Ltd. from Matocq 2002.

that also occur at higher elevations were particularly affected. *Phenacomys intermedius*, the heather vole (Cricetidae), is distributed across the mountains of the west with three major clades, Oregon/California (Cascades and Sierra Nevada), Washington, and northern/interior. Fossil evidence from 2.5 Ma in Siberia and Alaska and late Pleistocene/early Holocene fossils in lowland areas indicate the species was likely widespread during cooler periods throughout the Great Basin (Repenning et al. 1987; Rensberger and Barnosky 1993). A relaxed Bayesian method for estimating molecular divergence finds these lineages diverged 1.8 Ma, with first the Oregon/California clade and then the northern/interior lineage splitting from Washington

(Chavez and Kenagy 2010). These divisions have remained intact through-out several glaciation cycles. Subsequent divergence occurred within each major clade during the middle Pleistocene, with the Sierra Nevada and Oregon/Cascade clades diverging about 0.74 Ma. Cold-adapted high-elevation species such as *P. intermedius* likely have had declining popula-tions since Holocene warming restricted their habitats; elevational shifts led to smaller populations, and resulting bottlenecks increased the diver-gence among populations. A similar example is also found in *Ochotona princeps* (America pika), which has a genetically different, disjunct popula-tion in the Sierra Nevada within a major clade (Galbreath et al. 2009).

Microdipodops megacephalus, the Mono Lake kangaroo mouse (Heteromyidae), consists of populations disjunct from other populations of more than 100 km (Grinnell 1914; Hafner 1981). An analysis of genetic dif-ferences among two subspecies using mtDNA establishes that despite mor-phological differences, the two subspecies are not genetically differentiated (Hafner et al. 2008). Further analysis of other populations from California and eastward finds that Mono Lake populations are more closely related to populations in Nevada than those north in California. Biogeographical evi-dence suggests that *M. megacephalus* migrated through the Lahontan Trough and that a geographic break between the Wassuk Range and White Mountains provided a path. In general, *M. megacephalus* prefers sandy pluvial basin habitats found in the Great Basin created during the Pleistocene and Holocene dry periods, and the Mono Lake population diverged west and upward in elevation from eastern populations during cooling periods of the late Pleistocene and Holocene (Hafner et al. 2007). The Mono Lake region biota is conjectured to have been constructed in two waves, during the interlacustral period (130–35 ka) and the nonlacustral middle Holocene (7–5 ka).

The *Thomomys bottae-umbrinus* (pocket gopher, Geomyidae) complex sequenced for cyt *b* in 225 individuals from 108 locations establishes a within-complex variation of 13 percent, with eight monophyletic groups that do not correspond to current taxonomy or morphological variation (Álvarez-Castañeda 2010). The complex has a large ecological range, including deserts and alpine environments from southern Oregon to Veracruz, Mexico, and from the Pacific to the Rocky Mountains. Female pocket gophers are philopatric. Bayesian, maximum-likelihood, and maximum parsimony analyses revealed similarity to earlier studies based on allozymes (Patton and Smith 1990) but now identify eight geographic clades, five of which occur in California: Northern California group (Coast Ranges from San Francisco Bay north and into southern Oregon and the northeastern slope of the Sierra Nevada); Pacific group (Central Valley south through the Mojave, east to the Colorado River, and northwestern Mexico); Snake River group (Idaho and some individuals from California, Nevada, and Utah); Southwestern group (Colorado Desert); and Baja California Group (southern Salton Sea and Baja California). It is suggested that eight species should be recognized, including *T. laticeps* in Northern California and *T. bottae* for the rest of California; their population structure was clearly shaped by events in the Pliocene and Pleistocene, and local adaptations are a result of morphological changes in coloration.

Restricted to Baja California and southwestern California, *Chaetodipus fallax* (San Diego pocket mouse, Heteromyidae), is found in chaparral, coastal sage, and desert habitats. Cyt *b* analyses of six subspecies from 22 locations identify three major clades, the Los Angeles Basin north of the Salton Trough and northern populations in the southwestern Mojave Desert at mid-elevations, south of the Salton Trough and throughout Baja, and southern Baja populations west of the Vizarino Desert (Rios and Álvarez-Castañeda 2010). There is no support for current subspecies classification. Vicariant events thought to be important in the formation in these regions include the inundation of the Salton Trough (0.5 Ma), uplift of the Transverse Ranges (5 Ma), formation of the Pleistocene Lake Cahuilla in the Salton Trough (10 ka), and Pleistocene glacial cycles.

Ammospermophilus leucurus (antelope ground squirrel, Sciuridae) occurs in North American deserts and was widespread in the Pliocene and early Pleistocene based on fossil records (Miller 1980). A sample of 73 individuals from 13 locations from latitudes of 43°–22°N using 555 bp of cyt *b* and 510 bp of the mtDNA CR identifies one large clade from Oregon to central Baja with very little subclade structure and one clade from the southern end of Baja California (Whorley et al. 2004). Molecular data analyses and demographic modeling suggest rapid northward expansion from a northern Baja clade in the last warming cycle of the Pleistocene (Van Devender et al. 1987). A mid-Baja seaway that existed about 1.5 Ma was important for haplotypic divisions, disrupting gene flow from northern populations.

Endemic to the Mojave Desert, *Xerospermophilus mohavensis* (Mohave ground squirrel, Sciuridae) is listed as threatened under the California Endangered Species Act. Cyt *b* sequence data from 46 individuals in 11 localities show little phylogeographic structure and indicate that *X. mohavensis* likely expanded from a pluvial Mojave River Basin refugium (Bell et al. 2010). Similar analyses of the sister species, *X. tereticaudus*, from 38 individuals from 14 locations identify four subclades in the desert Southwest. Bayesian methods show there is reciprocal monophyly between the two species, and their divergence corresponds with uplift of the Transverse Ranges in the late Pliocene. This scenario is supported by positive F_S values in the southern end of the *X. mohavensis* range, with significantly negative values in the northern end. The Antelope Peak and Phelan Peak Basins likely provided refugia in the western part of the Mojave immediately east of the San Gabriel Mountains, with northward expansion about 11.7 ka (Bell et al. 2010).

Chaetodipus penicillatus (desert pocket mouse, Heteromyidae) is found in the Mojave and Sonoran Deserts. Bayesian inference from mtDNA COIII sequences and ecological niche modeling on 220 individuals from 51 locations identified Mojave and Sonoran Desert clades with 0.024–0.018 sequence divergence and significant differences between the clades (Mantel test, r = 0.3218, P<0.01). The divergence places their split in the late Pleistocene before the LGM, with some populations present in northern refugia during the LGM (21–18 ka) (Jezkova et al. 2009). Ecological niche models predict there were refugia in the northern Mojave in Death Valley for the Mojave clade; and for the Sonoran clade, a refugium in the northeastern Sonoran Desert. Based on three to four nucleotide substitutions, the Mojave clade of *C. penicillatus* has three shallowly divided subclades. Subclade 1 is north on both sides of the Colorado River and with one population in Death Valley, subclade 2 is in the western Mojave north of the San Bernardino mountains, and subclade 3 is along the lower Colorado River and in western Arizona. Little correspondence is found with the tree topology and existing subspecific classification.

Chaetodipus baileyi (Bailey's pocket mouse) is distributed throughout Baja California and in the Mojave and Sonoran Deserts. Sequences of 699 bp of mtDNA COIII and 450 bp of cyt *b* from 12 localities identified three mtDNA haplotypes (Riddle et al. 2000). Populations east and west of the

Colorado River are a separate species, *C. rudinoris*, which is also divided into lineages north and south in Baja California. The phylogeography of this species supports two vicariant events: the Sea of Cortéz extending north in the late Neogene 3 Ma and the peninsular seaway across Baja about 1 Ma (Upton and Murphy 1997).

SUMMARY

The mammal fauna of California was significantly shaped by Beringian migrations that occurred episodically from the Eocene through the late Pleistocene. Wide-dispersing mammals show evidence of continental divisions across North America, with western clades retreating variously to the coast or ice-free regions during glaciation periods.

For small mammals, deep phylogeographic divergence is common due to vicariant events, and due to their limited dispersal they often remain isolated (Avise 2000). The divergence of the two major clades of *Neotoma fuscipes* in California suggests that this occurred in the Sierra Nevada, followed by an east to west spread of genotypes, a pattern consistent with a number of other vertebrates (Rodríguez-Robles et al. 1999; Maldonado et al. 2001; Matocq 2002). During the Pleistocene, the San Joaquin and Kings River were glaciated and would have prevented long-distance migration, which explains the southeastern clades of *N. fuscipes*. *Martes pennanti* also shows reduced gene flow across the Kings River. Different species may have responded to different glaciation cycles depending on their distribution at the time of the glaciations. Severe habitat loss and degradation in the past few centuries has decimated many populations that occur in riparian corridors from east to west, further isolating historical contiguous populations and preventing migration.

The Coast Ranges and the Sierra Nevada underwent significant uplift about 5 Ma. Prior to this time, the seaway in the San Joaquin Valley would have provided the wetland habitat needed by habitat specialists like *Sorex ornatus* or their ancestors. Barriers to dispersal existed in the Monterey Bay region from about 8 to 2 Ma because the southern Sierra Nevada drained into the Pacific Ocean from this location and presented a significant obstacle for the dispersal of small mammals and amphibians (Wake

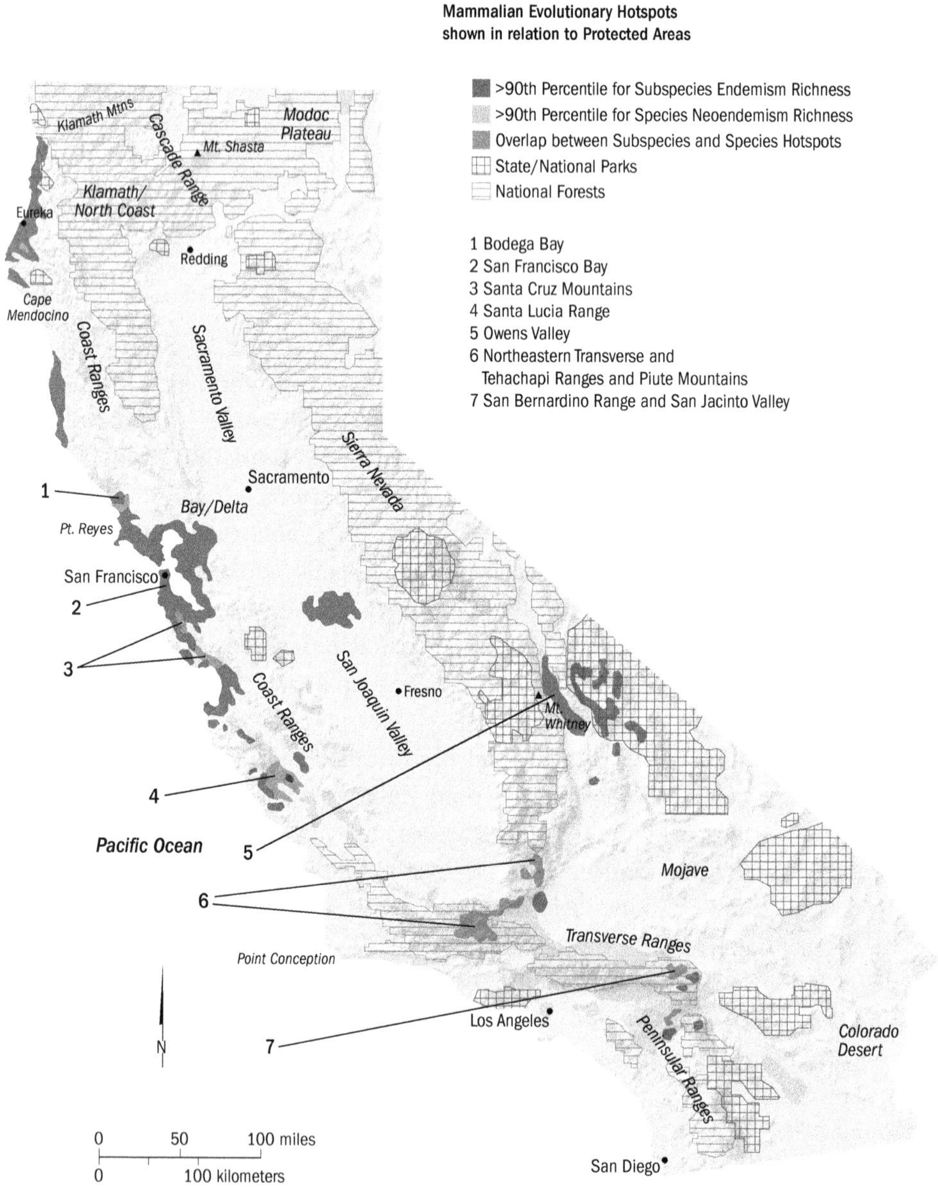

Mammalian Evolutionary Hotspots
shown in relation to Protected Areas
>90th Percentile for Subspecies Endemism Richness
>90th Percentile for Species Neoendemism Richness
Overlap between Subspecies and Species Hotspots
State/National Parks
National Forests
1 Bodega Bay
2 San Francisco Bay
3 Santa Cruz Mountains
4 Santa Lucia Range
5 Owens Valley
6 Northeastern Transverse and
Tehachapi Ranges and Piute Mountains
7 San Bernardino Range and San Jacinto Valley
Klamath Mtns
Cascade Range
Modoc Plateau
Mt. Shasta
Klamath/ North Coast
Eureka
Redding
Cape Mendocino
Coast Ranges
Sacramento Valley
Sierra Nevada
Sacramento
Bay/Delta
Pt. Reyes
San Francisco
San Joaquin Valley
Fresno
Mt. Whitney
Pacific Ocean
Mojave
Transverse Ranges
Point Conception
Los Angeles
Peninsular Ranges
Colorado Desert
San Diego
N
0 50 100 miles
0 100 kilometers

Map 12.1. Mammalian evolutionary hotspots shown in relation to protected areas. Used with permission from Wiley and Sons, from Davis et al. 2007.

1997; Vila et al. 1999). The central Sierra Nevada, the southern Sierra Nevada particularly around the San Joaquin River, the outer Coast Ranges particularly around the San Francisco Bay south to Monterey, the San Bernardino Mountains, the Tehachapi Mountains, the Peninsular Ranges, and Inyo Valley are identified as focal areas for endemic taxa based on mtDNA data for 27 species, with more than 75 percent of their range in California and MaxEnt estimations (Phillips et al. 2006; Davis et al. 2008) (Map 12.1).

The Transverse Ranges arose during the middle Pliocene, and because there is less rainfall in this area, there is less mesic habitat for those dependent on it (Yanev 1980; Vila et al. 1999). The Transverse Ranges probably played a role in the division of the southern and central clades of small vertebrates as found in *Ensatina* spp. (Wake 1997), *Taricha torosa* (Tan and Wake 1995), *Thomomys bottae* (Patton and Smith 1994), *Lampropeltis zonata* (Rodríguez-Robles et al. 1999), *Peromyscus californicus* (Smith 1979), and *Batrachoseps* spp. (Yanev 1980). All of these species would have been affected by expansion and contraction of mesic habitats during the Pleistocene.

The Mojave is recognized as a biotic transition region between the cooler Great Basin and the warmer Sonoran Desert (Axelrod 1983). A number of studies now verify that divergence for a number of taxa from the warm deserts began in the Miocene east of the Mojave (Riddle and Hafner 2006; Bell et al. 2010); however, the Mojave was likely influenced by later Pleistocene glaciation cycles. The Cordilleran ice sheet maximum 18 ka forced populations south, and during warming periods in the Pleistocene there was movement north (Booth 1987). Vegetation was dominated by scrub in xeric locations like the Armargosa River Basin and parts of the Death Valley (Spaulding 1990) that were likely refugia for many animal species during the LGM (Jaeger et al. 2005; Jezkova et al. 2009). Low-elevation areas around the lower Colorado River are thought to have remained warm deserts. Mojave vegetation similar to that found today was not present until 14 ka (Van Devender et al. 1987).

13 Marine Mammals

The Cetacea and Pinnipedia are well represented in the California Current System (Checkley and Barth 2009). Marine-dwelling members of the Carnivora include some members of the Mustelidae and Pinnipedia clades.

CARNIVORA/MUSTELIDAE

Within the Mustelidae, all otters are sister to *Mustela* (minks) and *Neovison* (true weasels). The genus *Enhydra* (sea otters) emerged from within the Old World river otter clade, but the relationship of *Enhydra* and *Hydrictis*, the spotted-neck otter native to sub-Saharan Africa, remains unresolved using Bayesian methods (Koepfli et al. 2008). *Enhydra lutris* is a densely furred animal historically ranging throughout the eastern Pacific from Japan, around the North Pacific Rim, and south along the western coast of North America to Baja California (Lidicker and McCollum 1997). The pre-European hunting population of sea otters is estimated to be as many as 300,000 individuals; however, they were virtually extirpated by hunters in the eighteenth through twentieth centuries, and it is estimated that 50 or fewer individuals survived on the California coast during this period (Riedman and Estes 1990). By 1980 the California population had reached a size of 1,400–1,600 individuals, and current population levels are estimated at 107,000 (Kornev and Korneva 2004). Amazingly, an analysis of 31 allozyme loci identifies 19.4 percent polymorphic loci and a mean expected heterozygosity level of 4.6 percent within the species, values expected for marine mammals (Lidicker and McCollum 1997). Although there is some support for a monophyletic origin of the California sea otter, there is no haplotype differentiation with the other two subspecies, *E. l. kenyoni,* which ranges from Aleutian Islands to Oregon, and *E. l. lutris,* in Japan.

CARNIVORA/PINNIPEDIA

Based on molecular evidence, Pinnipedia arose about 23 Ma during the late Oligocene or early Miocene (Flynn et al. 2005). The oldest fossil representative of this clade is *Puijila darwini* from 23 Ma in northern Canada, which supports an Arctic origin of this clade. Pinnipedia is represented by the Phocidae (earless seals) and the Otariidae (eared seals) in California. *Mirounga angustirostris* (Phocidae), the northern elephant seal, migrates 18,000–21,000 km and breeds two or three times a year onshore (Stewart and DeLong 1993). Their young are born in the late winter and remain primarily on land for about four months (Rick et al. 2011). The species is known from 28 archaeological sites from Baja California to Alaska ranging from 7 to 5 ka, but most are from 3.5 ka, suggesting that the species had a distribution similar to today, that is, from 28° to 49°N latitude (Rick et al. 2011). Hypotheses for variation in the distribution of populations through time include hunting by ancient humans who removed them from their preferred rookeries (sandy beaches) beginning about 13 ka. In addition, *M. angustirostris* was likely the preferred prey of saber-toothed cats, bears, and other predators.

Mirounga angustirostris was hunted to as few as eight individuals during the late nineteenth and early twentieth century, and as a result the species has extremely low levels of genetic variation. Pre-nineteenth-century DNA samples indicate much higher levels of diversity (Hoelzel et al. 2002). Covered by the Marine Mammal Protection Act enacted in 1972, their population is now about 124,000 (Carretta et al. 2009). *Mirounga angustirostris* provides an excellent cautionary tale as to the results of overhunting. Despite the increase in population numbers, the lack of genetic variability in northern elephant seals is reflected in low major histocompatibility complex (MHC) polymorphisms, greatly increasing its susceptibility to pathogens. An allozyme survey of 43 loci reveals zero polymorphic loci (Hoelzel et al. 1993). Lower birthrates during El Niño years combined with low MHC variation places this species at great risk for extinction (Le Beouf and Reiter 1991).

Eumetopias jubata (Otariidae), Steller's sea lion, has been present in California since the Pleistocene (Lyon 1941). This large animal can exceed 1,000 kg and is currently widely distributed, ranging from Japan, north to Alaska, and south to the Channel Islands in California (Orr and Helm 1989). *Eumetopias jubata* has experienced a sharp decline in population numbers of more than 80 percent in the past four decades, likely due to predation, hunting, and overfishing; it was listed as threatened under the Endangered Species Act in 1991 (Bickham et al. 1996; Horning and Mellish 2012). *Eumetopias jubatus* appears to have an Asian origin based on fossil evidence from an ancestral *Eumetopias*. An analysis of 668 individuals from throughout the species range for 13 polymorphic microsatellite loci establishes H_E levels of 0.237–0.843 and F_{ST} values of 0.121 among rookeries (Hoffman et al. 2006). An additional analysis of 238 AFLP loci from 285 individuals from 23 natal rookeries combined with mtDNA data substantiates relatively low genotype diversity. Most important, using Bayesian cluster analysis (STRUCTURE), *E. jubatus* is differentiated into an eastern Pacific stock (including Northern California) and Asian/western stocks at about 144°W, consistent with marking data of 8,500 individuals over a 24-year period. Different patterns of mtDNA and nucDNA diversity identify two main stocks, and the species is now recognized as having two subspecies, *E. j. jubatus* (the Asian/western stock) and *E. j. monteriensis* (the eastern stock) (Hoffman et al. 2009; Phillips et al 2011). A BEAST analysis of cyt *b* (1140 bp) and *ND1* (957 bp) mtDNA genes and the *HVRI* (238 bp) control region from 1,021 individuals representing >79 percent of the rookeries establish a most recent common ancestor for *E. jubatus* at 360 ka, with *E. j. monteriensis* emerging about 160 ka (Phillips et al. 2011).

Eumetopias jubatus and *Zalophus californianus* (California sea lion) split about 4.5 Ma (Repenning 1976; Schramm et al. 2009; Phillips et al. 2011). The population level of *Z. californianus* is about 355,000 (IUCNredlist.org). During the middle Pleistocene, the outer Pacific coast experienced significant and frequent changes in sea level and wave action (Jacobs et al. 2004). The Gulf of California is thought to have provided a refuge from these conditions for the marine fauna until they recolonized the Pacific coast in the late Pleistocene. Included in this scenario was *Zalophus californianus*. Forty-one mtDNA CR haplotypes in 299 individuals identify a latitudinally based genetic structuring of their populations (F_{ST} = 0.024–0.242; Schramm et al. 2009). The temperate region *Zalophus californianus* is enclosed within the Southern California Eddy. Adult female California sea lions are highly philopatric, which creates a strong ecological influence in population differentiation. The populations of *Z. californianus* along the eastern Pacific Coast should be considered management units for conservation needs (Mortiz 1994).

CETACEA OR "CETARTIODACTYLA"

Although traditionally classified under the Linnean system as within Cetacea, molecular data firmly place cetaceans within Artiodactyla, the even-toed ungulates and sister to the *Hippopotamus* (Shimamura et al. 1997; Price et al. 2005). Recent emphasis has been placed on the use of "Cetartiodactyla" to include both groups and is proposed as the new clade name (Agnarsson and May-Collado 2008); regardless, within the Cetacea, there are two primary lineages, the Mysticeti (baleen whales) and the Odontoceti (toothed whales).

The Balaenopteridae is estimated to have diverged from other members of the Mysticeti during the middle Miocene (Gingerich 2004). *Megaptera novaeangliae* (humpback whales, Balaenopteridae) has been studied extensively using an array of genetic tools. Prior to the decline of *M. novaeangliae* due to hunting, their numbers exceeded 100,000; although protected since 1966 under a moritorium, their populations are now estimated to be 80,000. *Megaptera novaeangliae* individuals migrate each year from high-latitude summer feeding grounds to their low-latitude winter breeding grounds, a distance of more than 9,000 km (Mackintosh 1965). Those individuals that summer off the central coast of California are part of the eastern Pacific population and winter in Baja California and the Gulf of Mexico (Baker and Palumbi 1996). The retention of some genetic diversity despite severe population decline is likely attributed to the relatively short time of the genetic bottleneck and their long generation times. A number of cetaceans are matrifocal, and thus the genetic structure of their populations should contain a matrilineal signature (Avise 2008). Combined with a high fidelity of females to their breeding sites, *M. novaeangliae* shows mtDNA population subdivision among three ocean basins (Baker et al. 1993).

Phocoena phocoena (harbor porpoise, Phocoenidae) is a continuously distributed coastal resident with circumpolar distribution in the temperate Northern Hemisphere that shows genetic structures consistent with those found in *Phoca vitulina* and *Eumetopias jubatus* (Bickham et al. 1996; Westlake and O'Corry-Crowe 2002; Harlin-Cognato et al. 2006). An analysis of 358 bp of the mtDNA CR shows differentiation between the northern Pacific and the eastern/southern group from Baja California to Monterey, with higher levels of genetic diversity near the southern genetic boundary near Baja (Taguchi et al. 2010). Mantel tests and an NCA support a range expansion from the south to the north and northwest following bottlenecks in the Pleistocene about 400–300 ka and 50–40 ka, consistent with interglacial periods. Early to middle Pleistocene glacial cycles correspond with a break in genetic variation around British Columbia that is inferred to have resulted from reduced gene flow from south to north and east to west. The AMOVA values of 11.87 and a Φ_{ST} value of 0.14 among groups support a hypothesis of rapid expansion from multiple refugia following the Pleistocene reductions in gene flow (Taguchi et al. 2010). A similar study using 402 bp of the mtDNA CR (n = 249) and nine nuclear and mitochondrial microsatellites (n = 194) (Chivers et al. 2002) also reveals genetic subdivision from north to south, for example, F_{ST} values of 0.0193 (P = 0.012) between British Columbia and San Francisco but a difference between San Francisco and Monterey of 0.0030 (P = 2.96) (Chivers et al. 2002).

Phocoenoides dalli (Dall's porpoise, Phocoenidae) occurs in coastal and pelagic populations from the eastern North Pacific and Alaska south to California and Baja California. One hundred thirteen and 119 individuals of *P. dalli* were analyzed using 379 bp of mtDNA CR and six microsatellite loci, respectively (Escorza-Treviño and Dizon 2000). Samples from Okhotsk Sea, North Pacific Ocean, Bering Sea, Alaskan Gyre, and along the Pacific Northwest coast of North America identify an overall Φ_{ST} of 0.1197 (P<0.0001) and distance between eastern and western regions of Φ_{ST} = 0.17. Based on mtDNA haplotype networks, the species likely originated in the Okhotsk Sea. Northern North America versus central North America differences show F_{ST} values of 0.0375 (P = 0.4653) for males and 0.1429 (P<0.0001) for females, which suggests differential dispersal. The mtDNA CR data suggest *Phocoenoides* diverged from *Phocoena* about 0.71 Ma based on a pairwise distance of 11.38 percent. Population differences among the western and eastern populations indicate they should be managed as different units.

SUMMARY

Marine mammals with some portion of their life cycle in California include 25 cetacean, 7 pinniped, and one mustelid taxa. *Hydrodamalis gigas* (Steller's sea cow), which formerly ranged throughout the North Pacific coast from Japan to California, became extinct in 1768. California populations of most marine mammals have been heavily affected by hunting and show a loss of genetic variation as measured by heterozygosity in comparison to more northern populations, which did not suffer as much exploitation (Lidicker and McCollum 1997) (Figure 13.1) (Map 13.1).

Low levels of genetic differentiation along the eastern Pacific Coast are present in *Enhydra lutris* and most cetaceans due to their high rates of dispersal. However, some taxa such as *Eumetopias jubatus* are differentiated into eastern Pacific stock and Asian/western Pacific stocks at about 144°W and should continue to be managed as separate units. Other species such as *Phocoenoides dalli* and *Zalophus califorianus* are latitudinally structured along the coast due to female philopatry and should continue to be managed as separate populations to conserve genetic diversity.

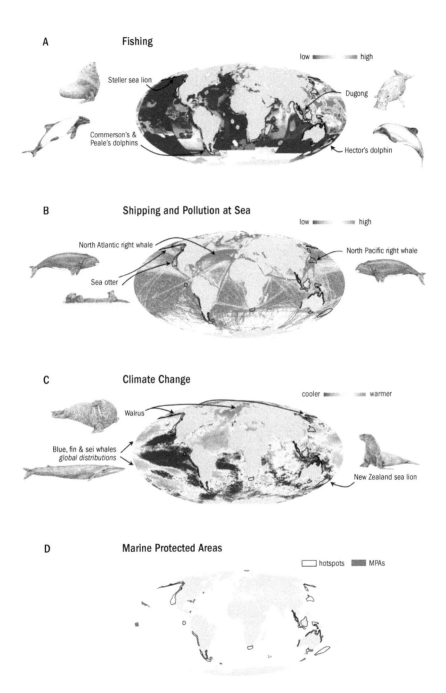

Figure 13.1. Geographic representation of global mammal extinction risks and human impacts on Marine Protected Areas. Based on models from Davidson et al. 2012. Used with permission from Ana Davidson. Illustration by Sharyn N. Davidson.

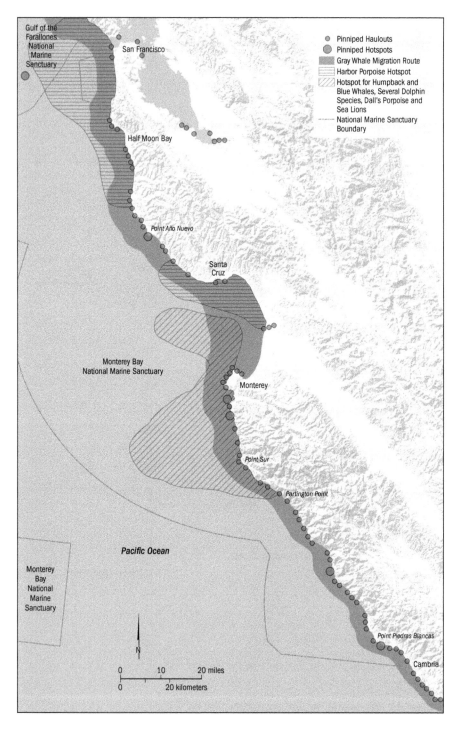

Legend:
- Pinniped Haulouts
- Pinniped Hotspots
- Gray Whale Migration Route
- Harbor Porpoise Hotspot
- Hotspot for Humpback and Blue Whales, Several Dolphin Species, Dall's Porpoise and Sea Lions
- National Marine Sanctuary Boundary

Gulf of the Farallones National Marine Sanctuary

San Francisco

Half Moon Bay

Point Año Nuevo

Santa Cruz

Monterey Bay National Marine Sanctuary

Monterey

Point Sur

Partington Point

Pacific Ocean

Monterey Bay National Marine Sanctuary

N

0   10   20 miles

0   20 kilometers

Point Piedras Blancas

Cambria

Map 13.1. Hotspots of marine mammals in Central California. Used with permission from the National Oceanic and Atmospheric Agency. Illustration by Sophie Beukelaer.

Unfortunately, there are inadequate data on cetaceans. The IUCN Red List of threatened species estimates that 56 percent of the members of the Dephinidae (oceanic dolphins) and 75 percent of Ziphiidae (beaked whales) are considered data deficient (www.redlist.org; Price et al. 2005).

PART III Summary

14 Consistent Phylogeographic Patterns across Taxa and Major Evolutionary Events

As most of California remained underwater until the Paleocene, most of the early organisms are marine. Insects were likely the first terrestrial organisms to colonize California during the Paleozoic, followed by non-flowering plant taxa in the Coniferophyta, Equistaceae, Pteridophyta, Lycophyta, and Pteridospermatophyta. In the Mesozoic, floristically, conifers and extinct lineages were dominant. But by the Eocene, climates were drier, and the flora was gradually replaced by broad-leaved deciduous forests, similar to those found in the southeastern United States today. North America and Eurasia were connected across the North Atlantic during the late Cretaceous, and some broad-leaved species then present in western North America are now found only in Asia. Adding to the complexity of the California flora, Anthophyta radiation was occurring in the area of Mexico during the Mesozoic, and many taxa migrated and diversified north during warmer climates. Many conifer taxa gradually became restricted to edaphic conditions, higher elevations, and cooler climates, where they remain today.

Late Mesozoic Southern California and Baja California delta sediments include archosaurs and some early mammals. The Triassic Hosselkus Formation in Shasta County includes a representative of the Sciuridae, but generally, because California was underwater and still forming throughout most of the Mesozoic, mammal diversification in North America occurred east of present-day western North America. Cretaceous mammals are known from incisors present in fossil beds of Alberta, Wyoming, and South Dakota.

Archaic Mammalia were present in California by the Paleocene but were replaced by modern mammalian migrations via Beringia in the Eocene when the first Artiodactyla are present in western North America. There was an increase in grasslands throughout western North America in the Paleogene, providing the habitat in which many mammal taxa evolved. The Central Valley contained a large sea with connections to the Pacific Ocean during the Oligocene and provided propagules for the icthyofauna to migrate and diversify into the forming streams of the Sierra Nevada. Modern mammal clades evolved by the Oligocene, and there was a general increase in body size due to global cooling. The late Oligocene to early Miocene was climatically transitional, from the warmer Paleogene to the cooler Neogene. Miocene California experienced significant climatic, geologic, and organismal transitions. Ancestors to the extant Felinae arose in the Miocene (~9 Ma) in western Eurasia and migrated to North America 8.5–8.0 Ma, coinciding with additional Eurasian carnivore migrations. The first members of the Mustelidae colonized North America in the late Miocene via Beringia and, substantiated by fossil evidence, included a number of now-extinct "paleomustelids." The earliest fossil Caudata in North America are from 25 Ma at the beginning of the Miocene (van Frank 1955).

The continued Miocene uplift of the Sierra Nevada was greater than 1,000 m, and sedimentation occurred into the Central Valley. The Coast Ranges were not yet formed, and the coastline was east of its current location. Miocene and Pliocene fossil floras indicate the gradual formation of a fairly stable Mediterranean-type climate over 23 million years that was fully developed by Pliocene. Upwelling conditions were poor at the beginning of the Miocene, but by 12 Ma they were quite rich, facilitating and stabilizing the Miocene marine mammal fauna. Members of the Sebastidae and other fish taxa radiated during upwelling peaks of the late Miocene and, depending on the taxon, in the Pleistocene (Jacobs et al. 2004). The Isthmus of Panama began forming 19–15 Ma and was complete by 3 Ma, resulting in the exchange of many species between North and South America. The Miocene rotation of blocks in Southern California resulted in the formation of the northern Channel Islands and the Los Angeles Basin. The continued formation of southern mountain ranges resulted in the development of rain shadows and of xeric desert species in southeastern California. By about 10 Ma, Death Valley was formed. Continued uplift of the Coast Ranges, the Transverse Ranges, and the Sierra Nevada occurred throughout the Pliocene. The White and Inyo Mountains separated from the Sierra Nevada about 4–3 Ma. The result of sedimentation from continued uplift and changing sea levels resulted in the Pliocene development of estuaries that had begun to form in the Miocene.

Most modern birds are present by the end of the Miocene; however, coalescence times among closely related taxa of bird taxa are 2.0–0.1 million years, indicating that diversification probably occurred in refugia over one or more glaciation cycles. Genetic signatures for many species indicate that in some cases speciation had occurred prior to Pleistocene glaciations; for example, haplotype divergences in *Parus* of 3–7 percent suggest speciation occurred about 5.2–3.7 Ma. More recent differentiation occurred in the late Pleistocene, and secondary contact after the last glaciations probably account for the different forms observed with zones of sympatry.

Pleistocene glaciation offered opportunities for colonization and retreat to refugia for a plethora of species. Genetic, paleontological, and archaeological data suggest that some mammals, such as *Martes pennanti*, colonized California within the last 5,000 years from eastern North America via British Columbia. *Vulpes vulpes* initially colonized North America over Beringia, and populations migrated south during the Illinoian glaciation (0.3–0.130 Ma) and then came back north, forming the Nearctic clade. Other populations of *Vulpes* went south during the Wisconsin glaciation (Péwé and Hopkins 1967) and gave rise to the North American part of the Holarctic clade.

Treelines dropped by about 1,000 m in the northwestern United States during the late Wisconsin glaciation (0.10–0.01 Ma) (Barnosky et al. 1987). However, the southern end of this region was more open and served as a refugium. As the glaciers retreated, many animal taxa retreated north but continue to in some areas of Northern California (Jacobsen et al. 1987; Perrine 2006). Boreal taxa with similar diversification patterns include *V. vulpes, Martes americana* (Stone et al. 2002), *Tamiasciurus* spp., *Glaucomys* spp. (Arbogast et al. 2001), *Myodes rutilus* (northern red-backed vole) (Runck and Cook 2005), and *Ursus americanus* (Stone and Cook 2000).

There is support for recent connectivity between the Cascades/Sierra Nevada and the Rocky Mountains for members of the butterfly complex, *Parnassius phoebus,* as also demonstrated with *Oeneis chryxus* and *Ochotona princeps* (Schoville and Roderick 2009; Nice and Shapiro 2001; Hafner and Sullivan 1995).

Evidence from pollen and isotope studies establishes the interglacial period (40–20 ka), a peak glacial period (20–15 ka), and the glacial-interglacial transition (15–10 ka). Regardless, 21–18 ka most of North America was covered in ice, and Eurasia and North America were connected via Beringia. Species in California experienced reduced and fragmented ranges, changes in distribution, and extinction. Although exact causes remain unclear, climate was undoubtedly a factor in the megafauna extinction of 15–10 ka. Phylogeographic methods, ecological niche modeling, fossils, and paleoclimatic reconstructions all contribute to the inference of refugia during the LGM that are now reflected as areas with high genetic diversity or as contact zones. Glacial periods restricted many habitats and likely forced the migration, extirpation, or extinction of many taxa. The many weather oscillations in the Pleistocene resulted in the filling and draining of estuaries important in the evolutionary and ecological sorting of aquatic taxa. Although earlier interpretations of the Pleistocene fossil record indicated that responses to climate change were idiosyncratic (Graham et al. 1996), recent evidence shows a number of patterns among groups with similar life history characteristics. Vagility, philopatry, isolation of habitat, reproductive output, and generation time all affect the ability of species to respond to climate change. Pleistocene glacial cycles were a significant factor in building the biota of California and shaping the genetic structure of populations that were already present. It is important to identify and understand glacial refugia because historical population structure influences the ability of a species to migrate and respond to current climatic change (Waltari et al. 2007).

KLAMATH/SISKIYOU REGION

The mountains of the remarkable Klamath-Siskiyou region are composed of a complex mosaic of Paleozoic rocks capped by Mesozoic volcanoes in the eastern section and subducted and uplifted oceanic substrates on the western margin that continued to form until the Paleogene. In addition, the region has had numerous intrusions of ultramafic and granitic material from the Mesozoic. Previously residing in more southern latitudes, the region did not reach its current location until the Paleocene. Throughout much of the Mesozoic, the eastern Klamath Range and the northern Sierra Nevada were the western edge of Pangaea. Many taxa in northwestern California have a coastal distribution, or their northern or southern limit of distribution occurs in the area (Soltis et al. 1997). The geologic complexity, influence of coastal climate, latitude, and relative lack of glaciation made this region a refugium for many species. Formerly separated lineages that later rejoined following climatic change resulted in the development of a contact zone hotspot (Hewitt 2000) from the southern Oregon Klamath-Siskiyou region through the Sierra Nevada in which many species hybridized. A high resolution (25 x 25 km) assessment documents phylogeographic breaks for a number of herpetofaunal taxa between the Trinity Mountains and the Cascade Range and high herptofaunal diversity on the Modoc Plateau (Rissler et al. 2006). A recent review of northwestern North American phylogeographic patterns emphasizes that much of the diversity and subsequent diversification within the Klamath-Siskiyou/Cascade/Sierra region is due to its historical importance as a refugium that was colonized from multiple directions (Shafer et al. 2010).

Far Northern California represents a confluence of the volcanic Cascades, the geologically ancient and complex Klamath-Siskiyou region, the northern Sierra Nevada, and the Warner Mountains, separated from the Sierra Nevada by the Modoc Plateau. The species that have hybridization zones or reach the northern or southern limit of their distribution show varying degrees of divergence. The area is a contact zone for northern and southern subspecies of *Picoides albolarvatus* (Alexander and Burns 2006) and Pacific Northwest and California lineages of *Strix occidentalis* (Barrowclough et al. 2005). Response to climatic change would have depended on specific adaptation to cooler climates, vagility, population

Table 14.1   Species with range limits, significant geographic structuring, or hybridization in the Klamath-Siskiyou region

Scientific Name	Significance of K-S Region	Reference
Ancient *Gila coerulea*	Split of Klamath and Sierra Nevada	Avise and Ayala 1976
Pinus balfouriana	Sierra Nevada/Klamath split 1 Ma	Eckert and Carstens 2008
Picea breweriana	Only present in Klamath-Siskiyous	Ledig et al. 2005
Boechera sparsiflora	Differentiation into *breweri/hoehleri*	Kiefer and Koch 2012
Collinsia linearis	Klamath R. divergence, 5.8–1.1 Ma	Baldwin et al. 2011
Ascaphus truei	Refugium	Nielson et al. 2006
Taricha granulosa	Refugium	Kuchta and Tan 2006
Anaxyrus boreas	Refugium	Goebel et al. 2009
Polystichum munitum	Split in K-S range between haplotypes	Soltis et al. 1997
Tolmiea menzieisii	cpDNA haplotype differentiation	Soltis et al. 1997
Calochortus spp.	Geographic structuring	Patterson and Givnish 2002, 2004
Notholithocarpus densiflorus	High cpDNA haplotype diversity, greater increase in K-S about 5.8 ka	Nettel et al. 2009
Grylloblatta spp.	Three new species	Schoville 2012
Acrotrichis xanthocera	High diversity	Moldenke 1999
Lycaeides complex	Strong admixture	Gompert et al. 2008
Greya politella	Southern subclade, secondary contact	Rich et al. 2008
Taricha t. sierrae	Diversified from *T. t. torosa* 3.4–2.6 Ma	Kuchta and Wake 2005
Plethodon stormi	Endemic to K-S, split along Klamath R.	Mahoney 2004
P. elongatus		
Dicamptodon tenebrosus	Southern refugium	Steele and Storfer 2006, 2007
Strix n. nebulosa	Split between SN/N Cascd pops 26.7 ka	Hull et al. 2010

size, connectivity of populations, and reproductive cycles. The Warner Mountains were dominated by conifer forests through the Pliocene, but Holocene drying resulted in the *Pinus-Juniperus* shrub steppe today. Habitat and climatic differences correspond to a split between the subspecies *Aphelocoma californica woodhouseii* and *A. c. californica* about 500 ka, but hybridization occurs in the Klamath-Siskiyou region, and similar scenarios exist between *Baeolophus inornatus* and *B. ridgwayi* and *Pica nuttalli* and *P. hudsonia* (Delaney et al. 2008).

In addition to the region's importance as a hybrid contact zone, its continued climatic fluctuations resulted in the presence of a number of relictual species to more recent arrivals affected by late Cenozoic climate change. Unlike the Sierra Nevada, the Wisconsin glaciation (0.11–0.01 ka) resulted in the formation of glaciers in northerly drainages and generally not on the ridges (Sawyer 2006) (Table 14.1). *Strix nebulosa nebulosa* is a good example of a species significantly affected by Pleistocene glaciations dependent on mature montane forests and almost exclusively in the Cascade Mountains north of the Klamath Basin with a disjunct population in the southern Sierra Nevada. Based on population differentiation statistics, the divergence of populations in the Sierra Nevada/Pacific Northwest split is estimated to have arisen 26.7 ka in the vicinity of the Klamath Basin. The populations remained isolated until postrefugial migration (Hull et al. 2010). Overall, the data suggest a number of recent bottlenecks and founder events and no evidence of hybridization between the Sierra Nevada population and others.

COAST RANGES

Geologic complexity, rapid uplift, and changing sea levels have been important in shaping the biota of the Coast Ranges. The northern Coast Ranges are the oldest and are composed of the Franciscan assemblage formed during the Mesozoic. Franciscan rocks are also present south of the San Francisco Bay Area, in the Santa Lucia Mountains of the inner central Coast Ranges, the San Luis Obispo region, and the Transverse Ranges. Beginning just north of San Francisco and continuing south along the Central Coast, the Salinian Block, composed of Cretaceous granitic rock, became joined with the North American Plate about 30 Ma and is the dominant substrate in the central and southern Coast Ranges. The San Francisco Bay Area is formed primarily of Franciscan material, with portions of the Salinian Block at Point Reyes and southwest of San Francisco that resulted from late Mesozoic/early Cenozoic plate movement and sedimentation. The inner Coast Ranges down the western edge of the Central Valley are composed primarily of the Great Valley sequence. The Franciscan complex along the Central Coast and the Channel Islands were present as islands during the Miocene, 17–13 Ma, and became part of the mainland during the Pliocene, with significant uplift about 5 Ma. The Channel Islands were resubmerged 500 ka and reemerged connected as one landmass from 30 to 15 ka. An important transition for the San Ynez connection of the Gabilan and Santa Lucia Mountains occurred 5–3 Ma. The marine seaway that existed in the Monterey canyon from 8 to 2 Ma was a significant barrier to the formation of different clades north and south. Continued uplift eventually blocked the exit of Central Valley waters from about 2 Ma to 600 ka when the Monterey seaway was blocked and until Central Valley waters eventually were diverted into the San Francisco Bay. The highest number of minimum-rank plant taxa per region in the California Floristic Province are found in the San Francisco Bay Area and in the central Coast Ranges, which also have the highest level of neoendemics (Kraft et al. 2010). The highest species richness for herpetofauna is found along the Central Coast around the San Francisco Bay and in the Peninsular and Transverse Ranges (Rissler et al. 2006).

There are phylogeographic breaks for a few species in the northern Coast Ranges. There is no obvious explanation for this other than changing sea levels, mountain uplift, and changing proximity to the coast, which likely resulted in barriers to dispersal at various points in geologic time. For example, *Rana boylii* has a phylogeographic break in Mendocino County consistent with deep divergence (Lind et al. 2011). Similarly, reciprocal monophyly exists for the northern *Rana aurora aurora* and the southern *R. a. draytonii* (Shaffer et al. 2004) and *Dicamptodon ensatus* and *D. tenbrosus* (Good 1989). Two very different clades for *Lavinia symmetricus* were identified from the Gualala River and the Pit River that were separated from other clades by 6–3 Ma. Different clades of *Lavinia* are also found in the Navarro River, Tomales Bay, the Red Hills, and the

Russian River–Clear Lake Basin. Gualala River *Lavinia* individuals are separated from other mtDNA haplotypes by 4.9 Ma (Dupré 1990).

The central Coast Ranges are the geographic limit of conifer forests, which are replaced by more xerically adapted chaparral and woodland species to the south. From the Miocene to the Pleistocene there were a number of marine embayments in the San Joaquin Valley and the central Coast Ranges, and at various times, the central Coast Ranges would have been isolated from the rest of California (Hall 2002). Monterey Bay and Sacramento–San Joaquin Delta lineages are the same for both *Lavinia symmetricus* and *L. exilicauda*, which probably resulted from hybridization, as the San Francisco and Monterey Bays were connected until as recently as 1.5 Ma (Dupré 1990). Monterey Bay was a barrier to dispersal and is of biogeographic importance for a number of taxa as it is the place where many reach their southern or northern limit of distribution. It was particularly significant for small mammals and amphibians (Wake 1997; Vila et al. 1999). It is the southern limit for at least six salamander species: *Ambystoma macrodactylum, Dicamptodon ensatus, Aneides flavipunctatus* (black salamander), *Ensatina eschscholtzii oregonensis, Batrachoseps attenuatus,* and *Taricha granulosa* (rough-skinned newt) (Kuchta and Tan 2006). Species that reach the northern limit of their distribution at Monterey Bay include *E. e. eschscholtzii, B. luciae* (Santa Lucia Mountains slender salamander), and *B. gabilanensis* (Gabilan Mountains slender salamander), and there are divergences between subspecies or varieties for at least three additional species (Kuchta and Tan 2006). The geographic barriers to dispersal would have been in place by 2 Ma between the northern clades of the *Sorex ornatus*. In southern Monterey County there is deep divergence for *Rana boylii*, with populations southwest of the Salinas River Valley (Lind et al. 2011), *Ambystoma californiense* (Shaffer et al. 2004), and *Actinemys marmorata* (Spinks and Shaffer 2005). Northern and west central Coast Range clades of *Neotoma fuscipes* diverged from the southern and east central clades about 2 Ma, which also coincides with the drainage of the Central Valley through the Monterey Bay (Matocq 2002).

Continued geologic changes in the Coast Ranges isolated the Santa Barbara clade of *Ambystoma californiense* around 2.5–2.0 Ma. The Santa Barbara and Sonoma clades of *A. californiense* were isolated

0.74 and 0.92 Ma, respectively. Vernal pools, which historically formed a ring around the Central Valley and generally to a greater extent in the northern end of the valley, provided habitat for *A. californiense*, which became isolated from southern populations about 1 Ma (Shaffer et al. 2004). When the Central Valley Embayment subsided in the Pliocene and early Pleistocene, gene flow occurred from the northern Coast Ranges and the San Joaquin Valley into the foothills of the southern Sierra Nevada and the inner eastern Coast Ranges. Molecular-based divergence times indicate that *Actinemys marmorata* may have started diverging 9–8 Ma, but divergence was more apparent by 4 Ma (Spinks and Shaffer 2005).

Taricha species divergence occurred about 15 Ma, subspecies diverged about 9 Ma, and within-haplotype clusters within subspecies diverged 5–2 Ma. Southern California populations are basal and remain from remnant populations that migrated and diverged north (Tan and Wake 1995). *Taricha torosa* subspecies indicate a split of 13–7 Ma between *T. t. torosa* and subclades, which diverged 2.3–1.9 Ma into coastal California, San Francisco and Monterey and Southern California. Genetic divergences among the taxa are similar to other Pacific Northwest amphibians in the area such as *Ensatina*, *Taricha*, and *Rana* (Moritz et al. 1992; Tan and Wake 1995; Macey et al. 2001). The minimum age for all species of *Batrachoseps attenuatus* is 7.6 Ma, with two major biogeographic barriers identified in the Sacramento–San Joaquin Delta and the San Andreas fault region that correspond with Miocene/Pliocene and Pleistocene waves of divergence. The phylogeography of *Sorex ornatus* provides a detailed and complicated picture of response to the changing riparian and marsh habitats of the coast and the Central Valley with admixture among clades in the northern Central Valley and northern delta region. Repeated cycles of cold throughout the Quaternary would have limited wetland habitats throughout the range of *S. ornatus* and *S. vagrans* and undoubtedly influenced the genetic structure of these populations (Maldonado et al. 2001).

Climatic fluctuations along the coast were a major factor in the deposition of estuarine sediments. Estuaries and streams were colonized following glaciations and then became isolated from others, resulting in the development of unique genotypes from Monterey Bay to Bodega Head for members of the Gobiidae (Dawson et al. 2002). Central and Southern

California were covered by the sea, and the Monterey/San Joaquin Valley seaway would have provided the wetland habitat needed by estuarine species. A divergence time of about 5 Ma in *Sebastes* from Monterey Bay is suggested to have occurred when upwelling was at its highest in this region, but an earlier radiation in the late Miocene, about 9 Ma, is also suggested. Los Angeles Basin marine incursions periodically from 2.0 to 0.100 Ma were important in the divergence of *Pseudacris cadverina* (Phillipsen and Metcalf 2009), *Lampropeltis zonata* (Rodríguez-Robles et al. 1999) and *Actinemys marmorata* (Shaffer et al. 2004; Spinks and Shaffer 2005).

SIERRA NEVADA

Formation of the Sierra Nevada began over 160 Ma in a tropical montane environment as Mesozoic granitic rocks capped by volcanoes, but it also experienced significant uplift at 25, 10, and 6–3 Ma with significant alterations by glaciations. The Cretaceous Sierra Nevada eroded by the Paleocene and Eocene into low hills with marine sedimentation. Pleistocene glaciations began in the Sierra Nevada about 2.5 Ma, and from 0.66 to 0.15 Ma major glaciation events lasted as long as 100,000 years. Continued geologic change, complex terrain, and recent glaciations have resulted in many examples of intraspecific variation, recent species divergences, and the shaping of genetic structures. The area along the northeastern Sierra Nevada was recognized as the Rocky-Pacific suture zone VI (Remington 1968), a center of differentiation (Austin and Murphy 1987) and a hybrid zone hotspot for the Sierra Nevada and Northern California (Swenson and Howard 2005). For example, *Hydromantes* has three species in California, *H. platycephalus* in the Sierra Nevada at high elevation and low elevation; *H. brunus* near Mount Shasta; and Owens Valley populations of putative *H. platycephalus*. However, *H. brunus* is nested within *H. platycephalus* and arose within the latter species' northern range from 0.7 to 0.3 Ma, probably from a small number of individuals. Directly north of the Sierra Nevada, the volcanically active Mount Lassen is the southernmost peak of the Cascade Range. *Charina bottae* has had repeated cycles of extirpation in this area due to volcanic disturbances and recolonization from both northern and southern clades that have an estimated separation time of 12.3–4.4 Ma. *Taricha taricha sierrae* populations in the Sierra Nevada and the southern Cascades are estimated to have diversified 3.4–2.6 Ma. This geographic separation is also thought to be important in California northern and southern clades of *Lampropeltis zonata* and *Diadophis punctatus* (Rodríguez-Robles et al. 1999).

The divergence of many northern and southern clades occurs in the central Sierra Nevada. Divergence between northern and southern lineages of *Hydromantes platycephalus* is estimated to have occurred between 4 and 3 Ma. Northern and southern clades of *Grylloblatta* species are estimated to have diverged about 2.51–1.94 Ma. There are two strongly supported clades of *Poecile gambeli* in the Sierra Nevada/Cascades and Great Basin/Rocky Mountains that are estimated to have diverged 0.61–0.53 Ma. Within the Sierra Nevada/Cascade clade, the subclade of *Poecile gambeli* that forms the northern populations has been isolated from the Transverse/Peninsular Ranges for 60,000 years. Recently derived species in the Sierra Nevada include *Lycaeides,* which colonized North America in the late Pliocene with rapid range expansion within each clade at the end of the Pleistocene from glacial refugia (Gompert et al. 2006; Schoville and Roderick 2009). In general, genetic variation in *Colias behrii* is very low but distinguishes northern, southern, and southwestern Sierra Nevadan groupings.

Models for the genetic histories of alpine species include expanding alpine insular habitat, alpine habitat as a refuge, and radiation from fragments (Schoville and Roderick 2009). The models include changes in population size and connectivity during changing warm and cold phases, with the intensity of gene flow dependent on species characteristics. For example, edaphic endemism in the *Collinsia* "metamorphica" clade on the quartzite-rich schist of the central high Sierra Nevada likely emerged 1.3–0.3 Ma. However, for some taxa, population differentiation is very low from southern refugia in the Sierra Nevada, as illustrated in *Cornus nuttallii* (Keir et al. 2011). River basins are demonstrated to provide important barriers to dispersal for both aquatic and terrestrial organisms. The southern Sierra Nevada in particular has evidence for clade separation across river basins for amphibians (Kuchta and Tan 2006; Kuchta 2007), mammals (Matocq 2002), reptiles (Rodríguez-Robles et al. 2001; Feldman and Spicer 2006), birds (Sgariglia and Burns 2003; Alexander and Burns 2006), and arthropods (Law and Crespi 2002; Starrett and Hedin 2007). Clade divergence within *Oncorhynchus mykiss* ranges from 2.0 to 0.6 Ma consistent with the Pleistocene, and mtDNA supports the origin of *Oncorhynchus mykiss aguabonita* (golden trout) from the Sacramento–San Joaquin Delta. Two separate colonization events established *O. m. aguabonita* in the Volcano Creek and Kern River populations of Kern County, but there is some evidence of gene flow through secondary contact.

Populations of *Rana boylii* south of the San Joaquin River in the Tulare Lake Basin are divergent from those in the north and are consistent with a number of other taxa that show similar deep phylogeographic splits with northern populations (Lind et al. 2011). The San Joaquin River serves as a separation for northern, southern, and eastern *Hydromantes platycephalus* populations, and their phylogeography supports a vicariant event from 4 to 3 Ma (Rovito 2010). There is a genetic discontinuity for three *Nebria* species near the headwaters of the San Joaquin River (Schoville et al. 2012). Two southern Sierra Nevada populations of *Martes pennanti* less than 100 km apart in contiguous forest have very low gene flow; however, these populations are separated by the Kings River, an established barrier to gene flow.

The Tehachapi Mountains provide an important connection between the central coast and the Sierra Nevada, as demonstrated in *Neotoma fuscipes* and other taxa (Table 14.2). Alternatively, examples of taxa with clade breaks across the Tehachapi Mountains are numerous. High genetic diversity with evidence of northern expansion into the Sierra Nevada or western expansion into the Coast Ranges is found in the Tehachapi Mountains for some taxa. Southern California consists of four subclades of *Neotoma lepida* that all overlap in the San Emigdio–Tehachapi Mountains. These mountains achieved their current height about 3 Ma, but during glacial periods they would have forced many taxa downslope, with subsequent divergence into the four subclades (Patton et al. 2008).

TRANSVERSE RANGES

The Transverse Ranges arose during the middle Pliocene. They are at the confluence of three major regional biotas, the Sierra Nevada, the Coast

Table 14.2   Taxa with significant evolutionary signatures in the Tehachapi
Mountains

Taxon	*Event*	*Reference*
Nyctoporis carinata	9.9–2 Ma with secondary contact	Chatzimanolis et al. 2010
Actinemys marmorata	W of Sierra Nevada from SF Delta to Tehachapi Mts. clade separate from Transverse Ranges clade 8–2 Ma	Spinks and Shaffer 2005
Puma concolor	Important contemporary conduit	Ernest et al. 2003
Ursus americanus	Important contemporary conduit	Brown et al. 2009
Sorex ornatus sinuosus	Split for northern clades and coastal clade	
Neotoma fuscipes	Conduit for eastern Sierra Nevada across Tehachapi Mts. to Baja California	Matocq 2002
Batrachoseps stebbinsi	Separate species in Tehachapi Mts. and Transverse Ranges	Jockusch and Wake 2002
Collinsia spp.	Splits between Transverse Ranges and Sierra Nevada 6.81–1.45 Ma	Baldwin et al. 2011

Ranges, and the Mojave Desert, and have long been recognized as an evolutionary hotspot, first proposed as the area for genetic connectivity with the coast (Peabody and Savage 1958) and later as an area for high hybridization between xeric and mesic lineages (Remington 1968). High levels of relictual plant species provide evidence of range expansion of Arcto-Tertiary species into southern latitudes, with concentrations in the San Jacinto and Santa Rosa Mountains (Stebbins and Major 1965; Raven and Axelrod 1978). The San Gabriel Mountains have been moving north over the past 12 million years, which likely has had an impact on less vagile species such as salamanders that are more dependent on mesic habitats (Jockusch and Wake 2002). The San Gorgonio Pass in the San Bernardino Mountains has a very strong cline of habitat types and is the location of many divergent clades (Patton et al. 2008).

A combined analysis of three bird taxa in California tested statistical congruence for the similarity of phylogeographic patterns for *Chamaea fasciata, Toxostoma redivivum*, and *Picoides albolarvatus gravirostris* in western North America. All three species showed intraspecific sequence divergences of less than one percent, and molecular clock data support respective intraspecific differentiation within the past one million years. The haplotype networks for *T. redivivum* and *P. albolarvatus* were similar in geographic distribution and separate Southern and Northern California haplotypes just north of the Transverse Ranges, although *P. albolarvatus* occurs at higher elevations. Separation of northern and southern haplotypes of *C. fasciata* occurred within the Transverse Ranges (Burns et al. 2007). NCA suggests that *C. fasciata* underwent range expansion, *T. redivivum* experienced allopatric separation, and *P. albolarvatus* underwent a number of long-distance colonization events that resulted in restricted gene flow.

Climatic fluctuations during the Pleistocene after allopatric speciation in *Xerospermophilus* about 1.6 Ma is hypothesized to be a result of rapid uplift of the Transverse Ranges and the later isolation of the basins in which these species occurred. This scenario is supported by positive F_S values in the southern end of the *X. mohavensis* range, with significantly negative values in the northern end. The Antelope Peak and Phelan Peak Basins likely provided refugia in the western part of the Mojave immediately east of the San Gabriel Mountains, with northward expansion about 11.7 ka (Bell et al. 2010).

The Transverse Ranges and Los Angeles Basin Pliocene Embayment break characterize an important phylogeographic break that often separates Central and Southern California for a number of clades (Calsbeek et al. 2003; Chatzimanolis and Caterino 2007; Polihronakis and Caterino 2010) (Table 14.3). Species distributions would have been affected by expansion and contraction of mesic habitats during the Pleistocene. Expansion in southern clades happened from southern refugia, and relationships between genetic divergence and geographic distance indicate a gradual movement north. Divergence times between, within, and around the Transverse Ranges range from 9.9 to 2.0 Ma (Segraves and Pellmyr 2001; Feldman and Spicer 2006; Polihronakis and Caterino 2010).

Table 14.3   Sample taxa that have significant evolutionary signatures in the
Transverse Ranges

*TR = Transverse Ranges, PR = Peninsular Ranges, SN = Sierra Nevada, LAB = Los
Angeles Basin*

Taxon	Event	Citation
Pinus lambertiana	Different haplotypes in TR, SN, and PR	Liston et al. 2007
Collinsia wrightii	3.9–0.5 Ma split	Baldwin et al. 2011
Collinsia heterophylla	Separate clade in TR and PR, 1.5–0.2 Ma	Baldwin et al. 2011
Nyctoporis carinata	East/west break in TR	Chatzimanolis et al. 2010
Sepedophilus castaneus	East/west break in TR	Chatzimanolis et al. 2010
Stenopelmatus "mahogani" complex	Divergence in TR	Vandergast et al. 2007
Pseudacris cadaverina	Divergence 0.6–0.25 Ma	Phillipsen and Metcalf 2009
Sceloporus magister	West TR, S of Cuyama River split	Leaché and Mulcahy 2007
Actinemys marmorata	Divergence around the TR	Spinks and Shaffer 2005
Poecile gambeli	Some isolation for ~ 60 ka, then divergence	Alexander and Burns 2006
Chamaea fasciata	Divergence within TR	Burns and Barhoum 2006
Toxostoma redivivum	N/S split around the TR	Burns et al. 2007
Picoides albolarvatus	Divergence within TR	Burns et al. 2007
Xerospermophilus mohavensis	1.6 Ma isolation from Mojave Desert	Bell et al. 2010
Thomomys bottae	S/central clade divergence	Patton and Smith 1994
Sorex ornatus	Divergence in TR and LAB	Maldonado et al. 2001
Lampropeltis zonata	Divergence	Feldman and Spicer 2006
Batrachoseps nigriventrus	Divergence in Sierra Pelona and NW TR	Jockusch and Wake 2002
Xantusia henshawi	Divergence in TR	Lovich 2001

Diversification for the phylogeography of *Sepedophilus castaneus* is consistent with other species in the Transverse Ranges, with three biogeographic areas: a western region (northwestern Transverse Ranges and Santa Ynez Mountains), a central region (central Transverse Ranges and Sierra Pelona), and an eastern region (San Gabriel, San Bernardino, and San Jacinto Mountains); however, no north-south break was identified for *S. castaneus*. A number of studies have placed the Sierra Pelona/San Gabriel divergence at 5.0–2.7 Ma (Rodríguez-Robles et al. 1999; Feldman and Spicer 2006; Spinks and Shaffer 2005), which corresponds to a Pliocene embayment that now corresponds to the Santa Clara River drainage (Hall 2002). The Transverse Ranges are thought to be the center of origin for present-day *Pseudacris cadaverina*. The northern, central, and southern groups were defined on a larger regional scale with divisions in the Transverse Ranges. The appropriate mesic habitat would have limited their occurrence in specific areas, but during wetter times some individuals did disperse, and divergence times of the major clades are about 0.6–0.259 Ma (Phillipsen and Metcalf 2009).

Chamaea fasciata is suggested to be a southern refugee from the Pleistocene that experienced a recent range expansion throughout the Holocene as it gradually migrated north and northwest along the coast and northeast through the Sierra Nevada (Burns and Barnham 2006). The San Gabriel and San Bernardino Mountains within the Transverse Ranges may have contained two refuge populations (Burns and Barhoum 2006). Similar to *C. fasciata* in life history traits and ecological and geographic distributions, *Toxostoma redivivum* displays mtDNA haplotype distributions that also indicate ancestry from southern refugia in the Transverse and Coast Ranges (Sgariglia and Burns 2003). However, *T. redivivum* shows a north/south split around the Transverse Ranges, unlike the *C. fasciata*, which shows a split within the Transverse Ranges (Burns et al. 2007). *Rana muscosa* shows divergences in the San Gabriel Mountains about 0.289 Ma; San Jacinto and San Bernardino populations diverged about 47 ka (Schoville et al. 2011).

DESERTS

The late Oligocene through mid-Miocene marks the development of major biogeographic regions in the western United States and likely is the time *Trimorphodon* (lyre snakes) diverged from other genera. Most Neotropical taxa are at their northern limit in Mexico. Consistent with this pattern, the Baja California clade diverged from the Sonoran/

Chihuahuan/western Mexico clade about 6.5 Ma, and the Sonoran/ Chihuahuan Desert clade diverged from western Mexico about 5.4 Ma (Devitt 2006).

The desert Southwest is thought to have been a continuous expanse of desert known as "Mojavia" (Axelrod 1958) from the Miocene through the early Pliocene. As a result, a number of taxa are hypothesized to have dispersed from east to west. Fossil records establish that *Crotalus atrox* spanned this area from 3.7 to 3.2 Ma; however, fossils of *C. atrox* are not present west of the Continental Divide until the late Pleistocene, and a split between eastern and western clades is present 1.36 Ma. Proposed Pleistocene refugial locations for the western clades include the southern Sonoran Desert and the lower Colorado River Valley (Castoe et al. 2007).

The San Gorgonio Pass in the Peninsular Ranges is an important transition zone for a number of taxa as it was covered by the Gulf of California 6.5–6.3 Ma (Dorsey et al. 2007). It has also been proposed, however, that vicariance between Mexico and Baja was 5.5–4 Ma, peaking at 3 Ma (Riddle et al. 2000). *Xantusia* started diversifying in southern Baja about 18.5 Ma, and Baja California began to split from Mexico about 13.2 Ma. Further diversification within the complex occurred when the Channel Islands split from the mainland in the middle Miocene with *X. riversiana* splitting from *X. vigilis*. *Xantusia vigilis* continued to diversify about 7.7 Ma, with population expansion and further divergence during Pleistocene uplifts 1.5 Ma. These findings are consistent with a fossil record that supports a late Miocene development for the biota of the desert Southwest. There is some disagreement about the exact timing of the Gulf of California changes in the Plio-Pleistocene, particularly when the Salton Trough extended into the Mojave. Deep sequence divergence within *Crotalus cerastes* and *C. mitchellii* (5.8–6.4 percent) supports their split in the late Miocene to early Pliocene. *Crotalus mitchellii* and *C. tigris* also diverged during the split of Baja California from Mexico. Shallow divergence was found between *C. tigris* (Sonoran Desert) and *C. ruber* (red sidewinder of southwestern California and all of Baja California), which have narrower distributions.

The ancestor of *Lampropeltis zonata* evolved 14.6–5.2 Ma during the cooler middle to late Miocene, and the northern and southern clades diverged 5.7–2.0 Ma when divided by inland seas that did not recede until

5–1.6 Ma (Rodríguez-Robles et al. 1999). *Corvus corax* (common raven, Corvidae) has an extensive range throughout the Northern Hemisphere. Separation of the California and Chihuahuan Desert clade is well supported with sequence divergences that suggest they split 2 Ma (Omland et al. 2000).

It is impossible to consider the evolution of the biota of Southern California deserts without also considering the geographic history of the surrounding areas of Baja California, Arizona, northern Mexico, and southern Nevada. The Bouse Embayment was a significant drainage that receded and formed the Colorado and Gila Rivers in the late Miocene. The Lahontan Trough between the Wassuk Range and the White Mountains has been described as an important post-Pleistocene migrational corridor and a possible route for range shifts for a number of species responding to climatic changes for east-west migration of Great Basin organisms (Reveal 1979; Riddle et al. 2000; Arbogast and Kenagy 2001). Resident fish species such as the *Gila intermedia* (Gila chub) and *Cyprinodon macularius* (desert pupfishes, Cyprinodontidae) from the Bonneville and Lahontan systems eventually became genetically isolated, as did species that migrated into these systems, such as members of the Cottidae (sculpins) and *Oncorhynchus clarkii* (Smith et al. 2002). *Phrynosoma mcallii* is endemic to the Salton Trough, a remnant of late Miocene to Pliocene inundation of the area. During the Pleistocene as the Colorado River developed, it provided a significant barrier to gene flow in this species, with higher genetic diversity retained in the southeastern extent of the range (Mulcahy et al. 2006). Two clades of *Charina trivirgata* separated by the Colorado River developed in the late Miocene to early Pliocene as the desert continued to dry from 30 to 15 Ma, and by 4.8 Ma they were distinct (Van Devender and Spaulding 1979; Wood et al. 2008).

Baja California began separating from mainland Mexico about 5.5 Ma and led to the development of the Gulf of California. Combined analyses of 12 desert taxa support a late Pliocene northward extension of the Gulf of California, a seaway across southern Baja, and the presence of a seaway across the Baja peninsula during the mid-Pleistocene (Riddle et al. 2000). A number of marine transgressions are evident from 5.5 to 1 Ma and resulted in the separation of regional biota through fragmentation. Established divergence within the *Sceloporus magister* group is estimated to have a posterior mean age of 6.23 Ma (Leaché and Mulcahy 2007). The transfer of Baja California from the North American Plate to the Pacific Plate began in the mid- to late Miocene and was the precursor to a number of divergence events that further separated clades in the *S. magister* complex (Carreño and Helenes 2002). The sister taxon of *S. magister*, *S. orcutti*, has a posterior mean age of about 4.108 with a great deal of admixture and gene flow between different desert regions along the Colorado River.

Separation of populations in the Great Basin and Death Valley are consistent with paleontological records. MtDNA substitution rates indicate that the ancestors to *Mylopharodon conocephalus* and *Ptychocheilus grandis* were introduced to North America from 4.7 to 2.4 Ma, and speciation between *Lavinia symmetricus* and the *L. exilicauda/Mylopharodon* clades was between 2.9 and 1.5 Ma.

Five morphologically distinctive pupfish species, *Cyprinodon macularius* from the Colorado Basin, *C. radiosus* from the Owens Valley, and *C. salinus*, *C. diabolis*, *C. nevadensis nevadensis*, *C. n. amargosae*, and *C. n. mionectes* from the Death Valley region, have very little allozyme differentiation (Turner 1974). In six remnant populations of *Cyprinodon macularius* from the Salton Sea in California, the Colorado River delta in Baja California, and the Sonoyta Basin in Mexico, overall differentiation is low. The low levels of genetic variation among these clades support the recent isolation of Death Valley populations of *C. salinus* and *C. diabolis* about 20 ka and the isolation of *C. nevadensis* about 5–4 ka (Soltz and Naiman 1978).

Mojave Desert vegetation was likely established by about 11 ka with the arrivals of current dominants, *Ambrosia dumosa* and *Larrea tridentata* (Axelrod 1979; Van Devender and Spaulding 1979). During warming periods, clades migrated northward; during cooling periods, clades migrated south, providing propagules to colonize the desert region. The youngest neoendemic plant taxa are present in the Mojave Desert and the Great Basin (Kraft et al. 2010). Established populations within ancient clades provided the source populations for colonizations, and admixture waxed and waned through vicariant events and climatic change until divergence into separate species or subspecies became complete. *Toxostoma redivivum* (Burns and Burnham 2006), *Neotoma fuscipes* (Matocq 2002),

Peromyscus californicus (Smith 1979), *Baeolophus inornatus* (Cicero 1996), and *Sorex ornatus* (Maldonado et al. 2001) all inhabited southern refugia. Mainland migrations were occasional from the east and the north and continued throughout the Quaternary.

COAST

For marine taxa, ice-free regions such as the Pacific coast, the Bering Sea, and the Aleutian Islands provided Pleistocene refugia for species such *Eumetopias jubatus*, *Phocoena phocoena*, and *Phoca vitulina*, all with similar genetic structure (Bickham et al. 1996; Westlake and O'Corry-Crowe 2002; Harlin-Cognato et al. 2006).

The northern Channel Islands experienced Miocene-Pliocene uplift (Ingersoll and Rumelhart 1999), and by the Plio-Pleistocene the Los Angeles area had transitioned from a marine to a terrestrial environment. In the late Pleistocene, 0.7 Ma, Los Angeles was submerged at varying extents, connecting a variety of estuaries and perhaps providing a conduit for gene flow. Air and sea surface temperatures of Southern California were about 6–10°C cooler during the Pleistocene glaciation events (Davis 1999). *Embiotoca lateralis* is a benthic invertebrate substrate feeder but competes with *E. jacksoni* for crustaceans on algae. Both species show strong genetic structuring into three major clades, the Northern Region, the Channel Islands, and the Southern Region (Bernardi 2005). Divergence distances between the northern and southern clades indicated a separation of about 328 ka for *E. jacksoni* and 150 ka for the *E. lateralis*.

SUMMARY

California, with its varied geologic history across space and time, has long provided a mecca for evolutionary biologists seeking to understand ancient and recent evolutionary events. Mediterranean to alpine climates combined with vicariant events associated with glaciations, changing sea levels, and rapid mountain formation have resulted in the formation of both deep and shallow divergences. Significant events include the second stage of the formation and movement of the Klamath-Siskiyou Ranges (450–65 Ma), the intermittent growth and changes of the Sierra Nevada (160, 50, 25, and 10 Ma), the Bering Land Bridge (35, 9, and 1 Ma), glaciation events in the Sierra Nevada (760 ka, 118 ka, 56 ka, 25–10 ka), the stabilization of the Mediterranean climate (8 Ma), the connection between North and South America (3 Ma), the connection between the Pacific Ocean and the Central Valley in the Monterey area (5–2 Ma), the formation of the Coast Ranges (~5 Ma), the formation of the Transverse and Peninsular Ranges (5–3 Ma), the flooding of the Los Angeles Basin (2 Ma), the inundation of the Salton Trough (0.5 Ma), the formation of the Channel Islands (3 Ma) and their connections to land (5–0.5 Ma), and the Bouse Embayment (0.5 Ma) (Yanev 1980; Hall 2002; Chatzimanolis and Caterino 2007). Refugia and high levels of diversity for multiple taxa are identified in the Klamath-Siskiyou region, northeastern California, the southern Sierra Nevada, the Central Coast, the Transverse Ranges, and the western Mojave Desert, reflecting both diversification and hybridization through secondary contact. Numerous questions remain about the extent of differentiation for plant and invertebrate species in alpine environments, and further exploration is warranted especially in areas identified with high levels of neoendemism, such as the southern Sierra Nevada. Regional analyses of species origins would provide insight into the development of the California biota, particularly in areas that are known suture zones such as northeastern California and the Tehachapi Range.

The supertree method and maximum agreement subtrees (MAST) in PAUP from nine different species with different ecological parameters and life histories from a data set with county distributions and 10 different climatic variables provide regional species groupings (LaPointe and Rissler 2005). The taxa included are all widely distributed across California: three amphibian, two mammal, one reptile, one bird, one insect, and one plant species. Although not a phylogeny, five groups were identified for which there was concordance among the taxa: the northern coast; Southern California and east of the Sierra Nevada; Stanislaus, Alameda, and Contra Costa Counties; Monterey, San Luis Obispo, and Santa Barbara Counties; and the Sierra Nevada except for Tuolumne County, which was allied with the northern coast. Five groups differed significantly by precipitation, temporal distribution of precipitation, radiation, and temperature and indicated that small spatial scales of <25 km may be important (LaPointe and Rissler 2005).

Pleistocene glaciation events resulted in the isolation of California populations of a number of species compared to their continental congeners (Arbogast 1999). Three major refugia existed during the LGM (~18 ka), the Pacific Northwest coast, Beringia, and areas south of the continental glaciers (Pielou 1991; Latch et al. 2009). California was colonized from all three refugia as the glaciation retreated, and distinct genetic signatures remain in those taxa that survived glaciation and that in some cases resulted in secondary contact (Hewitt 2000). Five to ten periods that occurred throughout the Quaternary would have limited the wetland habitats and forced a number of estuarine species to become similarly isolated along the coast or in river basins.

Post-Pleistocene examples of divergence in California plant taxa for nonconiferous taxa are limited to a few examples. The significant phylogeographic assessments of plant taxa that need to be completed will require the combination of rapidly evolving genetic markers such as microsatellites and gene spacer regions with less various nuclear and plastid markers. Examples from taxa that are widely distributed across ancient and recent geologic formations such as established in *Collinsia* will prove to be very useful in providing a deeper understanding of plant evolution in California.

The concordance of phylogeographic breaks across divergent groups of taxa supports the commonality of processes that shape the speciation of the biota of California. It is important, however, to consider how life history characteristics and ecological requirements affect lineages differently in time and space. Understanding the genetic structures of populations and how they have been influenced by vicariant events or human-mediated disturbance is important for the adaptive potential of highly fragmented populations.

15 Conservation Implications and Recommendations

Phylogeographic studies can be very useful in defining significant evolutionary units for conservation purposes (Roderick 1996). When concordance among divergent major clades is identified, phylogeographic studies can provide the basis for major conservation efforts in particular regions (Avise et al. 1987). Many regions of high biological diversity are in need of conservation efforts that will strengthen legal protection, establish or strengthen migratory corridors, and increase the size of parks and preserves. Anthropomorphic changes to the evolutionary process that can lead to loss of biological diversity include a loss of habitat or reduction in range size, habitat fragmentation, hybridization of nonlocal genotypes, reduction of genetic variation, disturbance, invasive species competition, changing fire regimes, and climate changes. Despite the seemingly daunting task of overcoming these negative changes to the environment, we are reminded that "profound social transformations are not impossible or unrealistic," as evidenced by the end of segregation, the fall of the iron curtain, the end of apartheid, and, perhaps, the end of anthropogenic extinction (Ehrlich and Pringle 2008, 11584).

California is first among the fifty states in number of endemic species and second only to Hawaii in number of species at risk of extinction.

Table 15.1   Number of plant and animal species at risk

Animals	
Total number of animals listed (includes subspecies & population segments)	157
Total number of candidate/proposed animals for listing	4
Number of animals state listed only	31
Number of animals federally listed only	71
Number of animals listed under both state and federal acts	55

Plants	
SE (state-listed endangered)	134
ST (state-listed threatened)	22
SR (state-listed rare)	64
SC (state candidate for listing)	0
FE (federally listed endangered)	139
FT (federally listed threatened)	47
Both state and federally listed	125

SOURCE: DFG website.

Although approximately 50 percent of California lands are managed by federal agencies, this management includes multiple uses, such as logging and heavy recreation. Determining which areas need further protection via acquisition or stricter protection requires coordination among local, state, and federal agencies and conservation organizations (Table 15.1).

In 1950, the population of California was approximately 10 million; and its current level of 37 million is projected to increase to 50 million by 2046 (Pitkin and Myers 2012). This projected population growth is equivalent to twenty-six additional Sacramento-sized cities or one additional Los Angeles County, which comprises eighty-eight cities. Of concern is not just the elimination of actual habitat in order to house and provide infrastructure for this population but also the concomitant introduction of invasive species, disturbance, and species extinction.

The specific nature of species-environment interactions sometimes appears idiosyncratic because of our lack of understanding of an ecosystem or the ecological requirements of particular species. Phylogeographic studies should focus on areas that contain the greatest number of endemic species or the greatest biodiversity, for example, the central Coast Ranges, the Klamath-Siskiyou region, the eastern Transverse Ranges, the Tehachapi Mountains, the Santa Lucia Mountains, and the eastern Mojave Desert. Among the issues that need to be considered in conservation are local, regional, and taxonomic rarity; ecological function; and threats to biodiversity. Areas with high levels of endemism that include species with the greatest range restrictions are the Central Coast, the Sierra Nevada, and the San Bernardino Mountains (Kraft et al. 2010). The Mojave Desert and the Great Basin have the highest level of neoendemics. The most important criterion for conservation is risk of extinction, which can be affected by genetic diversity, range size, demography, body size, and life history characteristics, especially fecundity. Isolated populations are demographically independent and therefore should be treated as separate management units.

The California Gap Analysis Project conducted between 1990 and 1998 mapped the distributions of the biota and identified those areas that are underprotected (Davis et al. 1998). The project identified about 18 percent of California as currently being managed for biodiversity but in areas that are generally less than 800 ha that primarily occur as clusters in the southern Sierra Nevada and the desert region. Land owned by government agencies covers about 30 percent of the state, but biodiversity is not the top management priority. More than 50 percent of the state's land area is in private ownership. The areas that are the least protected are the Coast Ranges and the Central Valley. About 73 of 194 community types mapped by the Gap Analysis Project have less than 10 percent representation and are primarily coastal scrub, prairie grasslands, Great Basin scrub, and hardwood woodlands, in addition to some conifer forests. Forty-one natural community types were ranked as the highest priority for conservation and generally occur around the perimeter of the Central Valley, with the highest areas of concern in Kern, Glenn, Colusa and Tehama Counties (Davis et al. 1998).

Arguments in favor of management at large spatial scales are well established. Habitat Conservation Plans usually occur at the county level; management plans, usually for a single forest; and statewide conservation plans, within artificial political boundaries. Continued efforts to establish regionally cohesive, ecologically sound plans are needed. Of particular concern is whether plans allow for range shifts and other adaptations that will occur under climate change.

Because of their habitat specificity and rarity, the most vulnerable species are endemic. And the most vulnerable ecosystems are those with high levels of spatial isolation, for example, alpine habitat and serpentine islands. Spatial isolation can mean different evolutionary dynamics in different populations of the same species in terms of gene flow, genetic drift, and strong selection from climate change and invasive species. Ecosystem function is dependent on biological diversity. Local declines in biodiversity can be more devastating than global declines since the beneficial impacts of organisms can be lost before a global extinction occurs. A change in the identity and abundance of species in an ecosystem can result in changes to the ecosystem processes: plant and nitrogen production and resistance to disturbance and disease.

Evolutionary process must be incorporated in conservation planning to conserve the ability of species to respond to evolutionary change, particularly with rapidly changing climatic scenarios. Maximizing genetic diversity alone does not necessarily conserve evolutionary processes; the conservation of genetic diversity must be accompanied by the conservation of the environment within which it evolved (Rissler et al. 2006). Although California State Parks has recently started to consider acquiring land in areas of high evolutionary diversity, in general, evolutionary potential is not part of regional or statewide planning (Davis et al. 2008). Conservation priorities should focus on areas of high diversity across major groups of organisms and require the conservation of large, heterogeneous landscapes.

Biosphere reserves are an excellent tool for conserving areas with high biodiversity and provide the greatest benefit when there is connectivity among habitats. The most beneficial reserves are supported by agreement among multiple stakeholders, the integration of biological and cultural diversity, and educational programs. California has created the following biosphere reserves: the Channel Islands and marine sanctuary waters for terrestrial and marine species; San Dimas, which contains mixed woodlands, chaparral, coastal scrub, and grasslands; San Joaquin, with its evergreen sclerophyllous woodland; Sequoia–Kings

Canyon, which includes the national parks and encompasses valley to alpine habitat; Stanislaus-Tuolumne, with mixed conifer forest north of Yosemite; the Coast Ranges, a complex of three areas; the Mojave and Colorado Deserts, also a cluster of areas; and the Golden Gate, a cluster of thirteen marine and terrestrial systems.

HYBRIDIZATION

Hybridization between historically allopatric lineages is a regular event during the course of evolution, and it can be the source of important adaptive variation, particularly in plants. Disturbance and other shifting environmental conditions can accelerate hybridization, however, and the human movement of species and genotypes can instigate hybridization that would not otherwise occur.

Hybridization among native and introduced fish species is particularly well documented and has been shown to have disastrous effects for native parental taxa (Edwards 1979; Siddiqui 1979; Dowling and Childs 1992; Muhlfeld et al. 2009). Specifically, hybridization between introduced *Oncorhynchus mykiss* and the 14 subspecies of native *Oncorhynchus* in the United States has resulted in the extinction of 2 of these subspecies, another 5 subspecies that are federally listed as threatened, and 7 that have been petitioned for listing as threatened (Muhlfeld et al. 2009).

In cities worldwide, non-native cultivars are hybridizing with their wild, native congeners. This is increasingly being recognized as a serious invasive species problem and a source of biotic homogenization (Ellstrand et al. 2010). Certainly more emphasis needs to be placed on the documentation of taxonomic homogenization by tracking taxa through space and time (Olden and Rooney 2006). Hybridization is occurring between cultivars and their native counterparts within the genera *Quercus, Pyrus, Acer, Malus, Platanus, Populus,* and *Cornus* (Culley and Hardiman 2007; Coart et al. 2003; Khasa et al. 2005; Petit 2004; Schierenbeck and Ellstrand 2010), often within or near riparian areas. However, genetic and population data on hybridizing riparian taxa are limited to just a few examples (Sala et al. 2000) within or outside the context of an ecosystem. There is a lack of models that take into account phylogeographic history, hybridization between historically allopatric species, and contemporary ecological factors such as varying population and habitat size.

Human activities such as introduction of non-natives and fragmentation or other means of degrading natural ecosystems promote hybridization of previously allopatric congeners (Vilà et al. 2000). When introgression accompanies hybridization, dilution of the native gene pool can occur (Abbott 1992), resulting in extinction of the native species (Anttila et al. 1998). Although it is often underrated, human-mediated hybridization is one of the leading causes of biodiversity loss (Muhlfeld et al. 2009). Through introgressive hybridization genetic diversity is reduced, limiting evolutionary flexibility and causing the loss of locally adapted genes along with the populations that once contained them (Rieseberg 1991; Ellstrand and Elam 1993).

Despite the prevalence of invasive species in our landscapes, little is known about how non-native taxa or the spread of their foreign alleles to native taxa may affect ecological relationships (Levine et al. 2003), even for the most dramatic invasions (Dukes and Mooney 2004). Recent experimentation on the loss of riparian ecosystem function has encompassed a number of factors—but not the role of gene flow from non-native populations (Sweeney et al. 2004). Some inroads have been made in the modeling of rates of gene flow that takes into account landscape diversity using conditional genetic distances, but examples are limited (Dyer et al. 2010; Sork and Waits 2010).

HABITAT LOSS AND FRAGMENTATION

Habitat destruction and fragmentation have the most direct effect on species. This effect does not entail just the destruction of habitat but also the aftermath of rural subdivisions, pipelines, power lines, and roads. The species most vulnerable to habitat fragmentation are those with large home ranges or requiring dispersal corridors among discontinuous patches. In addition, edge effects reduce habitat quality, and fragmentation increases the number of edges. Invasive species are often generalists that prefer these disturbed edges.

Even small, isolated, insular habitats can be important repositories for survival. For example, the moth *Argyrotaenia isolatissima* has survived for eighty years on 25 ha on Santa Barbara Island (Rubinoff and Powell 2004). Santa Barbara Island, at only 260 ha, is the most isolated of the Channel Islands and has seven endemic taxa.

Habitat types lost include 90 percent of Southern California coastal sage scrub, 99 percent of coastal prairie, and 95 percent of vernal pools. Fragmentation increases the vulnerability of species to invasion, changes in disturbance regions, changes in pollination, changes to dispersal, loss of genetic connectivity, and an ultimate loss of native species diversity (Crooks et al. 2004). Decreased levels of gene flow, especially among small populations, lead to genetic drift due to inbreeding and overall decreases in genetic diversity (Reed and Frankham 2003). Even in mobile species, fragmentation can have long-term effects on populations; for example, *Canis latrans* and *Lynx rufus* in Southern California have reduced gene flow across major highways (Riley et al. 2005).

Human effects on the population structures of native taxa may be obscured by recent changes in the landscape resulting from tectonic activity along fault lines, marine incursion, and, in many cases, Pleistocene glaciations. Long-lived species have a different vulnerability to extinction than do short-lived species. Nevertheless, critical to their survival and recolonization is the availability of migration corridors and source populations at a number of spatial scales.

GENOTYPE SPECIFICITY

Phylogenetic data provide a picture of the evolutionary history of taxa and their relationships with other taxonomic groups, but population genetic data can be more informative for management and should be utilized to determine whether genetic diversity is being lost (Funk et al. 2010). There is increasing reliance on genetic data to identify reservoirs of genetic diversity within species. Genetically divergent populations occurring in fragmented habitats need designation as evolutionarily significant management units. Examination of fitness and heterozygosity via a metaanalysis of multiple data sets substantiates that a loss of heterozygosity does indeed have a negative impact on fitness (Reed and Frankham 2003). The International Union for the Conservation of Nature (IUCN) specifically lists genetic diversity as one its conservation goals.

The long-term goal of genetic research in conservation is to preserve historical genetic variation that may be critical to the long-term evolutionary survival of a species. Usually this accomplished through the conservation of genetically distinct populations. For example, for *Ambystoma californiense*, older clades in Sonoma and Santa Barbara Counties were identified as critical to the recovery of the species (Shaffer et al. 2004). Quantitative trait loci analysis, which identifies traits of functional diversity, combined with molecular clock data and knowledge of vicariant events, may be crucial to survival of a species (Shaffer et al. 2004). As populations become at risk demographically, some reintroductions may be necessary and will need to be accomplished with genetically and ecologically similar populations.

Genotype specificity can be very important in site restoration. Species without quantitative or molecular differentiation suggest that propagule transfer should have a lower risk of maladaption. Every effort should be made to assess projected climatic shifts and whether the reproductive and phenotypic traits are suitable to future scenarios. The use of native genotypes in restoration is important for avoiding outbreeding depression, which can result in the decline in a native population (Erickson 2008). Local ecotypes are adapted to site conditions, and even though the use of natives may be more expensive, the outcome is greater restoration success (Lesica and Allendorf 1999). The U.S. Forest Service has developed protocols for the use of native genotypes in postfire restoration and other applications.

The conservation of evolutionary processes is dependent on available, appropriate habitat. Those habitats in which these processes are most likely to occur include unusual combinations of climate, geology, and strong environmental gradients (Davis et al. 2008). Surrounding habitats must be conserved in order to maintain ecosystem resilience and are critical to the conservation of evolutionary process at many spatial and temporal scales.

Recommendations to ensure the survival of species include the use of genetically appropriate choices of plant materials for ecosystem restoration in order to maintain biological diversity and the genetic diversity necessary for evolution to continue to occur. Appropriate genetic linkages among fragmented populations must be established, evolutionary interactions conserved, surrounding biological diversity maintained, and changes to ecosystem function minimized (Rogers and Montalvo 2004).

INVASIVE SPECIES

Urbanization reduces specialized habitats, allowing generalists to flourish and further contribute to the functional homogenization of habitats (Olden et al. 2004). Development, agricultural practices, and disturbance have resulted in the virtual disappearance of the pre-European Great Central Valley grasslands: approximately one percent of native grasslands remain (Barry et al. 2006). There are roughly 1,300 non-native plant species in California, of which more than 200 are considered major economic or ecological problems. Invasive species are projected to increasingly spread with climate change and replace natives that often have narrow physiological and dispersal limitations. For example, the invasive Argentine ant (*Linepithema humile*) has replaced virtually every native ant species that forages aboveground in seasonally dry Mediterranean habitats (Holway and Suarez 2006).

Native Lepidopterans are increasingly using non-native plants as hosts. In fact, at least 34 percent of California's native butterfly species use non-native plants (Graves and Shapiro 2003). Unfortunately, in general, specialists are less likely to adapt to new host plants than are generalists, and herbivores with large ranges are more likely to adapt to new hosts than are those with small ranges (Jahner et al. 2011). Of the more than 2,000 endemic plant species in California, over two-thirds are anticipated to lose more than 80 percent of their extant range in the next hundred years (Loarie et al. 2008), a process that will likely cause the co-extinction of many insect herbivores (Jahner et al. 2011).

The susceptibility of an ecosystem to invasion is significantly influenced by biodiversity; as species richness declines, an ecosystem becomes increasingly vulnerable to invasion. Biodiversity influences ecosystem properties and the ability of an ecosystem to respond to change (Hooper et al. 2005).

CLIMATE CHANGE

Species shifts in response to climate change are a serious concern (Warren et al. 2001; Zacherl et al. 2003). The velocity of temperature change is estimated globally to be 0.42 km per year under the A1B (alternative energy balanced with fossil fuel use) emission scenario, with the highest velocities in flooded grasslands and deserts and the lowest in temperate montane regions (Loarie et al. 2008). The key is to provide ecosystems with alternatives to a climatic and disturbance future that is somewhat uncertain. An argument has been made to prioritize the conservation of regions with high levels of evolutionary diversity in order to capitalize on genetic and functional diversity (Cadotte and Davies 2010). Montane areas will likely provide refuge for many species, and the creation of large desert reserves may reduce losses of xeric taxa (Loarie et al. 2008). Small reserves are problematic in that they provide only temporary protection. Measures of evolutionary diversity need to be developed to maximize ecosystem health and minimize losses to biological diversity (Cadotte and Davies 2010).

The life histories of insects are particularly susceptible to temperature and the availability of food resources, primarily the plants with which they evolved (Pelini et al. 2009). However, because insects and plants disperse and reproduce at varying rates, their respective responses to climate change could become a problem for either group (Schweiger et al. 2008). Butterfly species have demonstrated migration to higher elevations or latitudes of 240 km per 30 years (Parmesan et al. 1999), whereas tree migration responds to climate change at 20–40 km per 100 years (Davis and Shaw 2001). The context of the plant-animal interaction also needs to be taken into consideration, for example, the strength of the relationship or the availability of alternative host plants or pollinators. For example, *Erynnis propertius* (dusky-wing skipper) occurs on the Pacific coast from Baja California to British Columbia where the host plants, species of *Quercus*, occur (Guppy and Shepard 2001; Pelini et al. 2010). Southern and Central California populations of *E. propertius* feed on *Q. agrifolia*, and higher-elevation populations feed on *Q. kelloggii*. Host plant experiments have found that *E. propertius* have zero survival on other *Quercus* species. Southwestern Oregon populations are dependent on *Q. garryana*, but overlaps between *Q. agrifolia* and *Q. garryana* are very limited. If *E. propertius* has reduced fitness on alternative host plants and the host plants migrate at much slower rates than *E. propertius,* their survival is in question (Pelini et al. 2010).

Significant changes are projected to occur in the subalpine flora and fauna of California as climate change progresses (Hayhoe et al. 2004; Moritz et al. 2008). The plants with the smallest ranges are found in the high Sierra Nevada and Transverse Ranges and are projected to be the most heavily affected by climate change (Loarie et al. 2008). Coastal communities are particularly vulnerable to increasing fluctuations in temperature and wave action and increased coastal inundation.

LEGISLATION, GOVERNMENT, AND NONGOVERNMENTAL ORGANIZATIONS

A number of laws and regulations protect California's threatened and endangered species, among them, the California Endangered Species Act, the U.S. Endangered Species Act, the Native Plant Protection Act, and the National Environmental Protection Act (NEPA) and California Environmental Quality Act (CEQA), both of which require environmental review of proposed development projects. Yet with the projected population growth, losses will continue even with major efforts. The California Natural Diversity Database, a "natural heritage program," is a valuable source of at-risk species and part of a network of like programs nationwide that provide location and ecological information on special status plants and animals and communities to scientists, nongovernmental organizations (NGOs), and the public. Different conservation groups exist that concentrate their efforts on particular organismal groups; for example, the California Native Plant Society works to protect native plant species, whereas CalTrout focuses on the conservation of steelhead, trout, and salmon in rivers across the state in their historic ranges.

The Natural Community Conservation Planning Act (NCCPA) was instituted by the California Department of Fish and Wildlife (CDFW) in 1991 to identify and provide resources for regional awareness of the importance of biodiversity and works in concert with the California and federal Endangered Species Acts. The NCCPA uses a regional approach to conserve biological diversity but requires species- and community-specific data to promote the success of habitat conservation. As of December 2013, there were 25 Natural Community Conservation Plans in California; 20 counties with conservation land areas in their general plans; 29 additional authorities, districts, or companies as signatories to plans; and 52 cities as signatories to plans.

The Habitat Conservation Planning Branch (HCPB) of the CDFW is a critical component of ecosystem-level conservation. It works closely with the NCCPA, which is identifying and providing large-area protection of native ecosystems. Mitigation banking for conservation was approved in 2013 and received long-term funding; draft guidelines for the program are currently under review. Regional biodiversity plans must conserve the ability of diverse assemblages of species to respond to evolutionary and ecological processes on a dynamic landscape (Whittaker et al. 2005; Kraft et al. 2010). Impediments to species dispersal can have a great effect on endemic species, preventing them from responding to climate change, and thus the conservation or creation of migrational corridors is critical (Loarie et al. 2008).

A number of major international agreements are relevant to biodiversity. They include UNESCO's Man and Biosphere Program, the Convention on the International Trade in Endangered Species (wild fauna and flora) (CITES), and the Convention on Biological Diversity (CBD).

RESTORATION, RESTORATION, RESTORATION

Ecosystem restoration will be of increasing importance as the population of California grows. Two and a half centuries of overharvesting, dam building, levee construction, water diversion, overgrazing, invasive species introduction, and development have resulted not only in the direct loss of habitat but also in alterations to the landscape that prevent the ability of many taxa to respond to evolutionary change.

Habitat restoration should occur at every scale, regionally and incrementally, no matter how small the increment. This includes the restoration of lawns to native habitat or for growing food; and city and suburban landscapes at the urban-wildland interface should be planted with native species. Although conflicts between humans and native species will certainly become exacerbated, it is the human landscape that must provide the natural canvas on which evolution will occur.

Biodiversity is well established as key to ecosystem service, and, in turn, ecosystem degradation is a significant catalyst in biodiversity loss (Norris 2012). Intense urbanization has been documented to have a major impact on community species composition, population sizes, and genetic diversity (Delaney et al. 2010). Biodiversity is critical for the provision of ecosystem services such as pollination, timber, water quality, carbon sequestration, flood prevention, and nutrient cycling. Although seed banks are providing a mechanism to conserve genetic samples of species while there is time, the practice is clouded by climate change. Losses to biodiversity can be reversed through sound restoration practices; however, reintroductions of some taxa or the restoration of some ecosystems have proven to be very difficult under even the best circumstances. A metaanalysis of eighty-nine restoration projects in an array of ecosystems found that although reference ecosystems maintain higher biodiversity and ecosystem service values, on average, ecological restoration increases biodiversity by 44 percent and ecosystem services by 25 percent (Rey Benayas et al. 2009).

EDUCATION

Surprisingly, or perhaps not surprising at all given the limited number of population geneticists and systematic biologists currently being trained in California, phylogeographic studies are lacking in a number of critical regions. Unfortunately, land management and conservation decisions are hindered by a lack of taxonomic expertise or knowledge of biological diversity. A recent assessment led by the Chicago Botanical Garden and the U.S. Department of Education documented that undergraduate degrees in botany have declined nationwide by 50 percent and graduate degrees by 41 percent in the past twenty-five years, with similar trends in zoology (Kramer et al. 2010). In addition, in a survey of faculty at more than four hundred universities, about 40 percent indicated that courses with organismal emphases have declined in the past decade. The public universities of California have the responsibility to educate professionals trained in the evolution of its diverse biota, which includes incorporating evolutionary process in the management of its resources.

COLLECTIONS

Most collections in California are partly or entirely supported by the state and housed in public institutions. These collections have proven to be invaluable for predicting species responses to climate change, establishing historical species distributions, providing genetic information, and testing new hypotheses. In addition to their educational value, collections are an integral part of managing fish, wildlife, and plant communities and deserve state funding just as our land and species deserve protection. The more knowledge we have about the past and present biodiversity of California, the better we can protect it.

AGRICULTURE

Agriculture can play a major role in the conservation of biological diversity and evolutionary process. Sustainable agriculture practices have not only proven useful in terms of direct economic benefits; practiced properly, they can reduce soil erosion, reduce the need for fertilizers, improve groundwater recharge, increase forage for livestock, and improve flood control. Native plant hedgerows have been quite effective in strengthening pollinator populations and reducing invasive species. A metaanalysis of sixty-three studies that compared biodiversity of farms that were organically and conventionally managed found a significant positive effect of organic practices, with a mean increase in species richness of 30 percent and a mean increase in abundance of 50 percent (Bengtsson et al. 2005).

Carefully managed grazing can control non-native species and sensitive animal and plant species (Weiss 1999). Seasonal wetlands can be improved by grazing, which removes invasive annual grasses that reduce sensitive amphibian and plant habitat (Marty 2005; Wingo 2009). However, overgrazing, particularly in riparian areas, reduces appropriate habitat for a number of sensitive species, including *Actinemys marmorata* and *Rana aurora draytonii*, by destabilizing stream banks, trampling vegetation, increasing nutrients, and widening channels, which results in higher temperatures (Moyle 2002).

THE ROLE OF THE INDIVIDUAL

Each person has the ability to promote biological diversity by planting native species, supporting the funding and hiring of organismal biologists, supporting conservation organizations, and directly participating in educating the public about the importance of conservation. State and local jurisdictions should be encouraged to plant native vegetation along highways in order to increase local biodiversity and to build bird- and bat-friendly structures that attract beneficial species. In addition to improving food security by buying and growing local food and reducing waste by recycling and composting, the urban agriculturalist and gardener contributes to a lower carbon footprint and likely increases his or her knowledge of local ecosystems and local biological diversity.

Urban agriculture and native plant gardens are also a useful way to provide habitat, not only for the plants, but also for insects. The importance of using native plants in restoration and general landscaping cannot be overemphasized. A comparison of bird diversity on land landscaped primarily with native plants versus non-native ornamentals found that in an area with a mean size of 2.53 ha there were statistically significant increases in native insect and bird species abundance (Burghardt et al. 2009). Non-native species provide few ecological benefits to wildlife and pose the potential risk of invasion. Most insects are unable to use non-native species for growth and reproduction (Tallamy 2007).

Although there has been an increase in the use of native species by landscape architects and contractors, much work remains to be done in this area (Brzuszek et al. 2007). A number of communities nationwide are increasingly using native plants to improve water quality, manage invasive species, restore habitat, and even provide an alternative to sewage treatment (McMahan 2006). The Leadership in Energy and Environmental Design (LEED) Platinum status can only be achieved through the use of native plants in landscaping (new.usgbc.org/leed/).

REGIONAL STRATEGIES

Klamath-Siskiyou Region

The IUCN has identified the Klamath-Siskiyou ecoregion as an "area of global botanical significance" (Della Sala et al. 1999; Villa-Lobos 2003). The biodiversity of this region is a result, in part, of species from the north reaching their southern limit of distribution and species from the south reaching their northern limit. Geological complexity, especially in those areas that provide edaphic soils, further contribute to the botanical diversity of the area. The Klamath-Siskiyou region escaped oceanic submergence, major glaciation effects, and volcanic activity throughout its history and is considered to have retained botanical stability for over 100 million years (Whittaker 1961; Villa-Lobos 2003).

Forest restoration and restoration for conservation have different goals, but on the basis of molecular markers, breeding groups for *Notholithocarpus densiflorus*, one of the species most susceptible to sudden oak death (SOD), have been developed into conservation areas based on adaptive characteristics and resistance to SOD (Nettle et al. 2009). On the coast there are two clades of *N. densiflorus*, the central coastal clade and the Sierra Nevada/Klamath/southern Oregon clade, with some admixture on the northern coast. Deep divergence is present among the Sierra Nevada and Klamath populations, and the best strategy for the conservation of *N. densiflorus* is the maintenance of as much genetic diversity as possible (Nettle et al. 2009).

Based on the decline of *Strix occidentalis caurina*, the U.S. Forest Service has established a management policy for this region that includes habitat protection for a network of late successional reserves (Taylor and Skinner 1998). The Klamath-Siskiyou ecoregion is primarily contained within national forests, but only 10.5 percent of these forests are protected or managed for late successional reserves or old growth to conserve northern spotted owl habitat (Caesar et al. 2006). However, the management for exclusively old growth forests does not entail the conservation of genetic diversity of other species (Noss et al. 1999)

There is a great deal of political discussion about finding solutions for the water issues in California that will please both environmentalists and the thirsty water users of this state, but in reality the conservation of biological diversity will depend on economic sacrifices by many. The biggest need for biodiversity conservation in this area is better permanent protection, which will require a strong alliance between the financial and conservation communities. Of the 4.5 million ha Klamath River watershed, approximately two-thirds are public land, most of it federal, but habitat protection for the native fisheries cannot occur without the restoration of water to the Klamath River. In places where salmonid diversity is high, such as the Smith River and the Klamath Basin, the prevention of extinction may require the prioritization of fish conservation over other considerations (Katz et al. 2012). Salmonids are at the southern extent of their range in this region, and environmental degradation has already caused a significant decline in the quality of their habitat in the Klamath Basin (Quinones and Moyle 2012). Cold water habitat must be maintained and enhanced to ensure the survival of these fisheries.

Although there are a number of wilderness areas in the Klamath-Siskiyou region, increased habitat connectivity is required for the core areas. In particular, the region needs protection of rare species hotspots, old growth forests, key watersheds, representative physical and vegetative habitat sites in order to maintain viable populations of focal species. Scientifically based advocacy groups working toward these goals in the region include the Klamath Center for Conservation Research in Orleans, California; the Klamath-Siskiyou Wildland Center in Ashland, Oregon, which focuses on the Klamath and Rouge River watersheds; and the Conservation Biology Institute in Corvallis, Oregon.

The Sierra Nevada

The Sierra Nevada is home to 40 native fish taxa, 45 percent of which are either listed as threatened or in some state of decline because of watershed decline, dams and diversions, and invasive species (Moyle et al. 2012). Californians are at the point at which they need to decide how much they value native biodiversity and get very serious about water conservation and watershed management. Eighty-one percent of anadromous taxa and 73 percent of inland fish taxa are threatened with extinction, and some species are already extinct (Katz et al. 2012). The western Sierra Nevada historically provided major spawning grounds for anadromous fishes such as *Oncorhynchus tshawytscha, O. mykiss,* and *Lampetra tridentatus,* all of which are now in decline. Twenty-five of 32 species (78 percent) of salmonids native to California will be extinct in California within the next century. The detrimental genetic effects of hatchery fish on the population genetics and adaptive ability of wild populations is well established (Chilcote et al. 2011). Recommended actions for fish populations include enforcement of existing laws, enhancement of native populations rather than supplementing with hatchery fish, restoration of habitat connectivity, including dam removal, and restoration of riparian vegetation to protect cold water habitats.

The Sierra Nevada Conservancy is a state agency created by legislation in 2004 that involves 22 counties encompassing over 10 million ha. It has been charged with finding ways to conserve the environment that are both economically and socially acceptable. Funded by Proposition 84, it has contributed significantly to the conservation of the region through 175 projects that range from interpretive programs to developing recreational plans compatible with conservation goals to land acquisition through land trusts. Other conservation organizations include the Center for Sierra Nevada Conservation, which advocates for sound management practices for species conservation. The Truckee-Donner Land Trust and the Trust for Public Land recently purchased 1,200 ha at Donner Summit to save it from development.

Globally, amphibian decline has reached alarming levels. According to the IUCN, 41 percent of species are at risk of extinction. *Rana muscosa* decline in the Sierra Nevada has been linked to predation by invasive trout, habitat degradation, *Chytridiomycosis*, channelization, climate change, fire, and pollution (Briggs et al. 2005). Pesticide use has been strongly implicated in amphibian decline (Davidson 2004). Development levels as low as 8 percent are documented to have a significant impact on amphibian populations (Riley et al. 2005).

Coast Ranges

The Coast Ranges, which run the length of California, were formed variously in the early Paleozoic through the late Cenozoic from a variety of substrates, resulting in high geographic diversity. The numerous studies cited here underline the importance of this habitat for evolutionary diversification, yet due to the desirable climate and vegetation, human development along the coast is intense. Coastal Southern California has very high levels of biological diversity, with one-third of the native flora and 487 native vertebrate species in a six-county region that also contains 60 percent of the state's human population (Vandergast et al. 2008). Ninety percent of coastal sage shrub habitat and 99 percent of coastal prairies are gone (Hunter et al. 2003). Formerly in an almost continuous habitat along the coast from Oregon to Santa Cruz, *Sequoia sempervirens* forests have declined by 85 to 90 percent (Barbour et al. 2007). Continued threats to the Coast Ranges include habitat fragmentation due to human population growth, particularly in Southern and Central California. Increased urbanization and intensive agriculture have resulted in land conversion, erosion, and pollution. Overgrazing, invasive species, stream diversions, and recreational pressures continue to affect habitat quality.

Although the California Coastal Conservancy, along with a number of dedicated conservation organizations, focuses on the protection and restoration of coastal communities, there is at present no comprehensive plan for the protection of the Coast Range habitats. The northern Coast Ranges are the focus of the Coast Range Association, and the California Rangeland Conservation Coalition is dedicated to rangeland protection in the interior Coast Ranges.

The central coast includes the San Francisco Bay Area, a nine-county region that features high biological diversity. The National Park Service has a number of units in the region that total about 30,500 ha, but very few of these are in a natural state. The East Bay Regional Park District manages about 36,500 ha as natural lands (Trzyna 2001). Although these holdings are significant and important to habitat diversity, many units are small fragments and are highly disturbed. The Conservation Lands Network, a nonprofit consortium of scientists and policy makers, has made a number of important recommendations that emphasize the need for stream restoration and rangeland conservation and management (Bay Area Open Space Council 2011).

The newly designated Pinnacles National Park, east of Monterey, contains remnants of a volcano originally formed over 23 Ma along the San Andreas fault on a base of Santa Lucia granite (Chronic 1986). The park is located on about 9,713 ha in the Gabilan Mountains of the central Coast Ranges. Due to a variety of microclimates, the vegetation is diverse in the park: over 600 plant taxa support an equally diverse fauna, including purportedly the most diverse assemblage of bees on the planet (Messinger and Griswold 2002).

Estuarine/Coastal Habitat

The San Francisco Bay Delta was formerly a very large tidal marsh with a network of sloughs most of which has been converted to levees and islands surrounded by deep channels. Tidal fluctuations are responsible for most of the nutrient influx into the area (Lehman et al. 2010). Much of the San Francisco Bay Delta habitat has been converted. In the south bay, 80–90 percent of the shoreline habitat has been converted to salt production, urban development, or agriculture. However, efforts are under way to restore at least some of these areas. For example, about 6,000 ha of salt ponds are undergoing, or are proposed for, restoration (San Francisco Estuary Project 2007). Reductions in freshwater input, increased salinity due to rising sea levels, invasive species, pollution from agricultural and urban runoff, and warming waters further affect water quality in the delta, which is projected to become more similar to ocean rather than estuarine habitat. Species expected to undergo extinction in the delta in the next fifty years include *Spirinchus thaleichthys* (longfin smelt), *Hypomesus transpacificus* (delta smelt), *Pogonichthys macrolepidotus* (Sacramento splittail), and *Oncorhynchus tshawytscha* (Chinook salmon) (Moyle et al. 2012). *Gila crassicauda*, some *Oncorhynchus* clades, and *Archoplites interruptus* are already extirpated from the area (Moyle et al. 2012).

The Bay Delta Conservation Plan includes provisions for some restoration and habitat protection, but habitat degradation will continue to occur with water diversion. As currently proposed, spring run *Oncorhynchus tshawytscha*, *O. mykiss*, and *Acipenser medirostris* (green sturgeon) will continue to be affected by diversions and changing water temperatures. The Bay-Delta Interagency Ecological Program monitors and recommends action on the biological and physical aspects of the bay.

Central Valley

A former mosaic of wetlands, riparian forests, grassland, and oak woodlands, the Central Valley is unrecognizable after less than two hundred years of European expansion. Any species with a historical distribution in the Central Valley or with urbanization or suburbanization pressure is under threat due to the direct loss of habitat, fragmentation, and invasive species. The loss of chaparral along the flanks of the Central Valley due to development, as well as fire suppression and subsequent conflagrations, has led to the decline of a number of taxa, especially reptiles that are dependent on native insect species for their diet (Leaché et al. 2009).

Wetlands in California once covered more than 2 million ha, mostly in the Great Central Valley; today wetlands cover approximately 214,000 ha (Barbour et al. 2007). Riparian areas have been documented to increase species richness across taxonomic groups by more than 50 percent globally, yet less than 2 percent of historical riparian forests in the Central Valley remain intact (Sabo et al. 2005). Of the native grasslands in California, a major vegetation component of the Central Valley, less than 1 percent remains (Barbour et al. 2007). Approximately 90 percent of the plant species listed by the California Native Plant Society as rare and endangered occur in grasslands or oak savannas (Skinner and Pavlik 1994).

The Pacific Flyway, which runs roughly through the Central Valley, is used by approximately 14 million wild birds, and over 9 million birds winter in the wetlands (Garone 2011). Over 135 species of birds are completely dependent on streamside woodlands or need them at one stage in their life cycle, but over 95 percent of riparian habitat in the Sacramento Valley is lost or severely degraded (Katibah 1984). Special status species that depend on riparian habitat include *Coccyzus americanus occidentalis* (western yellow-billed cuckoo), *Buteo swainsoni*, *Riparia riparia* (bank swallow), and *Eumops perotis* (western mastiff bat) (CALFED 2000).

Vernal pools in the Central Valley and surrounding foothills have suffered devastating losses in the past 150 years. Almost 200 plant species are either endemic or associated with vernal pools, 55 percent of which are endemic to California (Holland 1976). Animals that depend on vernal pool habitat include a number of taxa from the Crustacea, Insecta, and Amphibia, and Aves. The presence of vernal pools in flat, easily accessible land has made them prone to development pressure, and alterations to surrounding areas can negatively affect their hydrological properties (Cheatham 1976). In addition to their direct loss, reduced connectivity among vernal pools has resulted in the extirpation and loss of gene flow in a number of taxa, both plant and animal. A CDFW report identified and prioritized vernal pool regions and their conservation needs throughout California (Keeler-Wolf et al. 1998).

Riparian habitat has multiple levels of conservation value. It prevents erosion and increases regional species richness by more than 50 percent globally, on average (Sabo et al. 2005). The clearing of riparian forest vegetation in the conterminous United States was already under way by 1730, and less than one-third of the original extent of these forests remains (Naiman et al. 1995). In California, less than 5 percent of the riparian forests present prior to European settlement remain, and of these forests, more than half are degraded, fragmented, or otherwise in poor condition (Barbour et al 1993). The degradation and decline of biodiversity in riparian ecosystems is exacerbated by the invasion of non-native vegetation that prevents the successful establishment and persistence of native species and communities and disrupts ecosystem processes (Griffin et al. 1989; Hood and Naiman 2000; Huenneke and Thompson 1995). Organisms thought to be formerly distributed throughout the Central Valley have experienced dramatic habitat loss due to conversion to farmland, drainage, and development.

Tehachapi Mountains/Transverse Ranges

The Tehachapi Mountains historically have been a migration corridor between the Transverse Ranges and the Sierra Nevada and have been long recognized as an area of high biological diversity and evolutionary divergence (Stebbins and Major 1965; Patton and Smith 1990). Conservation efforts have improved in recent years, but much of the area remains under threat. The Tejon Ranch, 109,265 ha at the juncture of the Tehachapi Mountains, the San Joaquin Valley, the Transverse Ranges, and the Mojave Desert, was recently designated a conservation easement (White et al. 2003). It is home to 20 state and federally listed species and 60 other rare taxa.

The Transverse Ranges have only about 30 percent of their original habitat intact, including the Ventana Wilderness Area in the Santa Lucia Range and the Ventura region. There are very high levels of fragmentation and isolation of contiguous habitat. Although much of the land falls within the Los Padres National Forest, the area is heavily impacted by logging, air pollution, disturbance to aquatic habitat, overgrazing, recreational use, and fire suppression (David Olson, TNC, https://worldwildlife.org/ecoregions/na1203).

A GIS analysis of 21 vertebrate and invertebrate species in Southern California was used to identify geographic areas with important phylogeographic breaks. It found five areas in which there was high genetic connectivity, six areas with high interpopulational divergence, and three areas with high intrapopulation genetic diversity for eight geographic regions, five of which are unprotected (Vandergast et al. 2008). The areas with the highest diversity include the northwestern and central Transverse Ranges, the Sierra Pelona, the San Gabriel Mountains, the San Bernardino Mountains, Warner Springs, and the southern Peninsular Ranges (Vandergast et al. 2008).

Mojave Desert

The Mojave Desert encompasses approximately 140,000 km^2 between two major metropolitan cities. It is under continuing threat from encroaching housing and building development, solar development, and invasive species. Although the recent national park monument designation has provided some protection, there are a number of state-endemic special status taxa in the Mojave Desert Region: 31 vertebrates, including *Gopherus agassizii*, *Uma scoparia*, *Xerospermophilus mohavensis*, *Cyprinodo*n spp. *Microtus californicus mohavensis* (Mojave River vole), *M. c. scirpensis* (Armargosa River vole), and 29 invertebrates (Bunn et al. 2007; Bell et al. 2010).

Conservation areas established include the Mojave National Preserve and Death Valley National Park and the Desert Tortoise Research Natural Area. The California Desert Protection Act of 2011 designated two new national monuments, Mojave Trails and Sand to Snow, and added land area to the Joshua Tree and Death Valley National Parks and the Mojave National Preserve. However, much of the desert remains under threat from groundwater overdrafting, solar and other kinds of development, mining, invasive species, overgrazing, military use, and inappropriate off-road vehicle use.

The Coast

Oceanographers predict that the CCS will become increasingly stratified and suffer reduced plankton production (Palacios et al. 2004). Circulation patterns will be altered by changing winds, which will increase upwelling (Checkley and Barth 2009; Snyder et al. 2003) and may offset increased stratification. Increased CO_2 production will raise acidification levels in the ocean. Global warming will result in the northward distribution of taxa, affecting the phenological development of organisms on both land and sea.

The California coast is 1,931 km long; its nearshore and offshore waters encompass 15,378,079 ha that are topographically and biologically diverse. Until the passage of the Marine Life Protection Act (MLPA) in 1999, only about one percent of these areas were under protection. As part of the CDFW code, the MLPA requires an evaluation and design or redesign of California's Marine Protected Areas in five regional areas (North Coast, San Francisco Bay, North Central Coast, Central Coast, and South Coast). The MLPA directs California to protect marine biodiversity and ecosystem integrity, rebuild depleted populations, and improve recreation and education in a way that is compatible with biodiversity. In collaboration with scientists and the public, the MLPA has resulted in the designation of twenty-nine new Marine Protected Areas.

There have been significant declines in the anadromous fisheries of California, most notably those of *Oncorhynchus tshawytscha, O. kisutch,* and *O. mykiss.* These declines have resulted in dramatic losses to populations of marine organisms and commercial fisheries. Of particular concern are disruptions to spawning habitat caused by dams, water diversion, urbanization, and logging, as well as introduced species in the streams that feed into the San Francisco Bay estuary. The most endangered fisheries are those that occur in coastal Southern California (Moyle et al. 1995). Aquatic organisms along the coast and throughout the waterways of California are threatened by invasive species, thermal and other forms of pollution, excessive recreational use, and channelization of streams and rivers. Increased collaboration among conservationists, fisheries management, and fishermen is essential to sustain economically viable fisheries and biological diversity off the coast of California (Salomon et al. 2011).

Bibliography

Abbott, R.J. 1992. Plant invasions, interspecific hybridization and the evolution of new plant taxa. *Trends in Ecology and Evolution* 7:401–5.

Ackerly, D. D. 2004. Adaptation, nice conservatism, and convergence: Comparative studies of leaf evolution in the California chaparral. *American Naturalist* 163:654–71.

———. 2009. Evolution, origin and age of lineages in the Californian and Mediterranean floras. *Journal of Biogeography* 36:1221–33.

Ackerly, D. D., C. A. Knight, S. B. Weiss, K. Barton, and K. P. Starmer. 2002. Leaf size, specific leaf area and microhabitat distribution of chaparral woody plants: Contrasting patterns in species and community level analyses. *Oecologia* 130:449–57.

Adam, D. P., J. P. Bradbury, H. J. Rieck, and A. M. Sarna-Wojcicki. 1990. Environmental changes in the Tule Lake Basin, Siskiyou and Modoc Counties, California, from 3 to 2 million years before present. *U.S. Geological Survey Bulletin* 1933:1–13.

Addicott, W. O. 1969. Tertiary climatic change in the marginal northeastern Pacific Ocean. *Science* 165:583–86.

Agnarsson, I., and L. May-Collado. 2008. The phylogeny of Cetartiodactyla: The importance of dense taxon sampling, missing data, and the remarkable promise of Cytochrome *b* to provide reliable species-level phylogenies. *Molecular Phylogenetics and Evolution* 48:964–85.

Aguilar, A., and W. J. Jones. 2009. Nuclear and mitochondrial diversification in two native California minnows: Insights into taxonomic identity and regional phylogeography. *Molecular Phylogenetics and Evolution* 51:373–81.

Aïnouche, A.-K., R. Bayer, and M.-T. Misset. 2004. Molecular phylogeny, diversification and character evolution in *Lupinus* (Fabaceae) with special attention to Mediterranean and African lupines. *Plant Systematics and Evolution* 246:211–22.

Alexander, M. P., and K. J. Burns. 2006. Intraspecific phylogeography and adaptive divergence in the white-headed woodpecker. *Condor* 108:489–508.

Álvarez-Castañeda, S. T. 2010. Phylogenetic structure of the *Thomomys bottae-umbrinus* complex in North America. *Molecular Phylogenetics and Evolution* 54:671–79.

Alvin, K. L.,1960. Further conifers of the Pinaceae from the Wealden Formation of Belgium. *Institut Royal des Sciences Naturelles de Belgique Memoires* 146:1-39.

Anacker, B. L., J. B. Whittall, E. E. Goldberg, and S. P. Harrison. 2011. Origins and consequences of serpentine endemism in the California flora. *Evolution* 65:365–76.

Anderson, P. M., A. V. Lozhkin, and L. B. Brubaker. 2002. Implications of a 24,000-yr palynological record for a Younger Dryas cooling and for boreal forest development in northeastern Siberia. *Quaternary Research* 57:325–33.

Andreasen, K., and B. G. Baldwin. 2003. Reexamination of relationships, habitat evolution, and phylogeography of checker mallows (*Sidalcea*; Malvaceae) based on molecular phylogenetic data. *American Journal of Botany* 90:436–44.

Anttila, C. K., C. C. Daehler, N. E. Rank, and D. R. Strong. 1998. Greater male fitness of a rare invader (*Spartina alterniflora*, Poaceae) threatens a common native (*Spartina foliosa*) with hybridization. *American Journal of Botany* 85:1597–1601.

Aranda-Manteca, F. J., D. Domning, and L. G. Barnes. 1994. A new middle Miocene Sirenian of the genus *Metaxytherium* from Baja California and California: Relationships and paleobiogeographic implications. *Proceedings of the San Diego Society of Natural History* 29:191–204.

Arbogast, B. S. 1999. Mitochondrial DNA phylogeography of the New World flying squirrels *Glaucomys*: Implications for Pleistocene biogeography. *Journal of Mammalogy* 80:142–55.

———. A brief history of the New World flying squirrels: Phylogeny, biogeography and conservation genetics. *Journal of Mammalogy* 88: 840–49.

Arbogast, B. S., R. A. Browne, and P. D. Weigl. 2001. Evolutionary genetics and Pleistocene biogeography of North American tree squirrels (*Tamiasciurus*). *Journal of Mammalogy* 82:302–19.

Arbogast, B. S., and G. J. Kenagy. 2001. Comparative phylogeography as an integrative approach to historical biogeography. *Journal of Biogeography* 28: 819–25.

Archie, J. W., and M. O. Quijano. 2010. Fine scale phylogeography of *Sceloporus occidentalis* in the Transverse Ranges of California reveals coincidence with geological complexity. Meeting abstract. Society for Integrative and Comparative Biology 37.9.

Arnason, U., and A. Gullberg. 1994. Relationship of baleen whales established by Cytochrome *b* gene sequence comparison. *Nature* 367:726–28.

Atwater, T. 1970. Implications of plate tectonics for the Cenozoic tectonic evolution of western North America. *Geological Society of America Bulletin* 81:3513–36.

Aubry, K. B. 1984. The recent history and present distribution of the red fox in Washington. *Northwest Science* 58:69–79.

Aubry, K. B., M. J. Statham, B. N. Sacks, J. D. Perrine, and S. Wisely. 2009. Phylogeography of the North American red fox: Vicariance in Pleistocene forest refugia. *Molecular Ecology* 18:2668–86.

Austin, G. T., and D. D. Murphy. 1987. Zoogeography of Great Basin butterflies: patterns of distribution and differentiation. *Great Basin Naturalist* 47:186-201.

Avise, J. C. 2000. Stability, equilibrium and molecular aspects of conservation in marine species. *In* A. M. Solé-Cava, C. A. M. Russo, and J. P. Thorpe [eds.], *Marine Genetics*, xi–xii. Kluwer, Dordrecht, The Netherlands.

———. 2008a. The history, purview, and future of conservation genetics. *In* S. P. Carroll and C. W. Fox [eds.], *Conservation Biology: Evolution in Action*, 5–15. Oxford University Press, Oxford.

———. 2008b. Three ambitious (and rather unorthodox) assignments for the field of biodiversity genetics. *Proceedings of the National Academy of Sciences USA* 105:11564–70.

———. 2009. Phylogeography: Retrospect and prospect. *Journal of Biogeography* 36:3–15.

Avise, J. C., R. T. Alisauskas, W. S. Nelson, and C. D. Ankney. 1992. Matriarchal population genetic structure in an avian species with female natal philopatry. *Evolution* 46:1084–96.

Avise, J. C., C. D. Ankney, and W. S. Nelson. 1990. Mitochondrial gene trees and the evolutionary relationship of mallard and black ducks. *Evolution* 44:1109–19.

Avise, J. C., and C. F. Aquadro. 1982. A comparative summary of genetic distances in the vertebrates: Patterns and correlations. *Evolutionary Biology* 15:151–85.

Avise, J. C., J. Arnold, R. M. Ball, E. Bermingham, T. Lamb, J. E. Neigel, C. A. Reeb, and N. C. Saunders. 1987. Intraspecific phylogeography: The mitochondrial DNA bridge between population genetics and systematics. *Annual Review of Ecology and Systematics* 18:489–522.

Avise, J. C., and F. J. Ayala. 1976. Genetic differentiation in speciose versus depauperate phylads: Evidence from the California minnows. *Evolution* 30:46–58.

Avise, J. C., R. A. Lansman, and R. O. Shade. 1979. The use of restriction endonucleases to measure mitochondrial DNA sequence relatedness in natural populations. I. Population structure and evolution in the genus *Peromyscus. Genetics* 92:279–95.

Avise, J. C., J. E. Neigel, and J. Arnold. 1984. Demographic influences on mitochondrial DNA lineage survivorship in animal populations. *Journal of Molecular Evolution* 20:99–105.

Avise, J. C., and D. Walker. 1998. Pleistocene phylogeographic effects on avian populations and the speciation process. *Proceedings of the Royal Society of London B* 265:457–63.

Axelrod, D. I. 1958. Evolution of the Madro-Tertiary Geoflora. *Botanical Review* 24:7.

———. 1964. The Miocene Trapper Creek flora of southern Idaho. *University of California Publications in Geological Sciences* 51:143–48.

———. 1966. The Pleistocene Soboba flora of Southern California. *University of California Publications in Geological Sciences* 60:1–109.

———. 1975. Evolution and biogeography of Madrean-Tethyan sclerophyll vegetation. *Annals of the Missouri Botanical Garden* 62:280–334.

———. 1976. History of the coniferous forests, California and Nevada. *University of California Publications in Botany* 70:1–62.

———. 1977. Outline history of California vegetation. *In* M. G. Barbour and J. Major [eds.], *Terrestrial Vegetation of California*, 140–93. John Wiley and Sons, New York.

———. 1979. Age and origin of Sonoran Desert vegetation. *Occasional Papers of the California Academy of Sciences* 132:1–74.

———. 1981. Holocene climatic changes in relation to vegetation disjunction and speciation. *American Naturalist* 117:847–70.

———. 1983. Paleobotanical history of the western deserts. *In* S. G. Wells and D. R. Haragan [eds.], *Origin and Evolution of Deserts*, 113–29. University of New Mexico Press, Albuquerque.

———. 1985. *Miocene Floras from the Middlegate Basin, West-Central Nevada*. University of California Press, Berkeley. 279 pp.

———. 1986. The Sierra redwood (*Sequoiadendron*) forest; end of a dynasty. *Geophytology* 16:25–36.

———. 1990. Ecologic differences have separated *Pinus remorata* and *P. muricata* since the early Pleistocene. *American Journal of Botany* 77:289–94.

———. 1995. *The Miocene Purple Mountain Flora of Western Nevada*. University of California Press, Berkeley. 62 pp.

Axelrod, D. I., and F. Govean. 1996. An early Pleistocene closed-cone pine forest at Costa Mesa, Southern California. *International Journal of Plant Sciences* 157:323–29.

Azevedo, J., and D. L. Morgan. 1974. Fog precipitation in coastal California forests. *Ecology* 55:1135–41.

Bagley, M. J., and G. A. E. Gall. 1998. Mitochondrial and nuclear DNA sequence variability among populations of rainbow trout (*Oncorhynchus mykiss*). *Molecular Ecology* 7:945–61.

Baker, C. S., D. A. Gilbert, M. T. Weinrich, R. H. Lambertsen, J. Calambokidis, B. McArdle, G. K. Chambers, and S. J. O'Brien. 1993. Population characteristics of DNA fingerprints in humpback whales (*Megaptera novaeangliae*). *Journal of Heredity* 84:281–90.

Baker, M., N. Nur, and G. R. Geupel. 1995. Correcting biased estimates of dispersal and survival due to limited study area: Theory and an application using wrentits. *Condor* 97:663–74.

Baker, C. S., and S. R. Palumbi. 1996. Population structure, molecular systematics and forensic identification of whales and dolphins. *In* J. C. Avise and J. L. Hamrick [eds.], *Conservation Genetics: Case Histories from Nature,* 10–49. Chapman and Hall, New York.

Baker, C. S., S. R. Palumbi, R. H. Lambertsen, M. T. Weinrich, J. Calambokidis, and S. J. O'Brien. 1990. Influence of seasonal migration on geographic distribution of mitochondrial DNA haplotypes in humpback whales. *Nature* 344:238–40.

Baker, C. S., A. Perry, J. L. Bannister, M. T. Weinrich, R. B. Abernethy, J. Calambokidis, J. Lien, R. H. Lambertsen, J. Urban R., O. Vasquez, P. J. Clapham, A. Alling, U. Arnason, S. J. O'Brien, and S. R. Palumbi. 1993. Abundant mitochondrial DNA variation and world-wide population structure in humpback whales. *Proceedings of the National Academy of Sciences* 90:8239–43.

Baldwin, B. G. 1992. Phylogenetic utility of the internal transcribed spacers of nuclear ribosomal DNA in plants: An example from the Compositae. *Molecular Phylogenetics and Evolution* 1:3–16.

Baldwin, B. G., D. H. Goldman, D. J. Keil, R. Patterson, T. J. Rosatti, and D. H. Wilken [eds.]. 2012. *The Jepson Manual: Vascular Plants of California.* 2nd ed. University of California Press, Berkeley.

Baldwin, B. G., S. Kalisz, and W. S. Armbruster. 2011. Phylogenetic perspectives on diversification, diversity, and phytogeography of *Collinsia* and *Tonella* (Plantaginaceae). *American Journal of Botany* 98:731–53.

Baldwin, B. G., and M. J. Sanderson. 1998. Age and rate of diversification of the Hawaiian silversword alliance (Compositae). *Proceedings of the National Academy of Sciences USA* 95:9402–6.

Ball, R. M., Jr., S. Freeman, F. C. James, E. Bermingham, and J. C. Avise. 1988. Phylogeographic population structure of red-winged blackbirds assessed by mitochondrial DNA. *Proceedings of the National Academy of Sciences USA* 85:1558–62.

Barbosa, O., and P. A. Marquet. 2002. Effects of forest fragmentation on the beetle assemblage at the relict forests of Fray Jorge, Chile. *Oecologia* 132:296–306.

Barbour, M. B., P. F. Drysdale, and S. Linstrom. 1993. *California's Changing Landscapes.* California Native Plant Society, Sacramento, CA. 244 pp.

Barbour, M. G., T. Keeler-Wolf, and A. A. Schoenherr. 2007. *Terrestrial Vegetation of California.* 3rd ed. University of California Press, Berkeley.

Barlow, M. 2002. Phylogeographic structure of the tidewater goby, *Eucyclogobius newberryi* (Teleostei: Gobiidae), in the San Francisco Bay Area and Ventura County: Implications for conservation management. MS thesis. University of California, Los Angeles. 77 pp.

Barnosky, C. W., P. M. Anderson, and P. J. Bartlein. 1987. The northwestern U.S. during deglaciation: Vegetational history and paleoclimatic implications. *In* W. F. Ruddiman and H. E.Wright Jr. [eds.], *North America and Adjacent Oceans during the Last Deglaciation,* Vol. K-3, 289–321. Geological Society of America, Boulder, CO.

Barron, J. A., L. E. Heusser, and C. Alexander. 2004. High resolution climate of the past 3,500 years of coastal northernmost California. *In* S. W. Starratt and N. L. Blomquist [eds.], *Proceedings of the Twentieth Annual Pacific Climate Workshop,* 13–22. Asilomar Conference Grounds, Pacific Grove, CA.

Barrowclough, G. F., J. G. Groth, L. A. Mertz, and R. J. Gutiérrez. 2005. Genetic structure, introgression, and a narrow hybrid zone between northern and California spotted owls (*Strix occidentalis*). *Molecular Ecology* 14:1109–20.

Barrowclough, G. F., R. J. Gutierrez, and J. G. Groth. 1999. Phylogeography of spotted owl (*Strix occidentalis*) populations based on mitochondrial DNA sequences: Gene flow, genetic structure, and a novel biogeographic pattern. *Evolution* 53:919–31.

Barry, S., S. Larson, and M. George. 2006. California native grasslands: A historical perspective. *Grasslands* 16:7–11.

Baskin, J. A. 1998. Mustelidae. *In* C. M. Janis, K. M. Scott, L. L. Jacobs [eds.], *Evolution of Tertiary Mammals of North America,* vol. 1, 152–73. Cambridge University Press, Cambridge.

Bassett, I. J., and B. R. Baum. 1969. Conspecificity of *Plantago fastigiata* of North America with *P. ovata* of the Old World. *Canadian Journal of Botany* 47:1865–68.

Bay Area Open Space Council. 2011. *The Conservation Lands Network: San Francisco Bay Area Upland Habitat Goals Project Report.* Berkeley, CA.

Beamish, R. J. 1980. Adult biology of the river lamprey (*Lampetra ayresi*) and the Pacific lamprey (*Lampetra tridentata*) from the Pacific coast of Canada. *Canadian Journal of Fish and Aquatic Science* 37:1906–23.

Becker, B., and B. Marin. 2009. Streptophyte algae and the origin of embryophytes. *Annals of Botany* 103: 999–1004.

Beidleman, R. G. 2006. *California's Frontier Naturalists*. University of California Press, Berkeley.

Bell, C. J., E. L. Lundelius, A. D. Barnosky, R. W. Graham, E. H. Lindsay, D. R. Ruez Jr., H. A. Semken Jr., S. D. Webb, and R. J. Zakrzewski. 2004. The Blancan, Irvingtonian, and Rancholabrean Mammal Ages. *In* M. O. Woodburne [ed.], *Late Cretaceous and Cenozoic Mammals of North America*, 232–314. Columbia University Press, New York.

Bell, K. C., D. J. Hafner, P. Leitner, and M. D. Matocq. 2010. Phylogeography of the ground squirrel subgenus *Xerospermophilus* and assembly of the Mojave Desert biota. *Journal of Biogeography* 37:363–78.

Bennett, K. D. 1997. *Evolution and Ecology: The Pace of Life*. Cambridge University Press, Cambridge.

———. 2004. Continuing the debate on the role of Quaternary environmental change for macroevolution. *Philosophical Transactions of the Royal Society London B* 359:295–303.

Benson, L., M. Kashgarian, R. Rye, S. Lund, F. Paillet, J. Smoot, C. Kester, S. Mensing, D. Meko, and S. Lindstrom. 2002. Holocene multidecadal and multicentennial droughts affecting Northern California and Nevada. *Quaternary Science Reviews* 21:659–82.

Benton, M. J., and D.A.T. Harper. 2009. *Introduction to Paleobiology and the Fossil Record*. Wiley Blackwell, London.

Bernardi, G. 2000. Barriers to gene flow in *Embiotoca jacksoni*, a marine fish lacking a pelagic larval stage. *Evolution* 54:226–37.

———. 2005. Phylogeography and demography of sympatric sister surfperch species, *Embiotoca jacksoni* and *E. lateralis*, along the California coast: Historic versus ecological factors. *Evolution* 59:386–94.

Bernatchez, L., and C. C. Wilson. 1998. Comparative phylogeography of Nearctic and Palearctic freshwater fishes. *Molecular Ecology* 7:431–52.

Betancourt, J. L., T. R. Van Devender, and P. S. Martin. 1990. Synthesis and prospectus. *In* J. L. Betancourt, T. R. Van Devender, and P. S. Martin [eds.], *Packrat Middens: The Last 40,000 Years of Biotic Change*, 435–47. University of Arizona Press, Tucson.

Bengtsson, J., J. Ahnstrom, and A. C. Weibull. 2005. The effects of organic agriculture on biodiversity and abundance: A meta-analysis. *Journal of Applied Ecology* 42:261–69.

Bermingham, E., and H. A. Lessios. 1993. Rate variation of protein and mitochondrial DNA evolution as revealed by sea urchins separated by the Isthmus of Panama. *Proceedings of the National Academy of Sciences USA* 90:2434–2738.

Bermingham, E., and C. Moritz. 1998. Comparative phylogeography: Concepts and applications. *Molecular Ecology* 7:367–69.

Bickham, J. W., J. C. Patton, and T. R. Loughlin. 1996. High variability for control-region sequences in a marine mammal: Implications for conservation and biogeography of Steller sea lions (*Eumetopias jubatus*). *Journal of Mammalogy* 77:95–108.

Booth, D. B. 1987. Timing and processes of deglaciation along the southern margin of the Cordilleran ice sheet. *In* W. F. Ruddiman and H. E. Wright Jr. [eds.], *North America and Adjacent Oceans during the Last Deglaciation*, vol. K-3, 71–90. Geological Society of America, Boulder, CO.

Bowen, B. W., and W. S. Grant. 1997. Phylogeography of the sardines (*Sardinops* spp.): Assessing biogeographic models and population histories in temperate upwelling zones. *Evolution* 51:1601–10.

Bowerman, N. D., and D. H. Clark. 2011. Holocene glaciation of the central Sierra Nevada, California. *Quaternary Science Reviews* 30:1067–85.

Bramble, D.M. 1982. *Scaptochelys*: Generic revision and evolution of gopher tortoises. *Copeia* 4:852-67.

Brattstrom, B. H. 1953. The amphibians and reptiles from Rancho La Brea. *Transactions of the San Diego Society of Natural History* 11:367–86.

———. 1955. A small herpetofauna from the Pleistocene of Carpinteria, California. *Copeia* 1967:138–39.

Briggs, C. J., V. T. Vredenburg, R. A. Knapp, and L. J. Rachowicz. 2005. Investigating the population-level effects of Chytridiomycosis: An emerging infectious disease of amphibians. *Ecology* 86:3149–59.

———. 1974. *Marine Zoogeography*. McGraw-Hill, New York.

———. 2003. Marine centres of origin as evolutionary engines. *Journal of Biogeography* 30:1–18.

Brito, P. H., and S. V. Edwards. 2008. Multilocus phylogeography and phylogenetics using sequence-based markers. *Genetica* 135:439–55.

Broderick, D. H. 1995. The biology of Canadian weeds. *Canadian Journal of Plant Science* 70:247–59.

Brown, S. K., J. M. Hull, D. R. Updike, S. R. Fain, and H. B. Ernest. 2009. Black bear population genetics in California: Signatures of population structure, competitive release, and historical translocation. *Journal of Mammalogy* 90:1066–74.

Brubaker, L. B., P. M. Anderson, M. E. Edwards, and A. V. Lozhkin. 2005. Beringia as a glacial refugium for boreal trees and shrubs: New perspectives from mapped pollen data. *Journal of Biogeography* 32:833–48.

Brunelli, J. P., J. M. Mallatt, R. F. Leary, M. Alfaqih, R. B. Phillips, and G. H. Thorgaard. 2013. Y chromosome phylogeny for cutthroat trout (*Oncorhynchus clarkii*) subspecies is generally concordant with those of other markers. *Molecular Phylogenetics and Evolution* 66:592–602.

Brunsfeld, S. J., J. Sullivan, D. E. Soltis and P. S. Soltis. 2001. Comparative phylogeography of northwestern North America: a synthesis. *In* J. Silvertown and J. Antonovics [eds.], *Integrating Ecological and Evolutionary Processes in a Spatial Context*, 319–39. Blackwell Science, Oxford.

Brzuszek, R. F., R. L. Harkess, and S. J. Mulley. 2007. Landscape architects use of native plants in the southeastern United States. *HortTechnology* 17:78–81.

Bunn, D., A. Mummert, M. Hoshovsky, K. Gilardi, and S. Shanks. 2007. *California Wildlife: Conservation Challenges California's Wildlife Action Plan*. Prepared by UC Davis Wildlife, California Department of Fish and Game.

Burge, D. O., D. M. Erwin, M. B. Islam, J. Kellerman, S. W. Kembel, D. H. Wilken, and P. S. Manos. 2011. Diversification of *Ceanothus* (Rhamnaceae) in the California Floristic Province. *International Journal of Plant Sciences* 172:1137–64.

Burge, D.O., and S. R. Manchester. 2008. Fruit morphology, fossil history, and biogeography of *Paliurus* (Rhamnaceae). *International Journal of Plant Sciences* 169:1066–85.

Burghardt, K. T., D. W. Tallamy, and W. G. Shriver. 2008. The impact of native plants on biodiversity in suburban landscapes. *Conservation Biology* 23:219–44.

Burnham, K. 2005. Point Lobos to Point Reyes: Evidence of approximately 180 km offset on the San Gregorio and northern San Andreas faults. *In* C. Stevens and J. Cooper [eds.], *Cenozoic Deformation in the Central Coast Ranges, California*, 1–28. San Jose, CA: Society for Sedimentary Geology.

Burns, K. J., M. P. Alexander, D. N. Barhoum, and E. A. Sgariglia. 2007. A statistical assessment of congruence among phylogeographic histories of three avian species in the California Floristic Province. *In* C. Cicero and J. V. Remsen Jr. [eds.], *Festschrift for Ned K. Johnson: Geographic Variation and Evolution in Birds*, 96–109. Ornithological Monographs, no. 63.

Burns, K. J., and D. N. Barhoum. 2006. Population-level history of the wrentit (*Chamaea fasciata*): Implications for comparative phylogeography in the California Floristic Province. *Molecular Phylogenetics and Evolution* 38:117–29.

Burton, R. S. 1998. Intraspecific phylogeography across the Point Conception biogeographic boundary. *Evolution* 52:734–45.

Cadotte, M. W., and J. Davies. 2010. Rarest of the rare: Advances in combining evolutionary distinctiveness and scarcity to inform conservation at biogeographical scales. *Diversity and Distributions* 16:376–85.

Caesar, R., N. Gillette, and A. I. Cognato. 2005. Population genetic structure of an edaphic beetle (Ptiliidae) among late successional reserves within the Klamath-Siskiyou ecoregion, California. *Annals of the Entomological Society of America* 98:931–40.

Caesar, R. M., M. Sörensson, and A. I. Cognato. 2006. Integrating DNA data and traditional taxonomy to streamline biodiversity assessment: An example from edaphic beetles in the Klamath ecoregion, California, USA. *Diversity and Distributions* 12:483–89.

CALFED. 2000. CALFED Bay Delta Program Record of Decision. CALFED Bay Delta Program, Sacramento, CA.

California Department of Fish and Game (CDFG). 2011. California Natural Diversity Database. Special Animals (898 taxa). www.dfg.ca.gov/biogeodata /cnddb/pdfs/spanimals.pdf.

Calsbeek, R., J. N. Thompson, and J. Richardson. 2003. Patterns of molecular evolution and diversification in a biodiversity hotspot: The California Floristic Province. *Molecular Ecology* 12:1021–29.

Camargo, A., B. Sinervo, and J. W. Sites Jr. 2010. Lizards as model organisms for linking phylogeographic and speciation studies. *Molecular Ecology* 19:3250–70.

Camp, C. L. 1952. *Earth Song, a Prologue to History*. University of California Press, Berkeley.

Campos, P. F., E. Willerslev, A. Sher, L. Orlando, E. Axelsson, A. Tikhonov, K. Aaris-Sørensen, A. D. Greenwood, R-D. Kahlke, P. Kosintsev, T. Krakhmal-naya, T. Kuznetsova, P. Lemey, R. MacPhee, C. A. Norris, K. Shepherd, M. A. Suchard, G. D. Zazula, B. Shapiro, and M. T. P. Gilbert. 2010. Ancient DNA analyses exclude humans as the driving force behind late Pleistocene musk ox (*Ovibos moschatus*) population dynamics. *Proceedings of the National Academy of Sciences USA* 107:5675–80.

Cane, M. A., P. Braconnot, A. Clement, H. Gildor, S. Joussaume, M. Kageyama, M. Khodri, D. Paillard, S. Tett, and E. Zorita. 2006. Progress in paleoclimate modeling. *Journal of Climate* 19:5031–57.

Carr, S. M., and G. A. Hughes. 1993. Direction of introgressive hybridization between species of North American deer (*Odocoileus*) as inferred from mitochondrial cytochrome-*b* sequences. *Journal of Mammalogy* 74:331–42.

Carreño, A. L., and J. Helenes. 2002. Geology and ages of the islands. *In* T. J. Case, M. L. Cody, and E. Ezcurra [eds.], *A New Island Biogeography of the Sea of Cortés*, 14–40. Oxford University Press, Oxford.

Carretta, J. V., K. A. Forney, M. S. Lowry, J. Barlow, J. Baker, D. Johnston, B. Hanson, R. L. Brownell Jr., J. Robbins, D. K. Mattila, K. Ralls, M. M. Muto, D. Lynch, and L. Carswell. 2009. NOAA Technical Memorandum National Marine Fisheries Service, U.S. Pacific Marine Mammal Stock Assessments.

Cassel, E. J., S. A. Graham, and C. P. Chamberlain. 2009. Cenozoic tectonic and topographic evolution of the northern Sierra Nevada, California, through stable isotope paleoaltimetry in volcanic glass. *Geology* 37:547–50.

Casteel, R. W., D. P. Adam, and J. D. Sims. 1975. Fish remains from Core 7, Clear Lake, Lake County, California. US Geological Survey Open File Rept. No. 75–173.

Castoe, T. A., C. L. Spencer, C. L. Parkinson. 2006. Phylogeographic structure and historical demography of the western diamondback rattlesnake (*Crotalus atrox*): A perspective on North American desert biogeography. *Molecular Phylogenetics and Evolution* 42:193–212.

Cavender, T. M., and R. R. Miller. 1972. *Smilodonichthys rastrosus*: A new Pliocene salmonid fish from western United States. *Museum of Natural History, University of Oregon, Bulletin* 18:1–44.

Chabot, L., and L. G. Allen. 2009. Global population structure of the tope (*Galeorhinus galeus*) inferred by mitochondrial control region sequence data. *Molecular Ecology* 18:545–52.

Chatzimanolis, S., and M. S. Caterino. 2007. Toward a better understanding of the "Transverse Range Break": Lineage diversification in Southern California. *Evolution* 61:2127–41.

Chatzimanolis, S., L. A. Norris, and M. S. Caterino. 2010. Multi-island endemicity: Phylogeography and conservation of *Coelus pacificus* (Coleoptera: Tenebrionidae) darkling beetles on the California Channel Islands. *Annals of the Entomological Society of America* 103:785–95.

Chavez, A. S., and G. J. Kenagy. 2010. Historical biogeography of western heather voles (*Phenacomys intermedius*) in montane systems of the Pacific Northwest. *Journal of Mammalogy* 91:874–85.

Cheatham, N. H. 1976. Conservation of vernal pools. *In* S. Jain [ed.], *Vernal Pools: Their Ecology and Conservation*, 86–89. Institute of Ecology Publication No. 9. University of California, Davis.

Checkley, D. M., and J. A. Barth. 2009. Patterns and processes in the California Current System. *Progress in Oceanography* 83:49–64.

Chen, C.-H., J-P. Huang, C.-C. Tsa and S.-M. Chaw. 2009. Phylogeny of *Calocedrus* (Cupressaceae), an eastern Asian and western North American disjunct gymnosperm genus, inferred from nuclear ribosomal nrITS sequences. *Botanical Studies* 50:425–33.

Cheng Y., R. G. Nicolson, K. Tripp, and S.-M. Chaw. 2000. Phylogeny of Taxaceae and Cephalotaxaceae genera inferred from chloroplast *mat*K gene and nuclear rDNA ITS region. *Molecular Phylogenetics and Evolution* 14:353-65.

Chesser, R. K. 1991. Gene diversity and female philopatry. *Genetics* 127:437–47.

Chilcote, M. W., K. W. Goodson, and M. R. Falcy. 2011. Reduced recruitment performance in natural populations of anadromous salmonids associated with hatchery-reared fish. *Canadian Journal of Fisheries and Aquatic Science* 68:511–22.

Chivers, S. J., A. E. Dizon, P. J. Gearin, and K. M. Robertson. 2002. Small-scale population structure of eastern North Pacific harbour porpoises (*Phocoena phocoena*) indicated by molecular genetic analyses. *Journal of Cetacean Research and Management* 4:111–22.

Christiansen, P., and J. M. Harris. 2005. Body size of *Smilodon* (Mammalia: Felidae). *Journal of Morphology* 266:369–84.

Chronic, H. 1986. *Pages of Stone: Geology of the Western National Parks and Monuments.* Mountaineers Press, Seattle, WA.

Cicero, C., and N. K. Johnson. 1995. Speciation in sapsuckers (*Sphyrapicus*): III. Mitochondrial-DNA sequence divergence at the cytochrome-B locus. *Auk* 112:547–63.

Clark, P. U., A. S. Dyke, J. D. Shakun, A. E. Carlson, J. Clark, B. Wohlfarth, J. X. Mitrovica, S. W. Hostetler, and A. M. McCabe. 2009. The last glacial maximum. *Science* 325:710–14.

Clarke, J. A. 2004. Morphology, phylogenetic taxonomy, and systematics of *Ichthyornis* and *Apatornis* (Avialae: Ornithurae). *Bulletin of the American Museum of Natural History* 286:1–179.

Clarke, J. A., C. P. Tambussi, J. I. Noriega, G. M. Erickson, and R. A. Ketcham. 2005. Definitive fossil evidence for the extant avian radiation in the Cretaceous. *Nature* 433:305–8.

Clementz, M. T., K. A. Hoppe, and P. L. Koch. 2003. A paleoecological paradox: The habitat and dietary preferences of the extinct tethythere *Desmostylus*, inferred from stable isotope analysis. *Paleobiology* 29:506–19.

Coart, E., X. Vekemans, M. J. M. Smulders, I. Wagner, J. Van Huylenbroeck, E. Van Bockstaele, and I. Roldan-Ruiz. 2003. Genetic variation in the endangered wild apple (*Malus sylvestris* (L.) Mill.) in Belgium as revealed by amplified fragment length polymorphism and microsatellite markers. *Molecular Ecology* 12:845–57.

Coates, A. G., and J. A. Obando. 1996. Geological evolution of the Central American Isthmus. *In* J. B. C. Jackson, A. F. Budd, and A. G. Coates [eds.], *Evolution and Environment in Tropical America*, 21–56. University of Chicago Press, Chicago, IL.

Cody, S., J. E. Richardson, V. Bull, E. Ellis, and R. T. Pennington. 2010. The Great American Biotic Interchange revisited. *Ecography* 33:326–32.

Cole, C. J., and T. R. Van Devender. 1976. Surface structure of fossil and recent epidermal scales from North American lizards of the genus *Sceloporus* (Reptilia, Iguanidae). *Bulletin of the American Museum of Natural History* 156:453–513.

Cole, K. L., K. Ironside, J. Eischeid, G. Garin, P. B. Duffy, and C. Toney. 2011. Past and ongoing shifts in Joshua tree distribution support future modeled range contraction. *Ecological Applications* 21:137–49.

Cole, K. L., K. Pohs, and J. A. Cannella. 2003. Digital range map of Joshua tree (*Yucca brevifolia*). U.S. Geological Survey, Forest and Rangeland Ecosystem Science Center, Corvallis, OR. http://sagemap.wr.usgs.gov/ftp/regional /USGS/YUBRRangemap.shp.xml.

Coleman, A. W., and V. D. Vacquier. 2002. Exploring the phylogenetic utility of ITS sequences for animals: A test case for abalone (*Haliotis*). *Journal of Molecular Evolution* 54:246–57.

Conroy, C. J., and J. A. Cook. 2000. Phylogeography of a post-glacial colonizer: *Microtus longicaudus* (Rodentia: Muridae). *Molecular Ecology* 9:165–75.

Conroy, C. J., and J. L. Neuwald. 2008. Phylogeographic study of the California vole, *Microtus californicus*. *Journal of Mammalogy* 89:755–67.

Cope, J. M. 2004. Population genetics and phylogeography of the blue rockfish (*Sebastes mystinus*) from Washington to California. *Canadian Journal of Fisheries and Aquatic Science* 61: 332–42.

Correa, E., C. Jaramillo, S. Manchester, and M. Gutierrez. 2010. A fruit and leaves of Rhamnaceae affinities from the late Cretaceous (Maastrichtian) of Columbia. *American Journal of Botany* 97:71–79.

Cox, B. F., J. W. Hillhouse, and L. A. Owen. 2003. Pliocene and Pleistocene evolution of the Mojave River, and associated tectonic development of the Transverse Ranges and Mojave Desert, based on borehole stratigraphy studies and mapping of landforms and sediments near Victorville, California. *In* Y. Enzel, S. G. Wells, and N. Lancaster [eds.], *Paleoenvironments and Paleohydrology of the Mojave and Southern Great Basin Deserts*, 1–42. Special Paper 368. Geological Society of America, Boulder, CO.

CPAD 1.8. July 2012, ©GreenInfo Network. www.calands.org.

Craig, M. T., F. J. Fodrie, L. G. Allen, L. A. Chartier, and R. J. Toonen. 2011. Discordant phylogeographic and biogeographic breaks in California halibut. *Bulletin of the Southern California Academy of Sciences* 110:141–51.

Crespi, E. J., L. J. Rissler, and R. A. Browne. 2003. Testing Pleistocene refugia theory: Historical biogeography of a high-elevation salamander, *Desmognathus wrighti*, in the southern Appalachians. *Molecular Ecology* 12:969–84.

Croizat, L. 1958. Panbiogeography or an introductory synthesis of zoogeography, phytogeography, and zoology. With notes on evolution, systematics, ecology, anthropology, etc. Vol. 1: *The New World*. Vol. 2: *The Old World*. Published by the author, Caracas, Venezuela.

Cronin, M. A., S. C. Amstrup, and G. W. Garner. 1991. Interspecific and intraspecific mitochondrial DNA variation in North American bears (*Ursus*). *Canadian Journal of Zoology* 69:2985–92.

Crooks, K. R., A. V. Suarez, and D. T. Bolger. 2004. Avian assemblages along a gradient of urbanization in a highly fragmented landscape. *Biological Conservation* 115:451–62.

Culley, T. M., and N. A. Hardiman. 2007. The beginning of a new invasive plant: A history of the ornamental Callery pear in the United States. *Bioscience* 57:956–64.

Culver, M., W. E. Johnson, J. Pecan-Slattery, and S. J. O'Brien. 2000. Genomic ancestry of the American puma *(Puma concolor)*. *Journal of Heredity* 91:186–97.

Cypher, B. L., G. D. Warrick, M. R. M. Otten, T. P. O'Farrell, W. H. Berry, C. E. Harris, T. T. Kato, P. M. McCue, J. H. Scrivner, and B. W. Zoellick. 2000. Population dynamics of San Joaquin kit foxes at the Naval Petroleum Reserves in California. *Wildlife Monographs* 45.

Daniels M. L., R. S. Anderson, and C. Whitlock. 2005. Vegetation and fire history since the late Pleistocene from the Trinity Mountains, northwestern California. *Holocene* 15:1062–71.

Davidson, A. D., A. G. Boyer, H. Kim, S. Pompa-Mansilla, M. J. Hamilton, D. P. Costa, G. Ceballos, and J. H. Brown. 2012. Drivers and hotspots of extinction risk in marine mammals. *Proceedings of the National Academy of Sciences* 109:3395–3400.

Davidson, C. 2004. Decline downwind: Amphibian population declines in California and historical pesticide use. *Ecological Applications* 14:1892–1902.

Davis, C. C., P. W. Fritsch, J. Li, and M. J. Donoghue. 2002. Phylogeny and biogeography of *Cercis* (Fabaceae): Evidence from nuclear ribosomal ITS and chloroplast ndhF sequence data. *Systematic Botany* 27:289–302.

Davis, E. B., M. S. Koo, C. Conroy, J. L. Patton, and C. Moritz. 2008. The California hotspots project: Identifying regions of rapid diversification of mammals. *Molecular Ecology* 17:120–38.

Davis, F. W., D. M. Stoms, A. D. Hollander, K. A. Thomas, P. A. Stine, D. Odion, M. I. Borchert, J. H. Thorne, M. V. Gray, and R. E. Walker. 1998. *The California Gap Analysis Project Final Report.* University of California, Santa Barbara.

Davis, M. B., and R. G. Shaw. 2001. Range shifts and adaptive responses to Quaternary climate change. *Science* 292:673–79.

Davis, O. K. 1999. Pollen analysis of a Holocene–late-Glacial sediment core from Mono Lake, Mono County, California. *Quaternary Research* 52:243–49.

Dawson, M. N. 2001. Phylogeography in coastal marine animals: A solution from California? *Journal of Biogeography* 28:723–36.

Dawson, M. N., K. D. Louie, M. Barlow, D. K. Jacobs, and C. C. Swift. 2002. Comparative phylogeography of sympatric sister species, *Clevelandia ios* and *Eucyclogobius newberryi* (Teleostei, Gobiidae), across the California Transition Zone. *Molecular Ecology* 11:1065–75.

Day, R. H., and D. A. Nigro. 2000. Feeding ecology of Kittlitz's and marbled murrelets in Prince William Sound, Alaska. *Waterbirds* 23:1–14.

Debruyne, R., G. Chu, C. E. King, K. Bos, M. Kuch, C. Schwarz, P. Szpak, D. R. Gröcke, P. Matheus, G. Zazula, D. Guthrie, D. Froese, B. Buigues, C. de Marliave, C. Flemming, D. Poinar, D. Fisher, J. Southon, A. N. Tikhonov, R. D. E. MacPhee, and H. N. Poinar. 2008. Out of America: Ancient DNA evidence for a New World origin of late Quaternary woolly mammoths. *Current Biology* 18:1320–26.

DeChaine, E. G., and A. P. Martin. 2004. Historic cycles of fragmentation and expansion in *Parnassius smintheus* (Papilionidae) inferred using mitochondrial DNA. *Evolution* 58:113–27.

———. 2005. Marked genetic divergence among sky island populations of *Sedum lanceolatum* (Crassulaceae) in the Rocky Mountains. *American Journal of Botany* 92:477–86.

DeCourten, F. L. 2008. *Geology of Northern California: A Regional Geology Supplement:* Brooks/Cole, Cengage Learning, Cambridge, MA. 45 pp.

Delaney, K. S., S. Zafar, R. K. Wayne. 2008. Genetic divergence and differentiation within the western scrub-jay (*Aphelocoma californica*). *Auk* 125:839–49.

Della Sala, D. A., S. B. Ried, T. J. Frest, J. R. Strittholt, and D. M. Olson. 1999. A global perspective on the biodiversity of the Klamath-Siskiyou ecoregion. *Natural Areas Journal* 19:300–319.

Devitt, T. J. 2006. Phylogeography of the western lyresnake (*Trimorphodon biscutatus*): Testing arid land biogeographical hypotheses across the Nearctic–Neotropical transition. *Molecular Ecology* 15:4387–4407.

Díaz-Jaimes, P., M. Uribe-Alcocer, A. Rocha-Olivares, F. J. García-de-León, P. Nortmoon, and J. D. Durand. 2010. Global phylogeography of the dolphinfish (*Coryphaena hippurus*): The influence of large effective population size and recent dispersal on the divergence of a marine pelagic cosmopolitan species. *Molecular Phylogenetics and Evolution* 57:1209–18.

Dickinson, W. R. 2008. Accretionary Mesozoic–Cenozoic expansion of the Cordilleran continental margin in California and adjacent Oregon. *Geosphere* 4:329–53.

Diester-Haass, L., P. A. Meyers, and L. Vidal. 2002. The late Miocene onset of high productivity in the Benguela Current upwelling system as part of a global pattern. *Marine Geology* 180:87–103.

Domning, D. P. 1978. Sirenian evolution in the northern Pacific Ocean. *University of California Publications in Geological Sciences* 118:1–176.

———. 1987. Sea cow family reunion. *Natural History,* April, 64–71.

Domning, D. P., C. E. Ray, and M. C. McKenna. 1986. Two new Oligocene Desmostylians and a discussion of Tethytherian systematics. *Smithsonian Contributions to Paleobiology* 59:1–56.

Dong, J., and D. B. Wagner. 1993. Taxonomic and population differentiation of mitochondrial diversity in *Pinus banksiana* and *Pinus contorta. Theoretical and Applied Genetics* 86:573–78.

Donley, M. W. 1979. *Atlas of California.* Pacific Book Center, Culver City, CA. 191 pp.

Donoghue, M. J., C. D. Bell, and J. Li. 2001. Phylogenetic patterns in Northern Hemisphere plant geography. *International Journal of Plant Sciences* 162: S41–S52.

Donoghue, M. J., and S. A. Smith. 2004. Patterns in the assembly of temperate forests around the Northern Hemisphere. *Philosophical Transactions of the Royal Society London B* 359:1633–44.

Dorsey, R. J., A. Fluette, K. McDougall, B. A. Housen, S. U. Janecke, G. J. Axen, and C. R. Shirvell. 2007. Chronology of Miocene–Pliocene deposits at Split Mountain Gorge, Southern California: A record of regional tectonics and Colorado River evolution. *Geology* 35:57–60.

Douglas, M. E., M. R. Douglas, G. W. Schuett, and L. W. Porras. 2006. Evolution of rattlesnakes (Viperidae; *Crotalus*) in the warm deserts of western North America shaped by Neogene vicariance and Quaternary climate change. *Molecular Ecology* 15:3353–74.

Dowling, T. E. and M. R. Childs. 1992. Impact of hybridization on a threatened trout the southwestern United States. *Conservation Biology* 6:355-64.

Doyle, J. A. 2012. Molecular and fossil evidence on the origin of angiosperms. *Annual Review of Earth and Planetary Sciences* 40:301–26.

Doyle, J. A., and L. J. Hickey. 1976. Pollen and leaves from the mid-Cretaceous Potomac Group and their bearing on early angiosperm evolution. *In* C. B. Beck [ed.], *Origin and Early Evolution of Angiosperms*, 139–206. Columbia University Press, New York.

Drummond, A. J., and A. Rambaut. 2007. BEAST: Bayesian evolutionary analysis by sampling trees. *BMC Evolutionary Biology* 7:214.

Ducea, M., S. Kidder, and G. Zandt. 2003. Arc composition at mid-crustal depths: Insights from the Coast Ridge Belt, Santa Lucia Mountains, California. *Geophysical Research Letters* 30:1703–7.

Dukes, J., and H. Mooney. 2004. Disruption of ecosystem processes in western North America by invasive species. *Revista Chilena de Historia Natural* 77:411–37.

Dupré, W. R. 1990. Quaternary geology of the Monterey Bay region, California. *In* R. E. Garrison, H. G. Greene, K. R. Hicks, G. E. Weber, and T. L. Wright [eds.], *Geology and Tectonics of the Central California Coastal Region, San Francisco to Monterey*, 185–92. Pacific Section of the American Association of Petroleum Geologists, Bakersfield, CA.

Dupré, W. R., R. B. Morrison, H. E. Clifton, K. R. Jajoic, D. J. Ponti, C. L. Powell II, S. A. Matheison, A. M. Sarna-Wojcicki, E. L. Leithold, W. R. Lettis, P. F. McDowell, T. K. Rockwell, J. R. Unruh, and R. S. Yeats. 1991. Quaternary geology of the Pacific margin. *In* R. B. Morrison [ed.], *Quaternary Nonglacial Geology, Conterminous U.S.*, 141–214. Geological Society of America, Boulder, CO.

Durham, J. W., and E. C. Allison. 1960. The geologic history of Baja California and its marine faunas. *Systematic Zoology* 9:47–91.

Duvernell, D. D., and B. J. Turner. 1998. Evolutionary genetics of Death Valley pupfish populations: Mitochondiral DNA sequence variation and population structure. *Molecular Ecology* 7:279–88.

Dyer, R. J., J. D. Nason, and R. C. Garrick. 2010. Landscape modelling of gene flow: Improved power using conditional genetic distance derived from the topology of population networks. *Molecular Ecology* 19:3746–59.

Echelle, A. A., and A. F. Echelle. 1993a. Allozyme perspective on mitochondrial DNA variation and evolution of the Death Valley pupfishes (Cyprinodontidae: *Cyprinodon*). *Copeia* 2:275–87.

———. 1993b. Allozyme variation and systematics of the New World Cyprinodotines (Teliostei: Cyprinodontidae). *Biochemical Systematics and Ecology* 21:583–90.

Eckert, A. J. 2011. Seeing the forest for the trees: Statistical phylogeography in a changing world. *New Phytologist* 189:894–97.

Eckert, A. J., and B. C. Carstens. 2008. Does gene flow destroy phylogenetic signal? The performance of three methods for estimating species phylogenies in the presence of gene flow. *Molecular Phylogenetics and Evolution* 49:832–42.

Eckert, A. J., and B. D. Hall. 2006. Phylogeny, historical biogeography, and patterns of diversification for *Pinus* (Pinaceae): Phylogenetic tests of fossil-based hypotheses. *Molecular Phylogenetics and Evolution* 40: 166–82.

Edwards, R. J. 1979. A report of Guadalupe bass (*M. treculi*) x smallmouth bass (*M. dolomieui*) hybrids from two localities in the Guadalupe River, Texas. *Texas Journal of Science* 31:231–38.

Edwards, S. V., and P. Beerli. 2000. Perspective: Gene divergence, population divergence, and the variance in coalescence time in phylogeographic studies. *Evolution* 54:1839–54.

Edwards, S. W. 2004. Paleobotany of California. *Four Seasons* 4(2):3–75.

Ehrlich, P. R. 1961. Intrinsic barriers to dispersal in checkerspot butterfly. *Science* 134:108–9.

Ehrlich, P. R., and R. M. Pringle. 2008. Where does biodiversity go from here? A grim business-as-usual forecast and a hopeful portfolio of partial solutions. *Proceedings of the National Academy of Sciences* 105, suppl. 1:11579–86.

Eizirik, E., J.-H. Kim, M. Menotti-Raymond, P. Crawshaw Jr., S. J. O'Brien, and W. E. Johnson. 2001. Phylogeography, population history and conservation genetics of jaguars (*Panthera onca*, Mammalia, Felidae). *Molecular Ecology* 10:65–79.

Elam, D. R., K. Lewiner, T. Keeler-Wolf, and S. A. Flint. May 1998. State of California Resources Agency, Department of Fish and Game, California Vernal Pool Assessment Preliminary Report. Prepared by the California Natural Diversity Database.

Ellstrand, N. C., D. Biggs, A. Kaus, P. Lubinsky, L. A. McDade, K. Preston, L. M. Prince, H. M. Regan, V. Rorive, O. A. Ryder, and K. A. Schierenbeck. 2010.

Got hybridization? A multidisciplinary approach for informing science policy. *BioScience* 60:384–88.

Ellstrand, N. C., and D. R. Elam. 1993. Population genetic consequences of small population size: Implications for plant conservation. *Annual Review of Ecology and Systematics* 24:217–42.

Emery, N. C., E. J. Forrestel, G. Jui, M. S. Park, B. G. Baldwin, and D. D. Ackerly. 2012. Niche evolution across spatial scales: Climate and habitat specialization in California *Lasthenia* (Asteraceae). *Ecology* 93:S151–S166.

Engel, M. S., and D. A. Grimaldi. 2004. New light shed on the oldest insect. *Nature* 427:627–30.

Enzel, Y., R. D. Cayan, R. Y. Anderson, and S. G. Wells. 1989. Atmospheric circulation during Holocene Lake stands in the Mojave Desert: Evidence of a regional climatic change. *Nature* 341:44–48.

Erickson, V. J. 2008. Developing native plant germplasm for national forests and grasslands in the Pacific Northwest. *Native Plant Journal* 9:3255–66.

Ernest, H. B., W. M. Boyce, V. C. Bleich, B. May, S. J. Stiver, and S. G. Torres. 2003. Genetic structure of mountain lion (*Puma concolor*) populations in California. *Conservation Genetics* 4:353–66.

Escobar García, P., M. Winkler, R. Flatscher, M. Sonnleitner, J. Krejčíková, J. Suda, K. Hulber, G.M. Schneeweiss, and P. Schonswetter. 2012. Extensive range persistence in peripheral and interior refugia characterizes Pleistocene range dynamics in a widespread Alpine plant species (*Senecio carniolicus*, Asteraceae). *Molecular Ecology* 21:1255–70.

Escorza-Treviño, S., and A. E. Dizon. 2000. Phylogeography, intraspecific structure and sex-biased dispersal of Dall's porpoise, *Phocoenoides dalli*, revealed by mitochondrial and microsatellite DNA analyses. *Molecular Ecology* 9:1049–60.

Escudero M., A. L. Hipp, and M. Luceño. 2010. Karyotype stability and predictors of chromosome number variation in sedges: A study in *Carex* section *Spriostachyae* (Cyperaceae). *Molecular Phylogenetics and Evolution* 57:353–63.

Estes, R. 1981. Gymnophiona, Caudata. *In* P. Wellnhofer [ed.], *Encyclopedia of Paleoherpetology*, Part 2, 1–115. Gustav-Fischer-Verlag, Stuttgart/New York.

Excoffier, L., P. E. Smouse, and J. M. Quattro. 1992. Analysis of molecular variance inferred from metric distances among DNA haplotypes: Application to human mitochondrial DNA restriction data. *Genetics* 131:479–91.

Farrar, C. D., and G. L. Bertoldi. 1988. Region 4, Central Valley and Pacific Coast Ranges. *In* W. Back, J. S. Rosenshein, and P. R. Seabier [eds.], *Hydrogeology: The Geology of North America*, vol. O-2. Geological Society of America, Boulder, CO.

Feduccia, A. 1995. Explosive evolution in Tertiary birds and mammals. *Science* 267:637–38.

———. 2003. Big Bang for Tertiary birds? *Trends in Ecology and Evolution* 18:172–76.

Fehlberg, S. D., and T. A. Ranker. 2009. Evolutionary history and phylogeography of *Encelia farinosa* (Asteraceae) from the Sonoran, Mojave, and Peninsular Deserts. *Molecular Phylogenetics and Evolution* 50:326–35.

Feldman, C. R., O. Flores-Villela, and T. J. Papenfuss. 2011. Phylogeny, biogeography, and display evolution in the tree and brush lizard genus *Urosaurus* (Squamata: Phrynosomatidae). *Molecular Phylogenetics and Evolution* 61:714–25.

Feldman, C. R., and G. S. Spicer. 2006. Comparative phylogeography of woodland reptiles in California: Repeated patterns of cladogenesis and population expansion. *Molecular Ecology* 15:2201–22.

Feng, Y., S-H. Oh, and P. S. Manos. 2005. Phylogeny and historical biogeography of the genus *Platanus* as inferred from nuclear and chloroplast DNA. *Systematic Botany* 30:786–99.

Fiedler, P. L. 2012. *Calochortus. In* B. G. Baldwin, D. H. Goldman, D. J. Keil, R. Patterson, T. J. Rosatti, and D. H. Wilken [eds.], *The Jepson Manual: Vascular Plants of California*, 2nd ed., 1378–84. University of California Press, Berkeley.

Fiedler, P. L., E. K. Crumb, and A. K. Knox. 2011. Reconsideration of the taxonomic status of Mason's *Lilaeopsis*—a state-protected rare species in California. *Madroño* 58:131–44.

Flagstad, O., and K. H. Roed. 2003. Refugial origins of reindeer (*Rangifer tarandus* L.) inferred from mitochondrial DNA sequences. *Evolution* 57:658–70.

Fleischer, R. C., C. E. McIntosh, and C. L. Tarr. 1998. Evolution on a volcanic conveyor belt: Using phylogeographic reconstructions and K–Ar-based ages of the Hawaiian Islands to estimate molecular evolutionary rates. *Molecular Ecology* 7:533–45.

Flynn, J. J., J. A. Finarelli, S. Zehr, J. Hsu, and M. A. Nedbal. 2005. Molecular phylogeny of the Carnivora (Mammalia): Assessing the impact of increased sampling on resolving enigmatic relationships. *Systematic Biology* 54:317–37.

Fontanella, F. M., C. R. Feldman, M. E. Siddall, and F. T. Burbrink. 2008. Phylogeography of *Diadophis punctatus:* Extensive lineage diversity and repeated patterns of historical demography in a trans-continental snake. *Molecular Phylogenetics and Evolution* 46:1049–70.

Fordyce, J. A., and C. C. Nice. 2003. Contemporary patterns in a historical context: Phylogeographic history of the pipevine swallowtail, *Bauus philenor* (Papilionidae). *Evolution* 57:1089–99.

Forister, M. L., J. A. Fordyce, and A. M. Shapiro. 2004. Geological barriers and restricted gene flow in the holarctic skipper *Hesperia comma* (Hesperiidae). *Molecular Ecology* 13:3489–99.

Friis, E. M., P. R. Crane, and K. Raunsgaard Pedersen. 2011. *Early Flowers and Angiosperm Evolution.* Cambridge University Press, Cambridge.

Fritsch, P. W., and B. C. Cruz. 2012. Phylogeny of *Cercis* based on DNA sequences of nuclear ITS and four plastid regions: Implications for transatlantic historical biogeography. *Molecular Phylogenetics and Evolution* 62:816–25.

Funk, W. C., E. D. Forsman, M. Johnson, T. D. Mullins, and S. M. Haig. 2010. Evidence for recent population bottlenecks in northern spotted owls (*Strix occidentalis caurina*). *Conservation Genetics* 11:1013–21.

Furnier, G. R., and W. T. Adams. 1986. Geographic patterns of allozyme variation in Jeffrey pine. *American Journal of Botany* 73:1009–15.

Galbreath, K. E., J. A. Cook, A. A. Eddings, and E. G. DeChaine. 2011. Diversity and demography in Beringia: Multilocus tests of paleodistribution models reveal the complex history of arctic ground squirrels. *Evolution* 65:1879–96.

Galbreath, K. E., D. J. Hafner, and K. R. Zamudio. 2009. When cold is better: Climate-driven elevation shifts yield complex patterns of diversification and demography in an alpine specialist (America pika, *Ochotona princeps*). *Evolution* 63:2848–63.

Garone, P. 2011. *The Fall and Rise of the Wetlands of California's Great Central Valley.* University of California Press, Berkeley.

Garza, J. C., and D. E. Pearse. 2008. Population genetic structure of *Oncorhynchus mykiss* in the California Central Valley. Report to the California Department of Fish and Game. 54 pp.

Gauthier, J., and K. de Queiroz. 2001. Feathered dinosaurs, flying dinosaurs, crown dinosaurs, and the name "Aves." *In* J. Gauthier and L. F. Gall [eds.], *New Perspective on the Origin and Evolution of Birds: Proceedings of the International Symposium in Honor of John H. Ostrom.* Peabody Museum, Yale University, New Haven, CT.

Ge, D., Z. Zhang, L. Xia, Q. Zhang, Y. Ma, and Q. Yang. 2012. Did the expansion of C_4 plants drive extinction and massive range contraction of micromammals? Inferences from food preference and historical biogeography of pikas. *Palaeogeography, Palaeoclimatology, Palaeoecology* 326–28:160–71.

Gernandt, D. S., S. Hernández-León, E. Salgado-Hernández, and J. A. Pérez de la Rosa. 2009. Phylogenetic relationships of *Pinus* subsection *Ponderosae* inferred from rapidly evolving cpDNA regions. *Systematic Botany* 34:481–91.

Gernandt, D. S., S. Magallón, G. Geada López, O. Zerón Flores, and A. Liston. 2008. Use of simultaneous analyses to guide fossil-based calibrations of Pinaceae phylogeny. *International Journal of Plant Sciences* 16:1086–99.

Gilbert, C., A. Ropiquet, and A. Hassanin. 2006. Mitochondrial and nuclear phylogenies of Cervidae (Mammalia, Ruminantia): Systematics, morphology, and biogeography. *Molecular Phylogenetics and Evolution* 40:101–17.

Gill, F. B., A. M. Mostrom, and A. L. Mack. 1993. Speciation in North American chickadees: I. Patterns of mtDNA genetic divergence. *Evolution* 47:195–212.

Gillespie, A. R., and P. H. Zehfuss. 2004. Glaciations of the Sierra Nevada, California, USA. *In* J. Ehiers and P. L. Gibbard [eds.], *Quaternary Glaciations—Extent and Chronology*, Part II, 51–62. Elsevier, Amsterdam.

Gillett, G. W., J. T. Howell, and H. Leschke 1995. *A Flora of Lassen Volcanic National Park, California.* Rev. V. H. Oswald, D. W. Showers, and M. Showers. California Native Plant Society, Sacramento, CA.

Gingerich, P. 2004. *Whale Evolution.* McGraw-Hill Yearbook of Science & Technology, Boston.

Goebel, A. M., T. A. Ranker, P. S. Corn, and R. G. Olmstead. 2009. Mitochondrial DNA evolution in the *Anaxyrus boreas* species group. *Molecular Phylogenetics and Evolution* 50:209–25.

Gompert, Z., J. A. Fordyce, M. L. Forister, and C. C. Nice. 2008. Recent colonization and radiation of North American *Lycaeides* (Plebejus) inferred from mtDNA. *Molecular Phylogenetics and Evolution* 48:481–90.

Gompert, Z., J. A. Fordyce, M. L. Forister, A. M. Shapiro, and C. C. Nice. 2006. Homoploid hybrid speciation in an extreme habitat. *Science* 314:1923–25.

Good, D. A. 1989. Hybridization and cryptic species in *Dicamptodon* (Caudata: Dicamptodontidae). *Evolution* 43:728–44.

Goodman, D. H., S. B. Reid, M. F. Docker, G. R. Haas, and A. P. Kinziger. 2008. Mitochondrial DNA evidence for high levels of gene flow among populations of a widely distributed anadromous lamprey *Entosphenus tridentatus* (Petromyzontidae). *Journal of Fish Biology* 72:400–417.

Gordon, S. P., C. M. Sloop, H. G. Davis, and J. H. Cushman. 2012. Population genetic diversity and structure of two rare vernal pool grasses in central California. *Conservation Genetics* 13:117–30.

Graham, A. 1993. History of the vegetation: Cretaceous (Maastrichtian)–Tertiary. *In* Flora of North America Editorial Committee [eds.], *Flora of North America*, vol. 1, 57–70. Oxford University Press, New York.

———. *Late Cretaceous and Cenozoic History of North American Vegetation.* Oxford University Press, Oxford.

Graham, R. W., E. L. Lundelius Jr., and M. A. Graham. 1996. Spatial response of mammals to late-Quaternary environmental fluctuations. *Science* 272:1601–6.

Graves, S. D., and A. M. Shapiro. 2003. Exotics as host plants of the California butterfly fauna. *Biological Conservation* 110:413–33.

Graves, W. R., and J. A. Schrader. 2008. At the interface of phylogenetics and population genetics, the phylogeography of *Dirca occidentalis* (Thymelaeaceae). *American Journal of Botany* 95:1454–65.

Griffin, G. F., D. M. Stafford-Smith, S. R. Morton, G. E. Allan, K. A. Masters, and N. Preece. 1989. Status and implications of the invasion of tamarisk

(*Tamarix aphylla*) on the Finke River, Northern Territory, Australia. *Journal of Environmental Management* 29:297–315.

Griffin, J. R. 1966. Notes on disjunct foothill species near Burney, California. *Leaflets of Western Botany* 10: 296–98.

———. 1988. Oak woodland. *In* M. G. Barbour and J. Major [eds.], Terrestrial vegetation of California, 383–415. John Wiley & Sons, New York.

Griffin, J. R., and W. B. Critchfield. 1972. The distribution of forest trees in California. USDA Forest Service, Berkeley, CA.

Grinnell, J. 1914. An account of the mammals and birds of the Lower Colorado Valley with especial reference to the distributional problems presented. *University of Colorado Publications in Zoology* 12:51–294.

Grivet, D. M.-F., R. J. Petit, and V. L. Sork. 2006. Contrasting patterns of historical colonization in white oaks (*Quercus* spp.) in California and Europe. *Molecular Ecology* 15:4085–93.

Gugger, P. F., A. González-Rodríguez, H. Rodríguez-Correa, S. Sugita, and J. Cavender-Bares. 2011. Southward Pleistocene migration of Douglas fir into Mexico: Phylogeography, ecological niche modeling, and conservation of "rear edge" populations. *New Phytologist* 189:1185–99.

Gugger, P. F., S. Sugita, and J. Cavender-Bares. 2010. Phylogeography of Douglas fir based on mitochondrial and chloroplast DNA sequences: Testing hypotheses from the fossil record. *Molecular Ecology* 19: 1877–97.

Guppy, C. S., and J. H. Shepard. 2001. *Butterflies of British Columbia: Including Western Alberta, Southern Yukon, the Alaska Panhandle, Washington, Northern Oregon, Northern Idaho, and Northwestern Montana.* University of British Columbia Press, Vancouver. 413 pp.

Hafner, D. J., and B. R. Riddle. 2008. Boundaries and barriers of North American warm deserts: An evolutionary perspective. *In* P. Upchurch, A, McGowan, and C. Slater [eds.], *Proceedings from Palaeogeography and Palaeobiogeography: Biodiversity in Space and Time,* 75–114. Cambridge University Press, Cambridge.

Hafner, D. J., and R. M. Sullivan. 1995. Historical and ecological biogeography of historical and ecological biogeography of Nearctic pikas (Lagomorpha: Ochotonidae). *Journal of Mammalogy* 76:302–21.

Hafner, J. C. 1981. Evolution, systematics, and historical biogeography of kangaroo mice, genus *Microdipodops*. PhD diss., University of California, Berkeley.

Hafner, J. C., D. J. Hafner, J. L. Patton, and M. F. Smith. 1983. Contact zones and the genetics of differentiation in the pocket gopher, *Thomomys bottae* (Rodentia: Geomyidae). *Systematic Zoology* 3:1–20.

Hafner, J. C., J. E. Light, D. J. Hafner, M. S. Hafner, E. Reddington, D. S. Rogers, and B. R. Riddle. 2007. Basal clades and molecular systematics of heteromyid rodents. *Journal of Mammalogy* 88:1129–45.

Hafner, J. C., N. S. Upham, E. Reddington, and C. W. Torres. 2008. Phylogeography of the pallid kangaroo mouse, *Microdipodops pallidus*: A sand-obligate endemic of the Great Basin, western North America. *Journal of Biogeography* 35:2102–18.

Hagstrum, J. T., and B. L. Murchey. 1993. Deposition of Franciscan complex cherts along the paleoequator and accretion to the American margin at tropical paleolatitudes. *Geological Society of America Bulletin* 105:766–68.

Haig, S. M., T. D. Mullins, and E. D. Forsman. 2004. Subspecific relationships and genetic structure in the spotted owl. *Conservation Genetics* 5:683–705.

Hall, C. A. 2002. Nearshore marine paleoclimate regions, increasing zoogeographic provinciality, molluscan extinctions, and paleoshorelines, California: Late Oligocene (27 Ma) to late Pliocene (2.5 Ma). *Special Paper* 357, v-489. Geological Society of America, Boulder, CO.

———. 2007. *Introduction to the Geology of Southern California and Its Native Plants.* University of California Press, Berkeley. 493 pp.

Hamrick, J. L., and M. J. W. Godt. 1996. Conservation genetics of endemic plant species. *In* J. C. Avise and J. L. Hamrick [eds.], *Conservation Genetics,* 281–304. Chapman and Hall, New York.

Hamrick, J. L., M. J. Godt, and S. L. Sherman-Broyles. 1992. Factors influencing levels of genetic diversity in woody plant species. *New Forests* 6:95–124.

Hamrick J. L., A. Schnabel, and P. V. Wells. 1994. Distributions of genetic diversity within and among populations of Great Basin conifers. *In* K. T. Harper, L. L. St. Clair, K. H. Thorne, and W. M. Hess [eds.], *Natural History of the Colorado Plateau and Great Basin,* 146–61. University Press of Colorado, Niwot.

Harden, D. R. 1998. *California Geology.* Prentice-Hall, Upper Saddle River, NJ. 477 pp.

Harlin-Cognato, A. J., W. Bickham, T. R. Loughlin, and R. L. Honeycutt. 2006. Glacial refugia and the phylogeography of Steller's sea lion (*Eumatopias jubatus*) in the North Pacific. *Journal of Evolutionary Biology* 19:955–69.

Harrison, S. 1991. Local extinction in a metapopulation context: An empirical evaluation. *Biological Journal of the Linnean Society* 42:73–88.

Harvey, H. T. 1986. Evolution and history of Giant Sequoia. *In* C. P. Weatherspoon, Y. R. Iwamoto, and D. D. Piirto [tech. coords.], *Proceedings of the Workshop on Management of Giant Sequoia.* General Technical Report PSW-GTR-95, 1–3. Pacific Southwest Forest and Range Experiment Station, Forest Service, U.S. Department of Agriculture, Berkeley, CA.

Hayhoe, K., D. Cayan, C. B. Field, P. C. Frumhoff, E. P. Maurer, N. L. Miller, S. C. Moser, S. H. Schneider, K. N. Cahill, E. E. Cleland, L. Dale, R. Drapek, R. M. Hanemann, L. S. Kalkstein, J. Lenihan, C. K. Lunch, R. P. Neilson, S. C. Sheridan, and J. H. Verville. 2004. Emissions pathways, climate change, and impacts on California. *Proceedings of the National Academy of Sciences USA* 101:12422–27.

Hennig, W. 1966. *Phylogenetic Systematics.* University of Illinois Press, Urbana. 268 pp.

Herman, A. B. 2002. Late early–late Cretaceous floras of the North Pacific region: Florogenesis and early angiosperm invasion. *Review of Palaeobotany and Palynology* 122:1–11.

Hewitt, G. 2000. The genetic legacy of the Quaternary ice ages. *Nature* 405:907–13.

———. 2004. Genetic consequences of climatic oscillations in the Quaternary. *Philosophical Transactions of the Royal Society of London B* 359:183–95.

Hickerson, M. J., B. C. Carstens, J. Cavender-Bares, K. A. Crandall, C. H. Graham, J. B. Johnson, L. Rissler, P. F. Victoriano, and A. D. Yoder. 2010. Phylogeography's past, present, and future: 10 years after Avise, 2000. *Molecular Phylogenetics and Evolution* 54:291–301.

Hickerson, M. J., G. Dolman, and C. Moritz. 2006. Comparative phylogeographic summary statistics for testing simultaneous vicariance. *Molecular Ecology* 15:209–23.

Hileman, L. C., and D. A. Baum. 2003. Why do paralogs persist? Molecular evolution of CYCLOIDEA and related floral symmetry genes in Antirrhineae (Veronicaceae). *Molecular Biology and Evolution* 20:591–600.

Hileman, L. C., M. C. Vasey, and V. T. Parker. 2001. Phylogeny and biogeography of the Arbutoideae (Ericaceae): Implications for the Madrean-Tethyan hypothesis. *Systematic Botany* 26:131–43.

Hillebrand, H. 2004. On the generality of the latitudinal diversity gradient. *American Naturalist* 163:192–211.

Hilton, R. P. 2003. *Dinosaurs and Other Mesozoic Reptiles of California.* University of California Press, Berkeley.

Hoelzel, A. R., R. C. Fleischer, C. Campagna, B. J. Le Boeuf, and G. Alvord. 2002. Impact of a population bottleneck on symmetry and genetic diversity in the northern elephant seal. *Journal of Evolutionary Biology* 15:567–75.

Hoelzel, A. R., J. Halley, S. J. O'Brien, C. Campagna, T. Arnbom, B. Le Boeuf, K. Ralls, and G. A. Dover. 1993. Elephant seal genetic variation and the use of simulation models to investigate historical population bottlenecks. *Journal of Heredity* 84:443–49.

Hollingsworth, B. D. 1998. The systematics of chuckwallas (*Sauromalus*) with a phylogenetic analysis of other Iguanid lizards. *Herpetological Monographs* 12:38–191.

Hoffman J. I., K. K. Dasmahapatra,W. Amos, C. D. Phillips, T. S. Gelatt, and J. W. Bickham. 2009. Contrasting patterns of genetic diversity at three different genetic markers in a marine mammal metapopulation. *Molecular Ecology* 18:2961–78.

Hoffman, J. I., C. W. Matson, W. Amos, T. R. Loughlin, and J. W. Bickham. 2006. Deep genetic subdivision within a continuously distributed and highly vagile marine mammal, the Steller's sea lion (*Eumetopias jubatus*). *Molecular Ecology* 15:2821–32.

Hoffman, P. F. 1999. The break-up of Rodinia, birth of Gondwana, true polar wander and the snowball earth. *Journal of African Earth Sciences* 17:17–33.

Hohmann, S., J. W. Kadereit, and G. Kadereit. 2006. Understanding Mediterranean-Californian disjunctions: Molecular evidence from Chenopodiaceae-Betoideae. *Taxon* 55:67–78.

Holland, R. F. 1976. The vegetation of vernal pools: A survey. *In* S. Jain [ed.], *Vernal Pools, Their Ecology and Conservation*, 11–15. Institute of Ecology Publication No. 9. University of California, Davis.

Holman, J. A. 2000. *Fossil Snakes of North America: Origin, Evolution, Distribution, and Paleoecology*. Indiana University Press, Bloomington.

———. 2003. *Fossil Frogs and Toads of North America*. Indiana University Press, Bloomington.

Holway, D., and A. Suarez. 2006. Homogenization of ant communities in mediterranean California: The effects of urbanization and invasion. *Biological Conservation* 127:319–26.

Hood, W. G., and R. Naiman. 2000. Vulnerability of riparian zones to invasion by exotic vascular plants. *Plant Ecology* 148:105–14.

Hooper, D. U., F. S. Chapin, J. J. Ewel, A. Hector, P. Inchausti, S. Lavorel, J. H. Lawton, D. M. Lodge, M. Loreau, S. Naeem, B. Schmid, H. Setälä, A. J. Symstad, J. Vandermeer, and D. A. Wardle. 2005. Effects of biodiversity on ecosystem functioning: A consensus of current knowldege. *Ecological Monographs* 75:3–35.

Horn, M. H., L. G. Allen, and R. N. Lea. 2006. Biogeography. *In* L. G. Allen, D. J. Pondella, and M. H. Horn [eds.], *The Ecology of Marine Fishes: California and Adjacent Waters*, 3–25. University of California Press, Berkeley.

Horning, M., and J.-A.-E., Mellish. 2012. Predation on an upper trophic marine predator, the Steller sea lion: Evaluating high juvenile mortality in a density dependent conceptual framework. *PLoS ONE* 7(1):e30173. doi:10.1371/journal.pone.0030173.

Hovanitz, W. 1940. Ecological color variation in a butterfly and the problem of "protective coloration." *Ecology* 21:371–80.

Howard, H. 1944. A Miocene hawk from California. *Condor* 46:236–37.

———. 1962. Fossil birds, with especial reference to the birds of Rancho La Brea, rev. ed. *Los Angeles County Museum of Natural History, Science Series no. 17, Paleontology* 10:3–44.

———. 1964. A fossil owl from Santa Rosa Island, California, with comments on the eared owls of Rancho La Brea. *Bulletin of the Southern California Academy of Sciences* 63:27–31.

Howard, J. L. 1992. *Torreya californica. In* Fire Effects Information System [online]. U.S. Department of Agriculture, Forest Service, Rocky Mountain

Research Station, Fire Sciences Laboratory (Producer). www.fs.fed.us /database/feis/.

Howard, R. D. 1979. *Geologic History of Middle California.* California Natural History Guide 43. University of California Press, Berkeley. 113 pp.

Howell, J. T. 1944. Certain plants of the Marble Mountains in California with remarks on the boreal flora of the Klamath area. *Wasmann Coll.* 6:13–20.

Huenneke, L. F., and J. K. Thomson. 1995. Potential interference between a threatened endemic thistle and an invasive nonnative plant. *Conservation Biology* 9:416–25.

Hull, J. M., J. J. Keane, W. K. Savage, S. A. Godwin, J. A Shafer, E. P. Jepsen, R. Gerhardt, C. Stermer, and H. B. Ernest. 2010. Range-wide genetic differentiation among North American great gray owls (*Strix nebulosa*) reveals a distinct lineage restricted to the Sierra Nevada, California. *Molecular Phylogenetics and Evolution* 56:212–21.

Hull, J. M., B. N. Strobel, C. W. Boal, A. C. Hull, C. R. Dykstra, A. M. Irish, A. M. Fish, and H. B. Ernest. 2008. Comparative phylogeography and population genetics within *Buteo lineatus* reveals evidence of distinct evolutionary lineages. *Molecular Phylogenetics and Evolution* 49:988–96.

Hunter, R. D., R. N. Fisher, and K. R. Crooks. 2003. Landscape-level connectivity in coastal Southern California, USA, as assessed through carnivore habitat suitability. *Natural Areas Journal* 23:302–14.

Ingersoll, R. V., and P. E. Rumelhart. 1999. Three-stage evolution of the Los Angeles Basin, Southern California. *Geology* 27:593–96.

Jaarola, M. N. Martínková, I. Gündüz, C. Brunhoff, J. Zima, A. Nadachowski, G. Amori, N. S. Bulatova, B. Chondropoulos, S., Fraguedakis-Tsolis, J. González-Esteban, M. José López-Fuster, A.S. Kandaurov, H. Kefelioğlu, M. da Luz Mathias, I. Villate, J.B. Searle. 2004. Molecular phylogeny of the speciose vole genus *Microtus* (Arvicolinae, Rodentia) inferred from mitochondrial DNA sequences. *Molecular Phylogenetics and Evolution* 3:647-63.

Ingles, L. G. 1965. *Mammals of the Pacific States.* Stanford University Press, Stanford, CA. 506 pp.

Jackman, T. R., and D. B. Wake. 1994. Evolutionary and historical analysis of protein variation in the blotched forms of salamanders of the *Ensatina* complex (Amphibia: Plethodontidae). *Evolution* 48:876–97.

Jacobs, D. K., T. A. Haney, and K. D. Louie. 2004. Genes, diversity, and geologic process on the Pacific Coast. *Annual Review of Earth and Planetary Sciences* 32:601–52.

Jacobson, G. L., Jr., T. Webb, and E. C. Grimm. 1987. Patterns and rates of vegetation change during the deglaciation of eastern North America. *In* W. F. Ruddiman and H. E. Wright Jr. [eds.], *The Geology of North America*, vol. k-3: *North America and Adjacent Oceans during the Last Deglaciation*, 277–88. Geological Society of America, Boulder, CO.

Jaeger, E. C., and A. C. Smith. 1966. *Introduction to the Natural History of Southern California.* University of California Press, Berkeley. 104 pp.

Jaeger, J. R., B. R. Riddle, and D. F. Bradford. 2005. Cryptic Neogene vicariance and Quaternary dispersal of the red-spotted toad (*Bufo punctatu*): Insights on the evolution of North American warm desert biotas. *Molecular Ecology* 14:3033–48.

Jagels, R., and M. A. Equiza. 2007. Why did *Metasequoia* disappear from North America but not from China? *Bulletin of the Peabody Museum of Natural History* 48:281–90.

Jahner, J. P., M. M. Bonilla, K. J. Badik, A. M. Shapiro, and M. L. Forister. 2011. Use of exotic hosts by Lepidoptera: Widespread species colonize more novel hosts. *Evolution* 65:2719–24.

Janis, C. M., J. Damuth, and J. M. Theodor. 2004. The species richness of Miocene browsers, and implications for habitat type and primary productivity in the North American grassland biome. *Palaeogeography, Palaeoclimatology, Palaeoecology* 207:371–98.

Janzen, F. J., J. G. Krenz, T. S. Haselkorn, E. D. Brodie Jr., and E. D. Brodie III. 2002. Molecular phylogeography of common garter snakes (*Thamnophis sirtalis*) in western North America: Implications for regional historical forces. *Molecular Eco*logy 11:1739–51.

Jarvis, K. J. and M. F. Whiting. 2006. Phylogeny and Biogeography of Ice Crawlers (Insecta: Grylloblattodea) based on six molecular loci: Designating Conservation Status for Grylloblattodea species. *Molecular Phylogenetics and Evolution.* 41:222-37.

Jezkova, T., J. R. Jaeger, Z. L. Marshall and B. R. Riddle. 2009. Pleistocene impacts on the phylogeography of the desert pocket mouse (*Chaetodipus penicillatus*). *Journal of Mammalogy* 90:306–20.

Jockusch, E. L., and D. B. Wake. 2002. Falling apart and merging: Diversification of slender salamanders (Plethodontidae: *Batrachoseps*) in the American West. *Biological Journal of the Linnean Society* 76:361–91.

Johnson, J. K. 1995. Speciation in vireos. I. Macrogeographic patterns of allozymic variation in the *Vireo solitarius* complex in the contiguous United States. *Condor* 97:903–19.

Johnson, J. K., and J. A. Marten. 1988. Evolutionary genetics of flycatchers. II. Differentiation in the *Empidonax difficilis* complex. *Auk* 105:177–91.

———. 1992. Macrogeographic patterns of morphometric and genetic variation in the sage sparrow complex. *Condor* 94:1–19.

Johnson, W. E., and S. J. O'Brien. 1997. Phylogenetic reconstruction of the Felidae using 16S rRNA and NADH-5 mitochondrial genes. *Journal of Molecular Evolution* (suppl. 1):S98–S116.

Johnson, N. K., and R. M. Zink. 1983. Speciation in sapsuckers (*Sphyrapicus*): I. Genetic differentiation. *Auk* 100:871–84.

Jorgensen, S. M., and J. L. Hamrick. 1997. Biogeography and population genetics of whitebark pine, *Pinus albicaulis*. *Canadian Journal of Forest Research* 27:1574–85.

Kadereit, J. W., and B. G. Baldwin. 2012. Western Eurasian–western North American disjunct plant taxa: The dry-adapted ends of formerly widespread north temperate mesic lineages and examples of long-distance dispersal. *Taxon* 61:3–17.

Kafton, D. L. 1976. Isozyme variability and reproductive phenology in Monterey cypress. PhD diss., University of California, Berkeley.

Karlstrom, K. E, M. L. Williams, J. McLelland, J. W. Geissman, and K.-I. Åhäll. 1999. Refining Rodinia: Geologic evidence for the Australia–western U.S. connection in the Proterozoic. *Geological Society of America Today* 9:1–7.

Käss, E., and M. Wink. 1997. Molecular phylogeny and phylogeography of *Lupinus* (Leguminosae) inferred from nucleotide sequences of the *rbc*L gene and ITS 1+2 regions of rDNA. *Plant Systematics and Evolution* 208:139–67.

Katibah, E. F. 1984. A brief history of the riparian forests in the central valley of California. *In* R. E. Warner and K. M. Hendrix [eds.], *California Riparian Systems: Ecology Conservation and Productive Management,* 23–29. University of California Press, Berkeley.

Katz, J., P. B. Moyle, R. M. Quiñones, J. Israel, and S. Purdy. 2012. Impending extinction of salmon, steelhead, and trout (Salmonidae) in California. *Environmental Biology of Fishes* doi:10.1007/s10641-012-9974-8.

Keator, G. 1998. *The Life of an Oak: An Intimate Portrait.* Heyday Books and California Oak Foundation, Berkeley, CA.

Keir, K. R., J. B. Bemmels, and S. N. Aitken. 2011. Low genetic diversity, moderate local adaptation, and phylogeographic insights in *Cornus nuttallii* (Cornaceae). *American Journal of Botany* 98:1327–36.

Kellogg, R. 1922. Pinnipeds from the Miocene and Pleistocene deposits of California. *University of California Publications in Geological Sciences* 13:23–132.

Kendall, A. W., Jr., and A. K. Gray. 2000. An historical review of *Sebastes* taxonomy and systematics. *Marine Fisheries Review* 62:1–15.

Kendrick, K. J., D. M. Morton, S. G. Wells, and R. W. Simpson. 2002. Spatial and temporal deformation along the northern San Jacinto fault, Southern California: Implications for slip rates. *Bulletin of the Seismological Society of America* 92:2782–92.

Kennedy, M. P., and G. W. Moore. 1971. Stratigraphic relations of Upper Cretaceous and Eocene formations, San Diego coastal area, California. *American Association of Petroleum Geologists Bulletin* 55:709–22.

Khasa, D., P. Pollefeys, A. Navarro-Quezada, P. Perinet, and J. Bousquet. 2005. Species-specific microsatellite markers to monitor gene flow between exotic poplars and their natural relatives in eastern North America. *Molecular Ecology Notes* 5:920–23.

Kiefer, C., and M. A. Koch. 2012. A continental-wide perspective: The gene pool of nuclear encoded ribosomal DNA and single-copy gene sequences in North American Boechera (Brassicaceae). *PLoS ONE* 7(5):e36491. doi:10.1371/journal.pone.0036491.

Kimsey, L. 1996. Status of terrestrial insects. *In Sierra Nevada Ecosystem Project: Final Report to Congress,* vol. 2, *Assessments and Scientific Basis for Management Options.* University of California, Davis, Centers for Water and Wildland Resources.

Kimura, M., S. M. Clegg, J. J. Lovette, K. R. Holder, D. J. Girman, B. Milá, S. P. Wade, and T. B. Smith. 2002. Phylogeographical approaches to assessing demographic connectivity between breeding and overwintering regions in a Nearctic-Neotropical warbler (*Wilsonia pusilla*). *Molecular Ecology* 11:1605–16.

Kirby, M. X., D. S. Jones, and B. J. MacFadden. 2008. Lower Miocene stratigraphy along the Panama Canal and its bearing on the Central American peninsula. *PLoS ONE* 3(7):e2791. doi:10.1371/journal.pone.0002791.NAL PAPER

Kishore, V. K., P. Velasco, D. K. Shintani, J. Rowe, C. Rosato, N. Adair, M. B. Slabaugh, and S. J. Knapp. 2004. Conserved simple sequence repeats for the Limnanthaceae (Brassicales). *Theoretical and Applied Genetics* 108:450–57.

Kleppe, J. A., D. S. Brothers, G. M. Kent, F. Biondi, S. Jensen, and N. W. Driscoll. 2011. Duration and severity of medieval drought in the Lake Tahoe Basin. *Quaternary Science Reviews* 30:3269–79.

Kluge, A. G. 1993. *Calabaria* and the phylogeny of erycine snakes. *Zoological Journal of the Linnean Society* 107:293–351.

Knowles, L. L. 2009. Statistical phylogeography. *Annual Review of Ecology and Systematics* 40:593–612.

Knowles, L. L., and W. P. Maddison. 2002. Statistical phylogeography. *Molecular Ecology* 11:2623–35.

Koepfli, K.-P., K. A. Deere, G. J. Slater, C. Begg, K. Begg, L. Grassman, M. Lucherini, G. Veron, and R. K. Wayne. 2008. Multigene phylogeny of the Mustelidae: Resolving relationships, tempo and biogeographic history of a mammalian adaptive radiation. *BMC Biology* 6:10. doi:10.1186/1741–7007.

Kornev, S. I., and S. M. Korneva. 2004. Population dynamics and present status of sea otters (*Enhydra lutris*) of the Kuril Islands and southern Kamchatka. *Marine Mammals of the Holarctic, Proceedings of 2004 Conference,* 273–78.

Korth, W. W., and R. E. Reynolds. 1991. A new heteromyid from the Hemingfordian of California and the first occurrence of sulcate incisors in heteromyids. *Journal of Vertebrate Paleontology* 11(Suppl.):41A.

Kraft, N. J. B., B. G. Baldwin, and D. D. Ackerly. 2010. Range size, taxon age and hotspots of neoendemism in the California flora. *Diversity and Distributions* 16:403–13.

Kramer, A. T., B. Zorn-Arnold, and K. Havens. 2010. Assessing botanical capacity to address grand challenges in the United States. www.bgci.org /usa/bcap. 64 pp.

Kreissman, B. 1991. *California: An Environmental Atlas and Guide.* Bear Klaw Press, Davis, CA. 255 pp.

Kruckeberg, A. R. 1984. *California Serpentines: Flora, Vegetation, Geology, Soils, and Management Problems.* University of California Press, Berkeley.

Kuchta, S. R., and A. M. Tan. 2006. Lineage diversification on an evolving landscape: Phylogeography of the California newt, *Taricha torosa* (Caudata: Salamandridae). *Biological Journal of the Linnean Society* 89:213–39.

Kuchta, S. R., D. S. Parks, R. L. Mueller, and D. B. Wake. 2009. Closing the ring: Historical biogeography of the salamander ring species *Ensatina eschscholtzii. Journal of Biogeography* 36:982–95.

Kuchta, S. R., and D. B. Wake. 2005. Ensatina *Ensatina eschscholtzii* Gray. *In* L. L. C. Jones, W. P. Leonard, and D. H. Olson [eds.], *Amphibians of the Pacific Northwest,* 110–13. Seattle Audubon Society, Seattle, WA.

Kurtén, B., and E. Anderson. 1980. *Pleistocene Mammals of North America.* Columbia University Press, New York. 442 pp.

Lacy, R. C. 1997. Importance of genetic variation to the viability of mammalian populations. *Journal of Mammalogy* 78:320–35.

Lamb, T., J. C. Avise, and J. Whitfield Gibbons. 1989. Phylogeographic patterns in mitochondrial DNA of the desert tortoise (*Xerobates agassizi*), and evolutionary relationships among the North American gopher tortoises. *Evolution* 43:76–87.

Lamb, T., T. R. Jones, and J. C. Avise. 1992. Phylogeographic histories of representative herpetofauna of the southwestern U.S.: Mitochondrial DNA variation in the desert iguana (*Dipsosaurus dorsalis*) and the chuckwalla (*Sauromalus obesus*). *Journal of Evolutionary Biology* 5:465–80.

Landini, W., G. Bianucci, G. Carnevale, L. Ragaini, C. Sorbini, G. Valleri, M. Bisconti, G. Cantalamessa, and C. DiCelma. 2002. Late Pliocene fossils of Ecuador and their role in the development of the Panamic bioprovince after the rising of the Central American Isthmus. *Canadian Journal of Earth Sciences* 39:27–41.

Lapointe, F. J., and L. J. Rissler. 2005. Notes and comments: Congruence, consensus, and the comparative phylogeography of codistributed species in California. *American Naturalist* 166:290–99.

Latch, E. K., J. R. Heffelfinger, J. A. Fike, and O. E. Rhodes Jr. 2009. Species-wide phylogeography of North American mule deer (*Odocoileus hemionus*): Cryptic glacial refugia and postglacial recolonization. *Molecular Ecology* 18:1730–45. l Publishing

Latta, R. G., Y. B. Linhart, D. Fleck, and M. Elliot. 1998. Direct and indirect estimates of seed versus pollen movement with a population of ponderosa pine. *Evolution* 52:61–67.

Latta, R. G., and J. B. Mitton. 1999. Historical separation and present gene flow through a zone of secondary contact in ponderosa pine. *Evolution* 53:769–76.

Law, J. H., and B. J. Crespi. 2002. Recent and ancient asexuality in *Timema* walkingsticks. *Evolution* 56:1711–17.

Lawver, L. A., and L. M. Gahagan. 2003. Evolution of Cenozoic seaways in the circum-Antarctic region. *Palaeogeography, Palaeoclimatology, Palaeoecology* 198:11–37.

Leaché, A. D., D.-S. Helmer, and C. Moritz. 2010. Phenotypic evolution in high-elevation populations of western fence lizards (*Sceloporus occidentalis*) in the Sierra Nevada Mountains. *Biological Journal of the Linnean Society* 100:630–41.

Leaché, A. D., M. S. Koo, C. L. Spencer, T. J. Papenfuss, R. N. Fisher, and J. A. McGuire. 2009. Quantifying ecological, morphological, and genetic variation to delimit species in the coast horned lizard species complex (*Phrynosoma*). *Proceedings of the National Academy of Sciences USA* 106:12418–23.

Leaché, A. D., and D. G. Mulcahy. 2007. Phylogeny, divergence times and species limits of spiny lizards (*Sceloporus magister* species group) in western North American deserts and Baja California. *Molecular Ecology* 16:5216–33.

Leavitt, D. H., R. L. Bezy, K. A. Crandall, and J. W. Sites Jr. 2007. Multi-locus DNA sequence data reveal a history of deep cryptic vicariance and habitat-driven convergence in the desert night lizard *Xantusia vigilis* species complex (Squamata: Xantusiidae). *Molecular Ecology* 16:4455–81.

Le Beouf, B. J., and J. Reiter. 1991. Biological effects associated with El Niño southern oscillation, 1982–83, on northern elephant seals breeding at Año Nuevo, California. *In* F. Trillmich and K. A. Ono [eds.], *Pinnipeds and El Niño: Responses to Environmental Stress*, 206–18. Springer-Verlag, New York.

Ledig, F. T. 2000. Founder effects and the genetic structure of Coulter pine. *Journal of Heredity* 91:307–15.

Ledig, F. T., and M. T. Conkle. 1983. Gene diversity and genetic structure in a narrow endemic, Torrey pine (*Pinus torreyana* Parry ex Carr.). *Evolution* 37:79–85.

Ledig, F.T., P. D. Hodgkiss, and D. R. Johnson. 2005. Genic diversity, genetic structure, and mating system and Brewer spruce (Pinaceae), a relict of the Arcto-Tertiary forest. *American Journal of Botany* 92:1975–86.

Lee, S. W., F. T. Ledig, and D. R. Johnson. 2002. Genetic variation at allozyme and RAPD markers in *Pinus longaeva* (Pinaceae) of the White Mountains, California. *American Journal of Botany* 89:566–77.

Lee, T. E., B. R. Riddle, and P. E. Lee. 1996. Speciation in the desert pocket mouse (*Chaetodipus penicillatus* Woodhouse). *Journal of Mammalogy* 77:58–68.

Lehman, P. W., S. Mayr, L. Mecum, and C. Enright. 2010. The freshwater tidal wetland at Liberty Island, CA, was both a source and sink of inorganic and organic material to the San Francisco Estuary. *Aquatic Ecology* 44:359–72.

Lesica, P., and F. W. Allendorf. 1999. Ecological genetics and the restoration of plant communities: Mix or match? *Restoration Ecology* 7:42–50.

Levine, J. M., M. Vilá, C. M. D'Antonio, J. S. Dukes, K. Grigulis, and S. Lavorel. 2003. Mechanisms underlying the impacts of exotic plant invasions. *Proceedings of the Royal Society of London, Series B* 270:775–81.

Lidicker, W. Z., Jr., and F. C. McCollum. 1997. Allozymic variation in California sea otters. *Journal of Mammalogy* 78:417–25.

Lijtmaer, D. A., K. C. R. Kerr, A. S. Barreira, P. D. N. Hebert, and P. L. Tubaro 2011. DNA barcode libraries provide insight into continental patterns of avian diversification. *PloS ONE* 6(7):e20744. doi:10.1371/journal. pone.0020744.

Lind, A. J., P. Q. Spinks, G. M. Fellers, and H. B. Shaffer 2011. Rangewide phylogeography and landscape genetics of the western U.S. endemic frog *Rana boylii* (Ranidae): Implications for the conservation of frogs and rivers. *Conservation Genetics* 12:269–84.

Lindberg, D. R., and J. H. Lipps. 1996. Reading the chronicle of Quaternary temperate rocky-shore faunas. *In* D. Jablonski, D. Erwin, and J. H. Lipps [eds.], *Evolutionary Paleobiology*, 161–82. University of Chicago Press, Chicago, IL.

Lindsay, S. L. 1981. Taxonomic and biogeographic relationships of Baja California chickarees (*Tamiasciurus*). *Journal of Mammalogy* 62: 673–82.

Liston, A. 1997. Biogeographic relationships between the Mediterranean and North American floras: Insights from molecular data. *Lagascalia* 19:323–30.

Liston, A., D. S. Gernandt, T. F. Vining, C. S. Campbell, and D. Piñero. 2003. Molecular phylogeny of Pinaceae and *Pinus*. *Acta Horticulturae* 615:107–14.

Liston, A., M. Parker-Defeniks, J. V. Syring, A. Willyard, and R. Cronn. 2007. Interspecific phylogenetic analysis enhances intraspecific phylogeographical inference: A case study in *Pinus lambertiana*. *Molecular Ecology* 16:3926–37.

Liston, A., L. H. Rieseberg, and T. S. Elias. 1989. Genetic similarity is high between intercontinental disjunct species of *Senecio* (Asteraceae). *American Journal of Botany* 76:383–88.

Livezey, B. C., and R. L. Zusi. 2007. Higher-order phylogeny of modern birds (Theropoda, Aves: Neornithes) based on comparative anatomy. II. Analysis and discussion. *Zoological Journal of the Linnean Society* 149:1–95.

Loarie, S. R., B. E. Carter, K. Hayhoe, S. McMahon, R. Moe, C. A. Knight, and D. D. Ackerly. 2008. Climate change and the future of California's endemic flora. *PLoS One* 3(6):e2502. doi:10.1371/journal.pone.0002502. e2502.

Longrich, N. R., T. Tokaryk, and D. J. Field. 2011. Mass extinction of birds at the Cretaceous–Paleogene (K–Pg) boundary. *Proceedings of the National Academy of Science USA* 108:15253–57.

Love, M. S., M. Yoklavich, and L. Thorsteinson. 2002. *The Rockfishes of the Northeast Pacific.* University of California Press, Berkeley. 416 pp.

Lovette, I. J. 2005. Molecular phylogeny and plumage signal evolution in a trans-Andean and circum-Amazonian avian species complex. *Molecular Phylogenetics and Evolution* 32:512–23.

Lovich, R. 2001. Phylogeography of the night lizard (*Xantusia henshawi*) in Southern California: Evolution across zones. *Herpetologica* 57:470–87.

Luikart, G., and F. W. Allendorf. 1996. Mitochondrial-DNA variation and genetic-population structure in Rocky Mountain bighorn sheep (*Ovis canadensis canadensis*). *Journal of Mammalogy* 77:109–23.

Lyon, G. M. 1941. A Miocene sea lion from Lomita, California. *University of California Publications in Zoology* 47:23–42.

Macdonald, D. 1984. *The Encyclopedia of Mammals* 1. Allen & Unwin, London. 446 pp.

Macey, J. R., J. L. Strasburg, J. A. Brisson, V. T. Vredenburg, M. Jennings, and A. Larson. 2001. Molecular phylogenetics of western North American frogs of the *Rana boylii* species group. *Molecular Phylogenetics and Evolution* 19:131–43.

Mackintosh, N. A. 1965. *The Stocks of Whales.* Fishing News (Books), London.

Mahoney, M. J. 2004. Molecular systematics and phylogeography of the *Plethodon elongates* species group: Combining phylogenetic and population genetic methods to investigate species history. *Molecular Ecology* 13:149–66.

Malamud-Roama, F. P., B. L. Ingram, M. Hughes, and J. L. Florsheim. 2006. Holocene paleoclimate records from a large California estuarine system and its watershed region: Linking watershed climate and bay conditions. *Quaternary Science Reviews* 25:1570–98.

Maldonado, J. E., C. Vila, and R. K. Wayne. 2001. Tripartite genetic subdivisions in the ornate shrew (*Sorex ornatus*). *Molecular Ecology* 10:127–47.

Mann, K., and J. Lazier. 2006. *Dynamics of Marine Ecosystems: Biological-Physical Interactions in the Oceans.* Wiley and Sons. 496 pp.

Manos, P. S., J. J. Doyle, and K. C. Nixon. 1999. Phylogeny, biogeography, and processes of molecular differentiation in *Quercus* subgenus *Quercus* (Fagaceae). *Molecular Phylogenetics and Evolution* 12:333–49.

Marincovich, L., Jr., and A. Y. Gladenkov. 1999. Evidence for an early opening of Bering Strait. *Nature* 397:149–51.

Marlow, J. R., C. B. Lange, G. Wefer, and A. Rosell-Melé. 2000. Upwelling intensification as part of the Pliocene-Pleistocene transition. *Science* 290:2288–91.

Martín-Bravo, S., and M. Escudero. 2012. Biogeography of flowering plants: A case study in mignonettes (Resedaceae) and sedges (*Carex*, Cyperaceae). *In*

L. Stevens [ed.], *Global Advances in Biogeography*, 257–90. Intech, Rijeka, Croatia.

Martínez-Solano, I., E. L. Jockusch, and D. B. Wake. 2007. Extreme population subdivision throughout a continuous range: Phylogeography of *Batrachoseps attenuatus* (Caudata: Plethodontidae) western North America. *Molecular Ecology* 20:4335–55.

Marty, J. T. 2005. Effects of cattle grazing on diversity in ephemeral wetlands. *Conservation Biology* 19:1626–32.

Mascarello, J. T. 1978. Chromosomal, biochemical, mensural, penile, and cranial variation in desert woodrats *(Neotoma lepida). Journal of Mammalogy* 59:477–95.

Mastrogiussepe, R. J., and J. D. Mastrogiuseppe. 1980. A study of *Pinus balfouriana* Grev. & Balf. (Pinaceae). *Systematic Botany* 5:86–104.

Matocq, M. D. 2002. Phylogeographical structure and regional history of the dusky-footed woodrat, *Neotoma fuscipes. Molecular Ecology* 11:229–42.

Maxwell, A. L. 1992. Mapping and habitat analysis of the California endemic tree, *Torreya californica,* in Marin County. MS thesis, San Francisco State University. 228 pp.

McCracken, G. F., and M. F. Gassel. 1997. Genetic structure in migratory and nonmigratory populations of Brazilian free-tailed bats. *Journal of Mammalogy* 78:348–57.

McKay, S. J., R. H. Devlin, and M. J. Smith. 1996. Phylogeny of Pacific salmon and trout based on growth hormone type-2 and mitochondrial NADH dehydrogenase subunit 3 DNA sequences. *Canadian Journal of Fisheries and Aquatic Sciences* 53:1165–76.

McMahan, L. R. 2006. Understanding cultural reasons for the increase in both restoration efforts and gardening with native plants. *Native Plants Journal* 7:31–34.

McNeal, D. W., and T. D. Jacobsen. 2002. *Allium* L. *In* Flora of North America Editorial Committee [eds.], *Flora of North America,* xx–xxx. New York Botanical Garden Press, New York.

Messinger, O., and T. Griswold. 2002. A pinnacle of bees. *Fremontia* 30(3–4):32–40.

Meyers, S., and A. Liston. 2008. The biogeography of *Plantago ovata* (Plantaginaceae). *International Journal of Plant Science* 169:954–62.

Michel, F., C. Rebourge, E. Cosson, and H. Descimon. 2008. Molecular philogeny of Parnassinae butterflies (Lepidoptera, Papilionidae) based on the sequences of four mitochondrial DNA segments. *Annales de la Societe Entomologique de France* 44:1-36.

Mila, B., J. E. McCormack, G. Castaneda, R. K. Wayne, and T. B. Smith. 2007. Recent postglacial range expansion drives the rapid diversification of a songbird lineage in the genus *Junco. Proceedings of the Royal Society B* 274:2653–60.

Miles, S., and C. Gouday [eds.]. 1997. *Ecological Subregions of California*. USDA, Forest Service, Pacific Southwest Division, R5-EM-TP-005, San Francisco.

Millar, C. I. 1983. A steepcline in *Pinus muricata. Evolution* 37:311–19.

———. 1996. Tertiary vegetation history. *In Sierra Nevada Ecosystem Project: Final Report to Congress*, vol. 2, *Assessments and Scientific Basis for Management Option*, 71–124. Wildland Resources Center Report No. 36, University of California, Davis.

———. 1998. Early evolution of pines. *In* D. M. Richardson [ed.], *Ecology and Biogeography of* Pinus, 69–94. Cambridge University Press, Cambridge.

———. 2012. Geologic, climatic, and vegetation history of California. *In* B. G. Baldwin, D. H. Goldman, D. J. Keil, R. Patterson, T. J. Rosatti, and D. H. Wilkin [eds.], *The Jepson Manual: Vascular Plants of California*, 2nd ed., 49–67. University of California Press, Berkeley.

Miller, C. N., Jr. 1973. Silicified cones and vegetative remains of *Pinus* from the Eocene of British Columbia. *Contributions of the University of Michigan Museum of Paleontology* 24:101–18.

Miller, R. R. 1950. Speciation in fishes of the genera *Cyprinodon* and *Empetrichtys*, inhabiting the Death Valley region. *Evolution* 4:155–63.

———. 1981. Coevolution of deserts and pupfishes (genus *Cyprinodon)* in the American Southwest. *In* R. J. Naiman and D. L. Soltz [eds.], *Fishes in North American Deserts*, 39–94. John Wiley & Sons, New York.

Miller, W. E. 1980. The late Pliocene Las Tunas local fauna from southernmost Baja, California, Mexico. *Journal of Paleontology* 54:762–805.

Milne, R. I. 2004. Phylogeny and biogeography of *Rhododendron* subsection *Pontica*, a group with a tertiary relict distribution. *Molecular Phylogenetics and Evolution* 33:389–401.

Milne, R. I., and R. J. Abbott. 2002. The origin and evolution of Tertiary relict floras. *Advances in Botanical Research* 38:281–314.

Milner, A. M., E. E. Knudsen, C. Soiseth, A. L. Robertson, D. Schell, I. T. Phillips, and K. Magnusson. 2000. Colonization and development of stream communities across a 200-year gradient in Glacier Bay National Park, Alaska, U.S.A. *Canadian Journal of Fisheries and Aquatic Sciences* 57:2319–35.

Modesto, S. P., and J. S. Anderson. 2004. The phylogenetic definition of Reptilia. *Systematic Biology* 53:815–21.

Mohr, J., A. C. Whitlock, and C. N. Skinner. 2000. Postglacial vegetation and fire history, eastern Klamath Mountains, California, USA. *Holocene* 10:587–601.

Moldenke, A. R. 1999. Soil-dwelling arthropods: Their diversity and functional roles. General Technical Report PNW-GTR-461, USDA Forest Service Pacific Northwest Research Station, Portland, OR.

Mönkkönen, M., and P. Viro. 1997. Taxonomic diversity of the terrestrial bird and mammal fauna in temperate and boreal biomes of the Northern Hemisphere. *Journal of Biogeography* 24:603–12.

Moore, D. M., and O. A. Chater. 1971. Studies on bipolar species. I. *Carex*. *Botaniska Notiser* 124:317–34.

Moore, M. J., P. S. Soltis, C. D. Bell, J. G. Burleigh, and D. E. Soltis. 2010. Phylogenetic analysis of 83 plastid genes further resolves the early diversification of eudicots. *Proceedings of the National Academy of Sciences USA* 107:4625–28.

Moratto, M. J. 1984. *California Archaeology*. Academic Press, New York.

Moritz, C. 1994. Defining "evolutionarily significant units" for conservation. *Trends in Ecology and Evolution* 9:373–75.

Moritz, C., J. L. Patton, C. J. Conroy, J. L. Parra, G. C. White, and S. R. Beissinger. 2008. Impact of a century of climate change on small-mammal communities in Yosemite National Park, USA. *Science* 322:261–64.

Mount, J. F., and P. W. Signor. 1989. Paleoenvironmental context of the metazoan radiation event and its impact on the placement of the Precambrian-Cambrian boundary: Examples from the southwestern Great Basin, USA. *In* N. Christie-Blick and M. Levy [eds.], *Late Proterozoic and Cambrian Tectonics, Sedimentation and Record of Metazoan Radiation in the Western U.S.*, 39–46. 28th International Geological Congress Field Trip Guidebook T331, American Geophysical Union.

Moyle, P. B. 1976. *Inland Fishes of California*. University of California Press, Berkeley. 405 pp.

———. 2002. *Inland Fishes of California*. Rev. and expanded. University of California Press, Berkeley. 517 pp.

Moyle, P. B., W. Bennett, J. Durand, W. Fleenor, B. Gray, E. Hanak, J. Lund, and J. Mount. 2012. *Where the Wild Things Aren't: Making the Delta a Better Place for Native Species*. Public Policy Institute of California, San Francisco. 53 pp.

Moyle, P. B., J. Hobbs, and T. O'Rear. 2012. Fishes. *In* A. Palaima [ed.], *Ecology, Conservation, and Restoration of Tidal Marshes: The San Francisco Estuary*, 161–73. University of California Press, Berkeley.

Moyle, P. B., R. M. Quiñones, and J. D. Kiernan. 2012. Effects of climate change on the inland fishes of California, with emphasis on the San Francisco Estuary region. California Energy Commission, Public Interest Research Program White Paper CEC-500–2011-xxx.

Moyle, P. B., R. M. Yoshiyama, J. E. Williams, and E. D. Wikramanayake. 1995. *Fish Species of Special Concern in California*. 2nd ed. California Department of Fish and Game, Sacramento.

Mueller, R. L. 2006. Evolutionary rates, divergence dates, and the performance of mitochondrial genes in Bayesian phylogenetic analysis. *Systematic Biology* 55:289–300.

Muhlfeld, C. C., T. E. McMahon, M. C. Boyer, and R. E. Gresswell. 2009. Local-habitat, watershed, and biotic factors in the spread of hybridization between native westslope cutthroat trout and introduced rainbow trout. *Transactions of the American Fisheries Society* 138:1036–51.

Muhs, D. R., K. R. Simmons, and B. Steinke. 2002. Timing and warmth of the Last Interglacial Period: New U-series evidence from Hawaii and Bermuda and a new fossil compilation for North America. *Quaternary Science Reviews* 21:1355–83.

Mulcahy, D. G. 2007. Molecular systematics of Neotropical cat-eyed snakes: A test of the monophyly of Leptodeirini (Colubridae: Dipsadinae) with implications for character evolution and biogeography. *Biological Journal of the Linnean Society* 92(3):483–500.

Mulcahy, D. G., A. W. Spaulding, J. R. Mendelson III, and E. D. Brodie Jr. 2006. Phylogeography of the flat-tailed horned lizard (*Phrynosoma mcallii*) and systematics of the *P. mcallii–platyrhinos* mtDNA complex. *Molecular Ecology* 15:1807–26.

Mulch, A., A. M. Sarna-Wojcicki, M. E. Perkins, and C. P. Chamberlain. 2008. A Miocene to Pleistocene climate and elevation record of the Sierra Nevada (California). *Proceedings of the National Academy of Science USA* 105:6819–24.

Murphy, R. W., and G. Aguirre-León. 2002. Non-avian reptiles: Origins and evolution. *In* T. J. Case, M. L. Cody, and E. Ezcurra [eds.], *A New Island Biogeography in the Sea of Cortés,* 181–220. Oxford University Press, New York.

Murphy, R. W., K. H. Berry, T. Edwards, A. E. Leviton, A. Lathrop, and J. D. Riedle. 2011. The dazed and confused identity of Agassiz's land tortoise, *Gopherus agassizii* (Testudines, Testudinidae) with the description of a new species and its consequences for conservation. *Zookeys* 113:39–71.

Murray, B. G., Jr. 1967. Grebes from the late Pliocene of North America. *Condor* 69:277–88.

Murray, K. F. and A. M. Barnes. 1969. Distribution and habitat of the woodrat, *Neotoma fuscipes,* in northeastern California. *Journal of Mammalogy* 50:43-48.

Myers, N., R. A. Mittermeier, C. G. Mittermeier, G. A. B. da Fonseca, and J. Kent. 2000. Biodiversity hotspots for conservation priorities. *Nature* 403:853–58.

Naiman, R. J., J. J. Magnuson, D. M. McKnight, and J. A. Stanford. 1995. *The Freshwater Imperative.* Island Press, Washington, D.C.

Near, T. J., T. W. Kassler, J. B. Koppelman, C. B. Dillman, and D. P. Philipp. 2003. Speciation in North American black basses, *Micropterus* (Actinopterygii: Centrarchidae). *Evolution* 57:1610–21.

Negrini, R. M., P. E. Wigand, S. Draucker, K. Gobalet, J. K. Gardner, M. Q. Sutton, and R. M. Yoh II. 2006. The Rambla highstand shoreline and the Holocene lake-level history of Tulare Lake, California, USA. *Quaternary Science Reviews* 25:1599–1618.

Nei, M. 1973. Analysis of gene diversity in subdivided populations. *Proceedings of the National Academy of Sciences USA* 70:3321–23.

———. 1978. Estimation of average heterozygosity and genetic distance from a small number of individuals. *Genetics* 89:583–90.

———. 1986. Stochastic errors in DNA evolution and molecular phylogeny. *In* H. Gershowitz, D. L. Rucknagel, and R.E. Tashian [eds.], *Evolutionary Perspectives and the New Genetics,* 133–47. Alan R. Liss, New York.

———. 1987. *Molecular Evolutionary Genetics.* Columbia University Press, New York.

Nei, M., and W. H. Li. 1979. Mathematical model for studying genetic variation in terms of restriction endonucleases. *Proceedings of the National Academy of Sciences USA* 76:5269–73.

Nei, M., and F. Tajima. 1981. DNA polymorphism detectable by restriction endonucleases. *Genetics* 97:145–63.

Nettel, A., R. S. Dodd, and Z. Afzal-Rafii. 2009. Genetic diversity, structure, and demographic change in tanoak, *Lithocarpus densiflorus* (Fagaceae), the most susceptible species to sudden oak death in California. *American Journal of Botany* 96:2224–33.

Nguyen, N. H., H. E. Driscoll, and C. D. Specht. 2008. A molecular phylogeny of the wild onions (*Allium*; Alliaceae) with a focus on the western North American center of diversity. *Molecular Phylogenetics and Evolution* 47:1157–72.

Nice, C. C., N. Anthony, G. Gelembiuk, D. Raterman, and R. French-Constant. 2005. The history and geography of diversification within the butterfly genus *Lycaeides* in North America. *Molecular Ecology* 14:1741–54.

Nice, C. C., and A. M. Shapiro. 2001. Patterns of morphological, biochemical, and molecular evolution in the *Oeneis chryxus* Complex (Lepidoptera: Satyridae): A test of historical biogeographical hypotheses. *Molecular Phylogenetics and Evolution* 20:111–23.

Niebling, C. R., and M. T. Conkle. 1990. Diversity of Washoe pine and comparisons with allozymes of ponderosa pine races. *Canadian Journal of Forest Research* 20:298–308.

Nielson, M., K. Lohman, C. Daugherty, F. W. Allendorf, K. L. Knudsen, and J. Sullivan. 2006. Allozyme and mitochondrial DNA variation in the tailed frog (Anura: Ascaphus): The influence of geography and gene flow. *Herpetologica* 62:235–58.

Nilsson, T. 1984. *The Pleistocene: Geology and Life in the Quaternary Ice Age.* D. Reidel, Boston, MA.

Norris, K. 2012. Biodiversity in the context of ecosystem services: The applied need for systems approaches. *Philosophical Transactions of the Royal Society B* 367:191–99.

Noss, R. F., J. R. Strittholt, K. Vance-Borland, C. Carroll, and P. Frost. 1999. A conservation plan for the Klamath-Siskiyou Ecoregion. *Natural Areas Journal* 19:392–411.

Novacek, M. J. 1994. The Great American Interchange. *Natural History* 103:40–44.

Nowak, R. M. 1979. North American Quaternary *canis. Monographs of the Museum of Natural History, University of Kansas* 6. 154 pp.

Olden, J. D., N. L. Poff, M. R. Douglas, M. E. Douglas, and F. D. Fausch. 2004. Ecological and evolutionnary consequences of biotic homogenization. *Trends in Ecology and Evolution* 19:18–24.

Olden, J. D., and T. P. Rooney. 2006. On defining and quantifying of biotic homogenization. *Global Ecology and Biogeography* 15:113–20.

Oline, D. K., J. B. Mitton, and M. C. Grant. 2000. Population and subspecific genetic differentiation in the foxtail pine (*Pinus balfouriana*). *Evolution* 54(5):1813–19.

Oliver, J. C., and A. M. Shapiro. 2007. Genetic isolation and cryptic variation within the *Lycaena xanthoides* species-group. *Molecular Ecology* 16:4308–20.

Olsen, A. M. 1954. The biology, migration and growth rate of the school shark, *Galeorhinus australis* (Macleay) (Carcharhinidae) in south-eastern Australian waters. *Australian Journal of Marine and Freshwater Research* 5:353–410.

Olson, S. L. 1985. The fossil record of birds. *In* D. S. Farner, J. R. King, and K. C. Parkes [eds.], *Avian Biology,* 79–238. Academic Press, New York.

Omland, K. E., J. Marzluff, W. Boarman, C. L. Tarr, and R. C. Fleischer. 2000. Cryptic genetic variation and paraphyly in ravens. *Proceedings of the Royal Society of London B* 267:2475–82.

Orme, A. R. 2008. Pleistocene pluvial lakes of the American West: A short history of research. *In* R. H. Grapes, D. Oldroyd, and A. Grigelis [eds.], *History of Geomorphology and Quaternary Geology,* 51–78. Special Publications, 301. Geological Society, London.

Orr, R. T. 1960. An analysis of the recent land mammals. *Systematic Zoology* 9:171–78.

Orr, R. T., and R. C. Helm. 1989. *Marine Mammals of California.* University of California Press, Berkeley.

Owen, J. G., and R. S. Hoffmann. 1983. *Sorex ornatus. Mammalian Species* 212:1–5.

Owen, L. A., R. C. Finkel, R. A. Minnich, and A. E. Perez. 2003. Extreme southwestern margin of late Quaternary glaciation in North America: Timing and controls. *Geology* 31:729–32.

Paetkau, D., L. P. Waits, and P. Clarkson. 1998. Variation in genetic diversity across the range of North American brown bears. *Conservation Biology* 12:418–29.

Palacios, D. M., S. J. Bograd, R. Mendelsshon, and F. B. Schwing. 2004. Long-term and seasonal trends in stratification in the California Current, 1950–1993. *Journal of Geophysical Research* 109(C10). doi:10.1029/2004JC002380.

Palumbi, S. R. 1994. Genetic divergence, reproductive isolation, and marine speciation. *Annual Review of Ecology and Systematics* 25:547–72.

Palumbi, S. R., F. Cipriano, and M. P. Hare. 2001. Predicting nuclear gene coalescence from mitochondrial data: The three-times rule. *Evolution* 55:859–68.

Parham, J. F., and T. A. Stidham. 1999. Late Cretaceous sea turtles from the Chico Formation of California. *PaleoBios* 19(3):1–7.

Parks, D. S. M. 2000. Phylogeography, historical distribution, migration, and species boundaries in the salamander *Ensatina eschscholtzii* as measured with mitochondrial DNA sequences. PhD diss., University of California, Berkeley.

Parmesan, C., N. Ryrholm, C. Stefanescu, J. K. Hill, C. D. Thomas, H. Descimon, B. Huntley, L. Kaila, J. Kullberg, T. Tammaru, W. J. Tennent, J. A. Thomas, and M. Warren. 1999. Poleward shifts in geographical ranges of butterfly species associated with regional warming. *Nature* 399:579–83.

Patterson, T. B., and T. J. Givnish. 2002. Phylogeny, concerted convergence, and phylogenetic niche conservatism in the core Liliales: Insights from *rbc*L and *ndh*F sequence data. *Evolution* 56:233–52.

———. 2004. Geographic cohesion and parallel adaptive radiations in *Calochortus* (Calochortaceae): Evidence from a cpDNA sequence phylogeny. *New Phytologist* 161:253–64.

Patton, J. L., D. G. Huckaby, and S. T. Álvarez-Castañeda. 2008. *The Evolutionary History and a Systematic Revision of the Woodrats of the* Neotoma lepida *Complex*. University of California Press, Berkeley.

Patton, J. L., and M. F. Smith. 1990. The evolutionary dynamics of the pocket-gopher *Thomomys bottae*, with emphasis on California populations. *University of California Publications in Zoology* 123.

———. 1994. Paraphyly, polyphyly, and the nature of species boundaries in pocket gophers (genus *Thomomys*). *Systematic Biology* 43:11–26.

Peabody, F. E., and J. M. Savage. 1958. Evolution of a Coast Range corridor in California and its effects on the origin and dispersion of living amphibians and reptiles. *In* C. L. Hubbs [ed.], *Zoogeography*, 159–86. American Association for the Advancement of Science, Washington, D.C.

Pease, K. M., A. H. Freedman, J. P. Pollinger, J. E. McCormack, W. Buermann, J. Rodzen, J. Banks, E. Meredith, V. C. Bleich, R. J. Schaefer, K. Jones, and R. K. Wayne. 2009. Ltd Landscape genetics of California mule deer (*Odocoileus hemionus*): The roles of ecological and historical factors in generating differentiation. *Molecular Ecology* 18:1848–62.

Pecon Slattery, J., and S. J. O'Brien. 1998. Patterns of Y and X chromosome DNA sequence divergence during the Felidae radiation. *Genetics* 148:1245–55.

Pelini, S. L., J. D. K. Dzurisin, K. M. Prior, C. M. Williams, T. D. Marsico, B. J. Sinclair, and J. J. Hellmann. 2009. Translocation experiments with butterflies reveal limits to enhancement of poleward populations under climate change. *Proceedings of the National Academy of Sciences USA* 106:11160–65.

Pellmyr, O., M. Balcázar-Lara, D. M. Althoff, K. A. Segraves, and J. Leebens-Mack. 2006. Phylogeny and life history evolution of *Prodoxus* yucca moths (Lepidoptera: Prodoxidae). *Systematic Entomology* 31:1–20.

Perrine, J. D. 2006. Ecology of the red fox (*Vulpes vulpes*) in the Lassen Peak region of California, USA. PhD diss., University of California, Berkeley.

Peterson, D., D. Cayan, J. Dileo, M. Noble, and M. Dettinger. 1995. The role of climate in estuarine variability. *American Scientist* 83:58–67.

Petit, R. J. 2004. Biological invasions at the gene level. *Diversity and Distributions* 10:159–65.

Petit, R. J., J. Duminil, S. Fineschi, A. Hampe, D. Salvini, and G. G. Vendramin. 2005. Comparative organization of chloroplast, mitochondrial and nuclear diversity in plant populations. *Molecular Ecology* 14: 689–701.

Petit, R. J., F. S. Hu, and C. Dick. 2008. Forests of the past: A window to future changes. *Science* 320:1450–52.

Petren, K., and T. J. Case. 1997. Phylogenetic analysis of body size evolution and biogeography in chuckwallas (*Sauromalus*) and other Iguanines. *Evolution* 51:206–19.

Péwé, T. L., and D. M. Hopkins. 1967. Mammal remains of pre-Wisconsin age in Alaska. *In* D. M. Hopkins [ed.], *The Bering Land Bridge*, 266–87. Stanford University Press, Palo Alto, CA.

Phillips, C. D., T. S. Gelatt, J. C. Patton, and J. W. Bickham. 2011. Phylogeography of Steller's sea lions: Relationships among climate change, effective population size, and genetic diversity. *Journal of Mammalogy* 92:1091–1104.

Phillips, S. J., R. P. Anderson, and R. E. Schapire. 2006. Maximum entropy modeling of species geographic distributions. *Ecological Modelling* 190:231–59.

Phillipsen, I. C., and A. E. Metcalf. 2009. Phylogeography of a stream-dwelling frog (*Pseudacris cadaverina*) in Southern California. *Molecular Phylogenetics and Evolution* 53:152–70.

Pielou, E. C. 1991. *After the Ice Age*. University of Chicago Press, Chicago, IL.

Pimentel, D., W. Westra, and R. F. Noss [eds.]. 2000. *Ecological Integrity: Integrating Environment, Conservation, and Health*. Island Press, Washington, D.C.

Pitkin, J., and D. Myers. 2012. Generational projections of the California population by nativity and year of immigrant arrival. Produced by the Population Dynamics Research Group, Sol Price School of Public Policy, University of Southern California. Text and supporting materials at www.usc.edu/schools/price/research/popdynamics.

Pittermann, J., S. A. Stuart, T. E. Dawson, and A. Moreau. 2012. Cenozoic climate change shaped the evolutionary ecophysiology of the Cupressaceae conifers. *Proceedings of the National Academy of Sciences USA* 109:9647–52.

Polihronakis, M., and M. S. Caterino. 2010. Contrasting patterns of phylogeographic relationships in sympatric sister species of ironclad beetles (Zopheri-
dae: *Phloeodes* spp.) in California's Transverse Ranges. *BMC Evolutionary
Biology* 10:195.

Polihronakis, M., M. S. Caterino, and S. Chatzimanolis. 2010. Elucidating the
phylogeographic structure among a mosaic of unisexual and bisexual
populations of the weevil *Geodercodes latipennis* (Coleoptera: Curculionidae) in the Transverse Ranges of Southern California. *Biological Journal of
the Linnean Society* 101:935–48.

Popp, M., V. Mirré, and C. Brochmann. 2011. A single mid-Pleistocene
long-distance dispersal by a bird can explain the extreme bipolar disjunction
in crowberries (*Empetrum*). *Proceedings of the National Academy of Sciences
USA* 108:6520–25.

Porinchu, D. F., G. M. MacDonald, A. M. Bloom, and K. A. Moser. 2003. Late
Pleistocene and early Holocene climate and limnological changes in the Sierra
Nevada, California, USA, inferred from midges (Insecta: Diptera: Chirono-
midae). *Palaeogeography, Palaeoclimatology, Palaeoecology* 198:403–22.

Posada, D., and K. A. Crandall. 2001. Evaluation of methods for detecting
recombination from DNA sequences: Computer simulations. *Proceedings of
the National Academy of Sciences USA* 98: 13757–62.

Potter, I. C. 1980. The Petromyzoniformes with particular reference to paired
species. *Canadian Journal of Fisheries and Aquatic Sciences* 37:1595–1615.

Premoli, A. A., S. Chischilly, and J. B. Mitton. 1994. Levels of genetic variation
captured by four descendant populations of pinyon pine (*Pinus edulis*
Engelm.). *Biodiversity and Conservation* 3:331–40.

Preston, R. E., and L. T. Dempster. 2012. Convolulaceae. *In* B. G. Baldwin, D.
H. Goldman, D. J. Keil, R. Patterson, T. J. Rosatti, and D. H. Wilken [eds.],
The Jepson Manual: Vascular Plants of California, 2nd ed., 654–64.
University of California Press, Berkeley.

Price, S. A., O. R. P. Bininda-Emonds, and J. L. Gittleman. 2005. A complete
phylogeny of the whales, dolphins and even-toed hoofed mammals (Cetartiodactyla). *Biological Review* 80:445–73.

Pritchard, J. K., M. Stephens, and P. Donnelly. 2000. Inference of population
structure using multilocus genotype data. *Genetics* 155: 945–59.

Quiñones, R. M., and P. B. Moyle. 2012. Integrating global climate change into
salmon and trout conservation: A case study of the Klamath River. *In* T. L.
Root, K. R. Hall, M. Herzog, and C. A. Howell [eds.], *Linking Science and
Management to Conserve Biodiversity in a Changing Climate,* xx–xxx.
University of California Press, Berkeley.

Ralls, K., J. Ballou, and R. L. Brownell. 1983. Genetic diversity in sea otters:
Theoretical considerations and management implications. *Biological
Conservation* 25:209–32.

Raven, P. H., and D. I. Axelrod. 1978. *Origin and Relationships of the California Flora*. California Native Plant Society, Sacramento.

Recuero, E., Í. Martínez-Solano, G. Parra-Olea, and M. García-París. 2006. Phylogeography of *Pseudacris regilla* (Anura: Hylidae) in western North America, with a proposal for a new taxonomic rearrangement. *Molecular Phylogenetics and Evolution* 39:293–304.

Redenbach, Z., and E. B. Taylor. 1999. Zoogeographical implications of variation in mitochondrial DNA of Arctic grayling (*Thymallus arcticus*). *Molecular Ecology* 8:23–35.

Reed, D. H., and R. Frankham. 2003. Correlation between fitness and genetic diversity. *Conservation Biology* 17:230–37.

Remington, C. L. 1968. Suture-zones of hybrid interaction between recently joined biotas. *In* T. Dobzhansky, M. K. Hecht, and W. C. Steere [eds.], *Evolutionary Biology,* 321–428. Plenum, New York.

Rensberger, J. M., and A.D. Barnosky. 1993. Short-term fluctuations in small mammals of the late Pleistocene from eastern Washington. *In* R. A. Martin and A. D. Barnosky [eds.], *Morphological Change in Quaternary Mammals of North America,* 299–342. Cambridge University Press, Cambridge.

Repenning, C. A. 1976. Adaptive evolution of sea lions and walruses. *Systematic Biology* 25:375–90.

Resh, V. H., and R. T. Carde. 2009. *Encyclopedia of Insects*. 2nd ed. Academic Press, New York.

Reveal, J. L. 1979. Biogeography of the Intermountain Region: A speculative appraisal. *Mentzelia* 4.

Rey Benayas, J. M., A. C. Newton, A. Diaz, J. M. Bullock. 2009. Enhancement of biodiversity and ecosystem services by ecological restoration: A meta-analysis. *Science* 325:1121–24.

Rich, K. A., J. N. Thompson, and C. C. Fernandez. 2008. Diverse historical processes shape deep phylogeographical divergence in the pollinating seed parasite *Greya politella*. *Molecular Ecology* 17:2430–48.

Rick, T. C., R. L. DeLong, J. M. Erlandson, T. J. Braje, T. L. Jones, J. E. Arnold, M. R. Des Lauriers, W. R. Hildebrandt, D. J. Kennett, R. L. Vellanoweth, and T. A. Wake. 2011. Where were the northern elephant seals? Holocene archaeology and biogeography of *Mirounga angustirostris*. *Holocene* 21:1159–66.

Riddle, B. R. 1996. The molecular phylogeographic bridge between deep and shallow history in continental biotas. *Trends in Ecology and Evolution* 11:207–10.

Riddle, B. R., M. N. Dawson, E. A. Hadly, D. J. Hafner, M. J. Hickerson, S. J. Mantooth, and A. D. Yoder. 2008. The role of molecular genetics in sculpting the future of integrative biogeography. *Progress in Physical Geography* 32:173–202.

Riddle, B. R., and D. J. Hafner. 2006. A step-wise approach to integrating phylogeographic and phylogenetic biogeographic perspectives on the history of a core North American warm deserts biota. *Journal of Arid Environments* 66:435–61.

Riddle, B. R., D. J. Hafner, and L. F. Alexander. 2000a. Comparative phylogeography of Baileys' pocket mouse (*Chaetodipus baileyi*) and the *Peromyscus eremicus* species group: Historical vicariance of the Baja California peninsular desert. *Molecular Phylogenetics and Evolution* 17:161–72.

Riddle, B. R., D. J. Hafner, L. F. Alexander, and J. R. Jaeger. 2000b. Cryptic vicariance in the historical assembly of a Baja California peninsular desert biota. *Proceedings of the National Academy of Sciences USA* 97:14438–43.

Riedman, M. L., and J. A. Estes. 1990. The sea otter (*Enhydra lutris*): Behavior, ecology, and natural history. *Biological Report* 90(14). U.S. Fish and Wildlife Service. 126 pp.

Rieseberg, L. H. 1991. Hybridization in rare plants: Insights from case studies in *Helianthus* and *Cercocarpus*. *In* D. A. Falk and K. E. Holsinger [eds.], *Conservation of Rare Plants: Biology and Genetics*, 171–81. Oxford University Press, Oxford.

Riley, S. P. D, G. T. Busteed, L. B. Kats, T. L. Van Dergon, L. F. S. Lee, R. G. Dagit, J. L. Kerby, R. N. Fisher, and R. M. Sauvajot. 2005. Effects of urbanization on the distribution and abundance of amphibians and invasive species in Southern California streams. *Conservation Biology* 19:1894–1907.

Rios, E., and S. T. Álvarez-Castañeda. 2010. Phylogeography and systematics of San Diego pocket mouse (*Chaetodipus fallax*). *Journal of Mammalogy* 91:293–301.

Rissler, L. J., R. J. Hijmans, C. H. Graham, C. Moritz, and D. B. Wake. 2006. Phylogeographic lineages and species comparisons in conservation analyses: A case study of California herpetofauna. *American Naturalist* 167:655–66.

Rissler, L. J., and W. H. Smith. 2010. Mapping amphibian contact zones and phylogeographical break hotspots across the United States. *Molecular Ecology* 19:5404–16.

Robbins, C. S., B. Bruun, and H. S. Zim. 1983. *Birds of North America*. Golden Press, New York. 360 pp.

Robinson, M. D., and T. R.Van Devender. 1973. Miocene lizards from Wyoming and Nebraska. *Copeia* 4:698–704.

Roderick, G. K. 1996. Geographic structure of insect populations: Gene flow, phylogeography, and their uses. *Annual Review of Entomology* 41:325–52.

Rodriguez, R. M., and L. K. Ammerman. 2004. Mitochondrial DNA divergence does not reflect morphological difference between *Myotis californicus* and *Myotis ciliolabrum*. *Journal of Mammalogy* 85:842–51.

Rodríguez-Robles, J. A., D. F. Denardo, and R. E. Staub. 1999. Phylogeography of the California mountain kingsnake, *Lampropeltis zonata* (Colubridae). *Molecular Ecology* 8:1923–34.

Rodríguez-Robles, J. A., G. R. Stewart, and T. J. Papenfuss. 2001. Mitochondrial DNA-based phylogeography of North American rubber boas, *Charina bottae* (Serpentes: Boidae). *Molecular Phylogenetics and Evolution* 18:227–37.

Rogers, D. L., C. I. Millar, and R. D. Westfall. 1999. Fine-scale genetic structure of whitebark pine (*Pinus albicaulis*): Associations with watershed and growth form. *Evolution* 53:74–90.

Rogers, D. L., and A. M. Montalvo. 2004. Genetically appropriate choices for plant materials to maintain biological diversity. USDA Forest Service Rocky Mountain Region, Lakewood, CO. www.fs.fed.us/r2/publications/botany /plantgenetics.pdf.

Ronquist, F., and J. P. Huelsenbeck. 2003. MrBayes 3: Bayesian phylogenetic inference under mixed models. *Bioinformatics* 19:1572–74.

Rose, K. D. 2006. *The Beginning of the Age of Mammals.* Johns Hopkins University Press, Baltimore, MD.

Rosenberg, N. A. 2003. The shapes of neutral gene genealogies in two species: Probabilities of monophyly, paraphyly, and polyphyly in a coalescent model. *Evolution* 57:1465–77.

Rovito, S. M. 2010. Lineage divergence and speciation in the web-toed salamanders (Plethodontidae: *Hydromantes*) of the Sierra Nevada, California. *Molecular Ecology* 19:4554–71.

Rubinoff, D., and J. A. Powell. 2004. Conservation of fragmented small populations: Endemic species persistence on California's smallest Channel Island. *Biodiversity and Conservation* 13:2537–50.

Ruegg, K. C., and R. J. Hijmans. 2006. Climate change and the origin of migratory pathways in the Swainson's thrush, *Catharus ustulatus. Journal of Biogeography* 33:1172–82.

Ruegg, K., and T. B. Smith. 2002. Not as the crow flies: A historical explanation for circuitous migration in Swainson's thrush (*Catharus ustulatus*). *Proceedings of the Royal Society of London B* 269:1375–81.

Runck, A. M., and J. A. Cook. 2005. Post-glacial expansion of the southern red-backed vole (*Clethrionomys gapperi*) in North America. *Molecular Ecology* 14:1445–56.

Rundel, P. W. 2011. The diversity and biogeography of the alpine flora of the Sierra Nevada, California. *Madroño* 58:153–84.

Sabo, J. L., R. Sponseller, M. Dixon, K. Gade, T. Harms, J. Heffernan, A. Jani, G. Katz, S. Candan, J. Watts, and J. Welter. 2005. Riparian zones increase regional species richness by harboring different, not more, species. *Ecology* 86:56–62.

Sage, R. D., and J. O. Wolff. 1986. Pleistocene glaciations, fluctuations, and low genetic diversity in a large mammal (*Ovis dalli*). *Evolution* 40:1092–95.

Sahney, S., and M. J. Benton. 2008. Recovery from the most profound mass extinction of all time. *Proceedings of the Royal Society B* 275:759–65.

Sala, O. E., F. S. Chapin III, J. J. Armesto, R. Berlow, J. Bloomfield, R. Dirzo, E. Huber-Sanwald, L. F. Huenneke, R. B. Jackson, A. Kinzig, R. Leemans, D. Lodge, H. A. Mooney, M. Oesterheld, N. L. Poff, M. T. Sykes, B. H. Walker, M. Walker, and D. H. Wall. 2000. Global biodiversity scenarios for the year 2100. *Science* 287:1770–74.

Salomon, A. K., S. K. Gaichas, O. P. Jensen, V. N. Agostini, N. A. Sloan, J. Rice, T. R. McClanahan, M. H. Ruckelshaus, P. S. Levin, N. K. Dulvy, and E. A Babcock. 2011. Bridging the divide between fisheries and marine conservation. *Bulletin of Marine Science* 87:251–74.

San Francisco Estuary Project. 2007. *Comprehensive Conservation and Management Plan*. San Francisco Estuary Project. www.sfestuary.org.

Sanmartín, I., H. Enghoff, and F. Ronquist. 2001. Patterns of animal dispersal, vicariance and diversification in the Holarctic. *Biological Journal of the Linnean Society* 73:345–90.

Sanmartín I., P. van der Mark, and F. Ronquist. 2008. Inferring dispersal: A Bayesian, phylogeny-based approach to island biogeography, with special reference to the Canary Islands. *Journal of Biogeography* 35:428–49.

San Mauro, D., M. Vences, M. Alcobendas, R. Zardoya, and A. Meyer. 2005. Initial diversification of living amphibians predated the breakup of Pangaea. *American Naturalist* 165:590–99.

Sarna-Wojcicki, A. M., C. E. Meyer, H. R. Bowman, N. T. Hall, P. C. Russell, M. J. Woodward, and J. L. Slate. 1985. Correlation of the Rockland ash bed, a 400,000-year-old stratigraphic marker in Northern California and western Nevada, and implications for middle Pleistocene paleogeography of Central California. *Quaternary Geology* 23:235–37.

Savage, D. E. 1951. Late Cenozoic vertebrates of the San Francisco Bay region. *University of California Publications, Bulletin of the Department of Geological Sciences* 28:215–314.

Savage, J. M. 1960. Evolution of a peninsular herpetofauna. *Systematic Zoology* 9:184–212.

Savage, W. K., A. K. Fremier, and H. B. Shaffer. 2010. Landscape genetics of alpine Sierra Nevada salamanders reveal extreme population subdivision in space and time. *Molecular Ecology* 16:3301–14.

Savolainen, O., T. Pyhäjärvi, and T. Knürr. 2007. Gene flow and local adaptation in trees. *Annual Review of Ecology, Evolution, and Systematics* 38:595–619.

Sawyer, J. O. 2006. *Northwestern California*. University of California Press, Berkeley.

Sawyer, J. O., and D. A. Thornburgh. 1977. Montane and sub-alpine vegetation of the Klamath Mountains. *In* M. G. Barbour and J. Major [eds.], *Terrestrial Vegetation of California*, 699–732. John Wiley & Sons, New York.

Scher, S. 1996 Genetic structure of natural *Taxus* populations in western North America. *In* T. B. Smith and R. K. Wayne [eds.], *Molecular Genetic Approaches in Conservation*, 424–41. Oxford University Press, New York.

Schierenbeck, K. A., and N. C. Ellstrand. 2009. Hybridization and the evolution of invasiveness in plants and other organisms. *Biological Invasions* 11:1093–1105.

Schierenbeck, K. A., and F. Phipps. 2010. Population genetics of *Howellia aquatilis* (Campanulaceae) in disjunct locations throughout the Pacific Northwest. *Genetica* 138:1161–69.

Schinske, J. N., B. Giacomo. K. J. David, and E. J. Routman. 2010. Phylogeography of the Diamond Turbot (*Hypsopsetta guttulata*) across the Baja California Peninsula. *Marine Biology* 157:123-34.

Schoenherr, A. A. 1992. *A Natural History of California*. University of California Press, Berkeley.

Schorn, H. E. 1984. Palynology of the late middle Miocene sequence, Stewart Valley. *Palynology* 8:259–60.

Schoville, S. D. 2012. Three new species of *Grylloblatta* Walker (Insecta: Grylloblattodea: Grylloblattidae), from southern Oregon and Northern California. *Zootaxa* 3412:42–52.

Schoville, S. D., A. W. Lam, and G. K. Roderick. 2010. A range-wide genetic bottleneck overwhelms contemporary landscape factors and local abundance in shaping genetic patterns of an alpine butterfly (Lepidoptera: Pieridae: *Colias behrii*). *Molecular Ecology* 21:4242–56.

Schoville, S. D., and G. K. Roderick. 2009. Alpine biogeography of a Parnassian butterfly during Quaternary climate cycles in North America. *Molecular Ecology* 15:3471–85.

———. 2010. Evolutionary diversification of cryophilic *Grylloblatta* species (Grylloblattodea: Grylloblattidae) in alpine habitats of California. *BMC Evolutionary Biology* 10:163.

Schoville, S. D., G. K. Roderick, and D. H. Kavanaugh. 2012. Testing the "Pleistocene species pump" in alpine habitats: Lineage diversification of flightless ground beetles (Coleoptera: Carabidae: *Nebria*) in relation to altitudinal zonation. *Biological Journal of the Linnean Society* 107:95–111.

Schoville, S. D., M. Stuckey, and G. K. Roderick. 2011. Pleistocene origin and population history of a neoendemic alpine butterfly. *Molecular Ecology* 20:1233–47.

Schoville, S. D., T. S. Tustall, V. T. Vredenburg, A. R. Backlin, E. Gallegos, D. A. Wood, and R. N. Fisher. 2011. Conservation genetics of evolutionary lineages of the endangered mountain yellow-legged frog, *Rana muscosa* (Amphibia: Ranidae), in Southern California. *Biological Conservation* 144:2031–40.

Schramm, Y., S. L. Mesnick, J. de la Rosa, D. M. Palacios, M. S. Lowry, D. Aurioles-Gamboa, H. M. Snell, and S. Escorza-Treviño. 2009. Phylogeography of California and Galápagos sea lions and population structure within the California sea lion. *Marine Biology* 156:1375–87.

Schwartz, M. K., K. Ralls, D. F. Williams, B. L. Cypher, K. L. Pilgrim, and R. C. Fleischer. 2005. Gene flow among San Joaquin kit fox populations in a severely changed ecosystem. *Conservation Genetics* 6:25–37.

Schweiger, O., J. Settele, O. Kudrna, S. Klotz, and I. Kühn. 2008. Climate change can cause spatial mismatch of trophically interacting species. *Ecology* 89:3472–79.

Scott, R. W. 1995. *The Alpine Flora of the Rocky Mountains.* Vol. 1, *The Middle Rockies.* University of Utah Press, Salt Lake City.

Segraves, K. A., and O. Pellmyr. 2001. Phylogeography of the yucca moth *Tegeticula maculata:* The role of historical biogeography in reconciling high genetic diversity with limited speciation. *Molecular Ecology* 10:1247–53.

Sgariglia, K. A., and K. J. Burns. 2003. Phylogeography of the California thrasher (*Toxostoma redivivum*) based on nested-clade analysis of mitochondrial-DNA variation. *Auk* 120:346–61.

Shafer, A. B. A., C. I. Cullingham, S. D. Côté, and D. W. Coltman. 2010. Of glaciers and refugia: A decade of study sheds new light on the phylogeography of northwestern North America. *Molecular Ecology* 19:4589–4621.

Shaffer, H. B., G. M. Fellers, S. R. Voss, J. C. Oliver, and G. B. Pauly. 2004. Species boundaries, phylogeography and conservation genetics of the red-legged frog (*Rana aurora/draytonii*) complex. *Molecular Ecology* 13:2667–77.

Shaffer, H. B., G. B. Pauly, J. C. Oliver, and P. C. Trenham. 2004. The molecular phylogenetics of endangerment: Cryptic variation and historical phylogeography of the California tiger salamander, *Ambystoma californiense. Molecular Ecology* 13:3033–49.

Sharp, R. P., C. R. Allen, and M. F. Meier. 1959. Pleistocene glaciers on Southern California mountains. *American Journal of Science* 257:81–94.

Sharsmith, C. W. 1940. A contribution to the history of the alpine flora of the Sierra Nevada. PhD diss., University of California, Berkeley.

Shimamura, M., H. Yasue, K. Ohshima, H. Abe, H. Kato, T. Kishiro, M. Goto, I. Munechika, and N. Okada. 1997. Molecular evidence from retroposons that whales form a clade with even-toed ungulates. *Nature* 388:666–70.

Siddiqui, A. Q. 1979. Reproductive biology of *Tilapia zillii* (Gervais) in Lake Naivasha, Kenya. *Environmental Biology of Fishes* 4:257–62.

Sigler, W. F., and J. W. Sigler. 1987. *Fishes of the Great Basin: A Natural History.* University of Nevada Press, Reno.

Simmons, N. B., K. L. Seymour, J. Habersetzer, and G. F. Gunnell. 2008. Primitive early Eocene bat from Wyoming and the evolution of flight and echolocation. *Nature* 451:818–21.

Sinclair, E. A., R. L. Bezy, K. Bolles, J. L. Camarillo, K. A. Crandall, and J. W. Sites Jr. 2004. Testing species boundaries in an ancient species complex with deep phylogeographic history: Genus *Xantusia* (Squamata: Xantusiidae). *American Naturalist* 164:396–414.

Sinervo, B., F. Méndez de la Cruz, D. B. Miles, B. Heulin, E. Bastiaans, M. Villagrán-Santa Cruz, R. Lara-Resendiz, N. Martínez-Méndez, M. L. Calderón-Espinosa, R. N. Meza-Lázaro, H. Gadsden, L. Avila, M. Morando, I. J. De la Riva, P. V. Sepulveda, C. Duarte Rocha, N. Ibargüengoytía, C. Aguilar Puntriano, M. Massot, V. Lepetz, T. A. Oksanen, D. G. Chapple, A. M. Bauer, W. R. Branch, J. Clobert, and J. W. Sites Jr. 2010. Erosion of lizard diversity by climate change and altered thermal niches. *Science* 328: 894–99.

Skinner, M. W., and B. M. Pavlik. 1994. *Inventory of Rare and Endangered Vascular Plants of California*. California Native Plant Society, Sacramento.

Slack, K. E., C. M. Jones, T. Ando, G. L. Harrison, R. E. Fordyce, U. Arnason, and D. Penny. 2006. Early penguin fossils, plus mitochondrial genomes, calibrate avian evolution. *Molecular Biology and Evolution* 23:1144–55.

Slatkin, M. 1995. Measure of population subdivision based on microsatellite allele frequencies. *Genetics* 139:457–62.

Slatkin, M., and R. R. Hudson. 1991. Pairwise comparisons of mitochondrial DNA sequences in stable and exponentially growing populations. *Genetics* 129:555–62.

Slauson, K. M., W. J. Zielinski, and K. D. Stone. 2009. Characterizing the molecular variation among American marten (*Martes americana*) subspecies from Oregon and California. *Conservation Genetics* 10:1337–41.

Sloop, C. M., C. Pickens, and S. P. Gordon. 2011. Conservation genetics of Butte County meadowfoam (*Limnanthes floccosa* ssp. *californica* Arroyo), an endangered vernal pool endemic. *Conservation Genetics* 12:211–323.

Smith, C. I., S. Tank, W. Godsoe, J. Levenick, E. Strand, T. Esque, and O. Pellmyr. 2011. Comparative phylogeography of a coevolved community: Concerted population expansions in Joshua trees and four yucca moths. *PLoS One* 6(10):e25628. doi:10.1371/journal.pone.0025628.

Smith, J. P., and J. O. Sawyer Jr. 1988. Endemic vascular plants of northwestern California and southwestern Oregon. *Madroño* 35:54–69.

Smith, M. F. 1979. Geographic variation in genic and morphological characters in *Peromyscus californicus*. *Journal of Mammalogy* 60:705–22.

Smith, S. A., and M. J. Donoghue. 2010. Combining historical biogeography with niche modeling in the *Caprifolium* clade of *Lonicera* (Caprifoliaceae, Dipsacales). *Systematic Biology* 59:322–41.

Smith, S. G., W. D. Muir, J. G. Williams, and J. R. Skalski. 2002. Factors associated with travel time and survival of migrant yearling Chinook salmon and steelhead in the lower Snake River. *North American Journal of Fisheries Management* 22:385–405.

Smith, S. V., and J. T. Hollibaugh. 1997. Annual cycle and interannual variability of ecosystem metabolism in a temperate climate embayment. *Ecological Monographs* 67:509–33.

Snyder, M. A., L. C. Sloan, N. S. Diffenbaugh, and J. L. Bell. 2003. Future climate change and upwelling in the California Current. *Geophysical Research Letters* 30:1823–27.

Soltis, D. E., M. A. Gitzendanner, D. D. Strenge, and P. S. Soltis. 1997. Chloroplast DNA intraspecific phylogeography of plant from the Pacific Northwest of North America. *Plant Systematics and Evolution* 206:353–73.

Soltz, D. L., and R. J. Naiman. 1978. The natural history of native fishes in the Death Valley System. *Natural History Museum of Los Angeles County Science Series* 30:1–76.

Sorensen, F. C., N. L. Mandel, and J. E. Aagaard. 2001. Role of selection versus historical isolation in racial differentiation of ponderosa pine in southern Oregon: An investigation of alternative hypotheses. *Canadian Journal of Forest Research* 31:1127–39.

Sork, V. L., F. W. Davis, R. Westfall, A. Flint, M. Ikegami, H. F. Wang, and D. Grivet. 2010. Gene movement and genetic association with regional climate gradients in California valley oak (*Quercus lobata* Née) in the face of climate change. *Molecular Ecology* 19:3806–23.

Sork, V. L., and L. Waits. 2010. Contributions of landscape genetics—approaches, insights, and future potential. *Molecular Ecology* 19:3489–3495.

Spellman, G. M., and J. Klicka. 2006. Testing hypotheses of Pleistocene population history using coalescent simulations: Phylogeography of the pygmy nuthatch (*Sitta pygmaea*). *Proceedings of the Royal Society Series B.* 273:3057–63.

Spinks, P. Q., and H. B. Shaffer. 2005. Range-wide molecular analysis of the western pond turtle (*Emys marmorata*): Cryptic variation, isolation by distance, and their conservation implications. *Molecular Ecology* 14:2047–64.

Spinks, P. Q., R. C. Thomson, and H. B. Shaffer. 2010. Nuclear gene phylogeography reveals the historical legacy of an ancient inland sea on lineages of the western pond turtle, *Emys marmorata*, in California. *Molecular Ecology* 19:542–56.

Starrett, J., and M. Hedin. 2007. Multilocus genealogies reveal multiple cryptic species and biogeographic complexity in the California turret spider *Antrodiaetus riversi* (Mygalomorphae, Antrodieatidae). *Molecular Ecology* 16:583–604.

Stearley, R. F., and G. R. Smith. 1993. Phylogeny of the Pacific trouts and salmons (*Oncorhynchus*) and genera of the family Salmonidae. *Transactions of the American Fisheries Society* 122:1–33.

Stebbins, G. L. 1942. The genetic approach to problems of rare and endemic species. *Madroño* 6:241–72.

———. 1949. Rates of evolution in plants. *In* G. L. Jepsen, G. G. Simpson, and E. Mayr [eds.], *Genetics, Paleontology, and Evolution*, 229–42. Princeton University Press, Princeton, NJ.

———. 1959. The role of hybridization in evolution. *Proceedings of the American Philosophical Society* 103:231–51.

———. 1974. Adaptive shifts and evolutionary novelty: A compositionist approach. *In* F. J. Ayala and T. Dobzhansky [eds.], *Studies in the Philosophy of Biology*, 285–338. University of California Press, Berkeley.

———. 1976. Ecological islands and vernal pools of California. *In* S. Jain [ed.], *Vernal Pools: Their Ecology and Conservation*, 1–4. Institute of Ecology Publication 9. University of California, Davis.

———. 1980. Rarity of plant species: A synthetic viewpoint. *Rhodora* 82:77–86.

Stebbins, G. L., and A. Day. 1967. Cytogenetic evidence for long continued stability in the genus *Plantago*. *Evolution* 21:409–28.

Stebbins, G. L., and J. Major. 1965. Endemism and speciation in the California flora. *Ecological Monographs* 35:1–35.

Steele, C. A., and A. Storfer. 2006. Coalescent-based hypothesis testing supports multiple Pleistocene refugia in the Pacific Northwest for the Pacific giant salamander (*Dicamptodon tenebrosus*). *Molecular Ecology* 15:2477–87.

———. 2007. Phylogeographic incongruence of codistributed amphibian species based on small differences in geographic distribution. *Molecular Phylogenetics and Evolution* 43:468–79.

Steinhoff, R. J., D. G. Joyce, and L. Fins. 1990. Isozyme variation in *Pinus monticola*. *Canadian Journal of Forest Research* 13:1122–32.

Steneck, R. S., J. Vavrinec, and A. V. Leland. 2004. Accelerating trophic-level dysfunction in kelp forest ecosystems of the western North Atlantic. *Ecosystems* 7:323–32.

Stewart, B. S., and R. L. Delong. 1995. Double migrations of the northern elephant seal, *Mirounga angustirostris*. *Journal of Mammalogy* 76:196–205.

Stewart, J. H., and C. A. Suczek. 1977. Cambrian and Pre-Cambrian paleogeography and tectonics in the western U.S. *In* J. H. L. Stewart, C. H. Stevens, and A. E. Fritsche [eds.], *Paleozoic Paleogeography of the Western U.S.*, 1–17. Symposium I: Pacific Section, Society of Economic Paleontologists and Mineralogists.

Stine, S. 1994. Extreme and persistent drought in California and Patagonia during mediaeval time. *Nature* 369:546–49.

Stock, C. 1992. *Rancho La Brea: A Record of Pleistocene Life in California*. 7th ed. Rev. J. M. Harris. Science Series no. 37. Natural History Museum of Los Angeles County. 113 pp.

Stockey, R. A., J. Kvacek, R. Hill, G. H. Rothwell, and Z. Kvacek. 2005. The fossil record of the Cupressaceae. *In* A. Farjon [ed.], *A Monograph of the Cupressaceae and* Sciadopitys, 54–68. Royal Botanic Gardens, Kew, Surrey.

Stone, K. D., and J. A. Cook. 2002. Molecular evolution of the Holarctic genus *Martes*. *Molecular Phylogenetics and Evolution* 24:169–79.

Stone, K. D., R. W. Flynn, and J. A. Cook. 2002. Post-glacial colonization of northwestern North America by the forest associated American marten (*Martes americana*). *Molecular Ecology* 11:2049–63.

Storer, T. I., and R. L. Usinger. 1964. *Sierra Nevada Natural History.* University of California Press, Berkeley.

Storfer, A., M. A. Murphy, J. S. Evans, C. S. Goldberg, S. Robinson, S. F. Spear, R. Dezzani, E. Delmelle, L. Vierling, and L. P. Waits. 2007. Putting the "landscape" in landscape genetics. *Heredity* 98:128–42.

Strauss, S. H., Y. P. Hong, and V. D. Hipkins. 1993. High levels of population differentiation for mitochondrial DNA haplotypes in *Pinus radiata*, *muricata*, and *attenuata*. *Theoretical and Applied Genetics* 86:605–11.

Street, J. H., R. S. Anderson, and A. Paytan. 2012. An organic geochemical record of Sierra Nevada climate since the LGM from Swamp Lake, Yosemite. *Quaternary Science Reviews* 40:89–106.

Suh, A., M. Paus, M. Kiefmann, G. Churakov, F. A. Franke, J. Brosius, J. O. Kriegs, and J. Schmitz. 2011. Mesozoic retroposons reveal parrots as the closest living relatives of passerine birds. *Nature Communications* 2:443–48.

Sutcliffe, A. J. 1985. *On the Track of Ice Age Mammals.* Harvard University Press, Cambridge, MA.

Sweeney, B. W., T. L. Bott , J. K. Jackson, L. A. Kaplan, J. D. Newbold, L. J. Standley, W. C. Hession, and R. J. Horwitz. 2004. Riparian deforestation, stream narrowing, and loss of stream ecosystem services. *Proceedings of the National Academy of Sciences* 101:14132–37.

Swenson, N. G., and D. J. Howard. 2005. Clustering of contact zones, hybrid zones, and phylogeographic breaks in North America. *American Naturalist* 166:581–91.

Syring, J., K. Farrell, R. Businský, R. Cronn, and A. Liston. 2007. Widespread genealogical 557 nonmonophyly in species of *Pinus* subgenus *Strobus*. *Systematic Biology* 56:163–81.

Syring, J., A. Willyard, R. Cronn, and A. Liston. 2005. Evolutionary relationships among *Pinus* (Pinaceae) subsections inferred from multiple low-copy nuclear loci. *American Journal of Botany* 92:2086–2100.

Syverson, V. J., and D. R. Prothero. 2010. Evolutionary patterns in late Quaternary California condors. *Palarch's Journal of Vertebrate Palaeontology* 7:1–18.

Taguchi, M., S. J. Chivers, P. E. Rosel, T. Matsuishi, and A. Syuiti. 2010. Mitochondrial DNA phylogeography of the harbour porpoise *Phocoena phocoena* in the North PaciÔ¨Åc. *Marine Biology* 157:1489–98.

Tajima, F. 1989. Statistical method for testing the neutral mutation hypothesis by DNA polymorphism. *Genetics* 123:585–95.

Talbot, S. L., and G. F. Shields. 1996. Phylogeography of brown bears (*Ursus arctos*) of Alaska and paraphyly within the Ursidae. *Molecular Phylogenetics and Evolution* 5:477–94.

Tallamy, D. W. 2007. *Bringing Nature Home: How You Can Sustain Wildlife with Native Plants*. Timber Press, Portland, OR.

Tan, A.-M., and D. B. Wake. 1995. MtDNA phylogeography of the California newt, *Taricha torosa* (Caudata: Salamandridae). *Molecular Phylogenetics and Evolution* 4:383–94.

Taylor, E. B. 1991. A review of local adaptation in Salmonidae, with special reference to Pacific and Atlantic salmon. *Aquaculture* 98:185–207.

Taylor, T. N., E. L. Taylor, and M. Krings. 2009. *The Biology and Evolution of Fossil Plants*. 2nd ed. Prentice-Hall, Englewood Cliffs, NJ.

Tedford, R. H., L. B. Albright II, A. D. Barnosky, I. Ferrusquia-Villafranca, R. M. Hunt Jr., J. E. Storer, C. C. Swisher III, M. R. Voorhies, S. D. Webb, and D. P. Whistler. 2004. Mammalian biochronology of the Arikareean through Hemphilian interval (late Oligocene through early Pliocene epochs). *In* M. O. Woodburne [ed.], *Late Cretaceous and Cenozoic Mammals of North America*, 169–231. Columbia University Press, New York.

Tedford, R. H., T. Galusha, M. F. Skinner, B. E. Taylor, R. W. Fields, J. R. Macdonald, J. M. Rensberger, S. D. Webb, and D. P. Whistler. 1987. Faunal succession and biochronology of the Arikareean through Hemphilian interval (late Oligocene through earliest Pliocene epochs) in North America. *In* M. O. Woodburne [ed.], *Cenozoic Mammals of North America: Geochronology and Biostratigraphy*, 153–210. University of California Press, Berkeley.

Templeton, A. R. 2004. Statistical phylogeography: Methods of evaluating and minimizing inference errors. *Molecular Ecology* 13:789–809.

Testa, J. W. 1986. Electromorph variation in Weddell seals (*Leptonychotes weddelli*). *Journal of Mammalogy* 67:606–10.

Thompson, C. E., E. Taylor, and J. D. McPhail. 1997. Parallel evolution of lake-stream pairs of threespine sticklebacks (*Gasterosteus*) inferred from mitochondrial DNA variation. *Evolution* 51:1955–65.

Thompson, R. S., and K. H. Anderson. 2000. Biomes of western North America at 18,000, 6000 and 0 ^{14}C yr bp reconstructed from pollen and packrat midden data. *Journal of Biogeography* 27:555–84.

Thompson, R. S., C. Whitlock, P. J. Bartlein, S. P. Harrison, and W. G. Spaulding. 1993. Climatic changes in the western United States since 18,000 yr B.P. *In* H. E. Wright Jr., J. E. Kutzbach, T. Webb III, W. F. Ruddiman, F. A. Street-Perrott, and P. J. Bartlein [eds.], *Global Climates since the Last Glacial Maximum*, 468–513. University of Minnesota Press, Minneapolis.

Tiffney, B. H. 1985. Perspectives on the origin of the floristic similarity between eastern Asia and eastern North America. *Journal of the Arnold Arboretum* 66:73–94.

Tiffney, B. H., and S. R. Manchester. 2001. The use of geological and paleontological evidence in evaluating plant phylogeographic hypotheses in the Northern Hemisphere Tertiary. 162 (6, suppl.): S3–S17.

Todisco, V., P. Gratton, E. V. Zakharov, C. W. Wheat, V. Sbordoni, and F. A. Sperling. 2012. Mitochondrial phylogeography of the Holarctic *Parnassius phoebus* complex supports a recent refugial model for alpine butterflies. *Journal of Biogeography* 39:1058–62.

Townsend, T. M., D. G. Mulcahy, B. P. Noonan, J. W. Sites Jr., C. A. Kuczynski, J. J. Wiens, and T. W. Reeder. 2011. Phylogeny of iguanian lizards inferred from 29 nuclear loci, and a comparison of concatenated and species-tree approaches for an ancient, rapid radiation. *Molecular Phylogenetics and Evolution* 61:363–80.

Tripodi, A. D., J. W. Austin, A. L. Szalanski, J. McKern, M. K. Carroll, R. K. Saran, and M. T. Messenger. 2006. Phylogeography of *Reticulitermes* termites (Isoptera: Rhinotermitidae) in California inferred from mitochondrial DNA sequences. *Annals of the Entomology Society of America* 99:697–706.

Truesdale, H. D., and L. R. McClenaghan Jr. 1998. Population genetic structure of tecate cypress (Cupressaceae). *Southwestern Naturalist* 43:363–73.

Trzyna, T. 2001. California's urban protected areas, progress despite daunting pressures. *Parks: International Journal for Protected Area Managers* 11(3).

Turner, B. J. 1974. Genetic divergence of Death Valley pupfish species: Biochemical versus morphological evidence. *Evolution* 28:281–94.

———. 1983. Genic variation and differentiation of remnant natural populations of the desert pupfish, *Cyprinodon macularius*. *Evolution* 37:690–700.

Upton, D. E., and R. W. Murphy. 1997. Phylogeny of the sideblotched lizards (Phrynosomatidae: *Uta*) based on mtDNA sequences: Support for a midpeninsular seaway in Baja California. *Molecular Phylogenetics and Evolution* 8:104–13.

Vandergast, A. G., A. J. Bohonak, S. A. Hathaway, J. Boys, and R. N. Fisher. 2008. Are hotspots of evolutionary potential adequately protected in Southern California? *Biological Conservation* 141:1648–64.

Vandergast, A.G., A. J. Bohonak, D. B. Weissman, and R. N. Fisher. 2007. Understanding the genetic effects of recent habitat fragmentation in the context of evolutionary history: Phylogeography and landscape genetics of a Southern California endemic Jerusalem cricket (Orthoptera: Stenopelmatidae: *Stenopelmatus*). *Molecular Ecology* 16:977–92.

Van Devender, T. R. 1987. Holocene vegetation and climate in Puerto Blanco Mountains, southwestern Arizona. *Quaternary Research* 27:51–72.

Van Devender, T. R., and J. I. Mead. 1978. Early Holocene and late Pleistocene amphibians and reptiles in Sonoran Desert packrat middens. *Copeia*, 464–47.

Van Devender, T. R., and W. G. Spaulding. 1979. Development of vegetation and climate in the southwestern United States. *Science* 204:701–10.

Van Frank, R. 1955. *Paleotaricha oligocenica*, new genus and species, and Oligocene salamander from Oregon. *Breviora, Museum of Comparative Zoology, Cambridge, MA* 45:1–12.

Vermeij, G. J. 1993. *Evolution and Escalation: An Ecological History of Life.* Princeton University Press, Princeton, NJ. 537 pp.

Vilà, C., I. R. Amorim, J. A. Leonard, D. Posada, J. Castroviejo, F. Petrucci-Fonseca, K. A. Crandall, H. Ellegren, and R. K. Wayne. 1999. Mitochondrial DNA phylogeography and population history of the grey wolf *Canis lupus. Molecular Ecology* 8:2089–2103.

Vila, R., C. D. Bell, R. Macniven, B. Goldman-Huertas, R. H. Ree, C. R. Marshall, Z. Bálint, K. Johnson, D. Benyamini, and N. E. Pierce. 2011. Phylogeny and paleoecology of *Polyommatus* blue butterflies show Beringia was a climate-regulated gateway to the New World. *Proceedings of the Royal Society B.* doi:10.1098/rspb.2010.2213.

Villa-Lobos, J. 2003. Klamath-Siskiyou: Jewel of the Pacific Coast. *Plant Talk* 31:29–33.

Wahlert, J. H. 1993. The fossil record. *In* H. H. Genoways and J. H. Brown [eds.], *Biology of the Heteromyidae*, 1–37. Special Publication 10. American Society of Mammalogists.

Wakabayashi, J., and T. L. Sawyer. 2001. Stream incision, tectonics, uplift, and evolution of topography of the Sierra Nevada, California. *Journal of Geology* 109:539–62.

Wake, D. B. 1997. Incipient species formation in salamanders of the *Ensatina* complex. *Proceedings of the National Academy of Sciences USA* 94:7761–67.

———. 2006. Problems with species: Patterns and processes of species formation in salamanders. *Annals Missouri Botanical Garden* 93:8–23.

Wake, T. A., and M. A. Roeder. 2009. A diverse Rancholabrean vertebrate microfauna from southern California includes the Ô¨Ärst fossil record of ensatina (*Ensatina eschscholtzii*: Plethodontidae). *Quaternary Research* 72:364–70.

Waltari, E., R. J. Hijmans, A. T. Peterson, A. S. Nyári, S. L. Perkins, and R. P. Guralnick. 2007. Locating Pleistocene refugia: Comparing phylogeographic and ecological niche model predictions. *PLoS ONE* 2(7):e563. doi:10.1371/journal.pone.0000563.

Wang, I. J. 2010. Recognizing the temporal distinctions between landscape genetics and phylogeography. *Molecular Ecology* 19:2605–8.

Waples, R. S., T. Beechie, and G. R. Pess. 2009. Evolutionary history, habitat disturbance regimes, and anthropogenic changes: What do these mean for

resilience of Pacific salmon populations? *Ecology and Society* 14(1):3. www
.ecologyandsociety.org/vol14/iss1/art3/.

Ward, S. N., and G. Valensise. 1996. Progressive growth of San Clemente Island,
California, by blind thrust faulting: Implications for fault slip partitioning in
the California continental borderland. *Geophysical Journal International*
126:712–34.

Warheit, K. I. 1992. A review of the fossil seabirds from the Tertiary of the
North Pacific: Plate tectonics, paleoceanography, and faunal change.
Paleobiology 18:401–24.

Warren, M. S., J. K. Hill, J. A. Thomas, J. Asher, R. Fox, B. Huntley, D. B. Roy,
M. G. Felfer, S. Jeffcoate, P. Harding, G. Jeffcoate, S. G. Willis, J. N. Greato-
rex-Davies, D. Moss, and C. D. Thomas. 2001. Rapid responses of British
butterflies to opposing forces of climate and habitat change. *Nature*
414:65–69.

Weinstock, J., E. Willerslev, A. Sher, T. Wenfei, S. Y. Ho, D. Rubenstein, J.
Storer, J. Burns, L. Martin, C. Bravi, A. Prieto, D. Froese, E. Scott, X. Lai,
and A. Cooper. 2005. Evolution, systematics, and phylogeography of Pleistocene horses in the New World: A molecular perspective. *PLoS Biology*
3(8):e241. doi:10.1371/journal.pbio.0030241.

Weir, B. S., and C. C. Cockerham. 1984. Estimating F-statistics for the analysis
of population structure. *Evolution* 38:1358–70.

Weir, J. T., and D. Schluter. 2004. Ice sheets promote speciation in boreal birds.
Proceedings of the Royal Society of London Series B 271:1881–87.

Weiss, S. 1999. Cars, cows, and checkerspot butterflies: Nitrogen deposition and
management of nutrient-poor grasslands for a threatened species. *Conservation Biology* 13:1476–86.

Wen, J., and S. M. Ickert-Bond. 2009. Evolution of the Madrean-Tethyan
disjunctions and the North and South American amphitropical disjunctions
in plants. *Journal of Systematics and Evolution* 47:331–48.

Wendling, B. M., K. E. Galbreath, and E. G. DeChaine. 2011. Resolving
the evolutionary history of *Campanula* (Campanulaceae) in western
North America. *PLoS ONE* 6(9):e23559. doi:10.1371/journal.pone
.0023559.

West, G. J. 2004. A late Pleistocene-Holocene pollen record of vegetation
change from Little Willow Lake, Lassen Volcanic National Park. *In* S. W.
Starratt and N. L. Blomquist [eds.], *California Proceedings of the Twentieth
Annual Pacific Climate Workshop,* 65–80. Asilomar Conference Grounds,
Pacific Grove, CA.

Westergaard, K. B., M. H. Jorgensen, T. M. Gabrielsen, I. G. Alsos, and C.
Brochmann. 2010. The extreme Beringian/Atlantic disjunction in *Saxifraga
rivularis* (Saxifragaceae) has formed at least twice. *Journal of Biogeography*
37:1262–76.

Westlake, R. L., and G. M. O'Corry-Crowe. 2002. Macrogeographic structure and patterns of genetic diversity in harbor seals (*Phoca vitulina*) from Alaska to Japan. *Journal of Mammalogy* 83:1111–26.

White, M. D., J. A. Stallcup, W. D. Spencer, J. R. Strittholt, and G. E. Heilman. 2003. *Conservation Significance of Tejon Ranch, a Biogeographic Cross-roads*. Environment Now, 2515 Wilshire Blvd., Santa Monica, CA.

Whitmeyer, S. J., and K. E. Karlstrom. 2007. Tectonic model for the Proterozoic growth of North America. *Geosphere* 3:220–59.

Whittaker, R. H. 1961. Vegetation history of the Pacific coast states and the "central" significance of the Klamath region. *Madroño* 16:5–23.

Whittaker, R. J., M. B. Araujo, J. Paul, R. J. Ladle, J. E. M. Watson, and K. J. Willis. 2005. Conservation biogeography: Assessment and prospect. *Diversity and Distributions* 11:3–23.

Whorley, J. R., S. Alvarez-Castañeda, J. S. Ticul, and G. J. Kenagy. 2004. Genetic structure of desert ground squirrels over a 20-degree-latitude transect from Oregon through the Baja California peninsula. *Molecular Ecology* 13:2709–20.

Wiley, E. O. 1988. Parsimony analysis and vicariance biogeography. *Systematic Biology* 37:271–90.

Willson, M. F. 1983. *Plant Reproductive Ecology*. John Wiley & Sons, New York.

Willyard, A., R. Cronn, and A. Liston. 2009. Reticulate evolution and incomplete lineage sorting among the ponderosa pines. *Molecular Phylogenetics and Evolution* 52:498–511.

Willyard, A., J. Syring, D. S. Gernandt, A. Liston, and R. Cronn. 2007. Fossil calibration of molecular divergence infers a moderate mutation rate and recent radiations for *Pinus*. *Molecular Biology and Evolution* 24:90–101.

Wilson, A. B. 2006. Genetic signature of recent glaciation on populations of a near-shore marine fish species (*Syngnathus leptorhynchus*). *Molecular Ecology* 15:1857–71.

Wingo, S. 2009. Targeted grazing as an effective control for non-native *Lolium multiflorum* among the rare species *Cordylanthus palmatus* in alkali meadow habitats. MS thesis, California State University, Chico.

Wisely, S. M., S. W. Buskirk, G. A. Russell, K. B. Aubry, and W. J. Zielinski. 2004. Genetic diversity and strructure of the fisher (*Martes pennanti*) in a peninsular and peripheral metapopulation. *Journal of Mammalogy* 85:640–48.

Wolf, A. T., R. W. Howe, and J. L. Hamrick. 2000. Genetic diversity and population structure of the serpentine endemic *Calystegia collina* (Convolvulaceae) in Northern California. *American Journal of Botany* 87:1138–46.

Wolfe, J. A. 1978. A paleobotanical interpretation of Tertiary climates in the Northern Hemisphere. *American Scientist* 66:694–703.

———. 1995. Paleoclimatic estimates from Tertiary leaf assemblages. *Annual Review of Earth and Planetary Sciences* 23:119–42.

Wolfe, J. A., and H. E. Schorn. 1994. Fossil floras indicate high altitude for west-cental Nevada at 16 Ma and collapse to about present altitudes by 12 Ma. *Geological Society of America Abstracts,* 521.

Wood, D. A., R. N. Fisher, and T. W. Reeder. 2008. Novel patterns of historical isolation, dispersal, and secondary contact across Baja California in the rosy boa (*Lichanura trivirgata*). *Molecular Phylogenetics and Evolution* 46:484–502.

Woolfenden, W. B. 1996. Quaternary vegetation history. *In Sierra Nevada Ecosystem Project: Final Report to Congress, vol. 2, Assesments and Scientific Basis for Management Options,* 47–70. Center for Water and Wildland Resources, University of California, Davis. Wildland Resources Center Report 37.

Wright, T. L. 1991. Structural geology and tectonic evolution of the Los Angeles Basin, California. *In* K. T. Biddle [ed.], *Active Margin Basins: American Association of Petroleum Geologists Memoir* 52:35–134.

Xiang, Q. Y., D. S. Soltis, and P. S. Soltis. 1998. Phylogenetic relationships of Cornaceae and close relatives inferred from *matK* and *rbc*L sequences. *American Journal of Botany* 85:285–97.

Xue, M., and R. M. Allen. 2010. Mantle structure beneath the western United States and its implications for convection processes. *Journal of Geophysical Research* 115:B07303. doi:10.1029/2008JB006079.

Yanev, K. P. 1980. Biogeography and distribution of three parapatric salamander species in coastal and borderline California. *In* D. M. Power [ed.], *The California Islands: Proceedings of a Multidisciplinary Symposium,* 531–50. Santa Barbara Museum of Natural History, Santa Barbara, CA. 787 pp.

Zacherl, D., S. D. Gaines, and S. I. Lonhart. 2003. The limits to biogeographical distributions: Insights from the northward range extension of the marine snail *Kelletia kelletii* (Forbes, 1852). *Journal of Biogeography* 30:913–24.

Zanazzi, A., M. J. Kohn, B. J. MacFadden, and D. O. Terry. 2007. Large temperature drop across the Eocene-Oligocene transition in central North America. *Nature* 445:639–42.

Zhanxiang, Q. 2003. Dispersal of Neogene carnivorans between Asia and North America. *Bulletin of the American Museum of Natural History* 279:18–31.

Zielinski, W., K. Slauson, C. Carroll, C. Kent, and D. Kudrna. 2001. Status of American marten populations in the coastal forests of the Pacific States. *Journal of Mammalogy* 82:478–90.

Zink, R. M., and R. C. Blackwell. 1988. Molecular systematics of the scaled quail complex (genus *Callipepla*). *Auk* 115:394–403.

Zink, R. M., and N. K. Johnson. 1984. Evolutionary genetics of flycatchers. I. Sibling species in the genera *Empidonax* and *Contopus*. *Systematic Zoology* 33:205–16.

Zink, R. M., A. E. Kessen, T. V. Line, and R. C. Blackwell-Rago. 2001. Comparative phylogeography of some arid land bird species. *Condor* 103:1–10.

Zink, R. M., J. Klicka, and B. R. Barber. 2004. The tempo of avian diversification during the Quaternary. *Philosophical Transactions of the Royal Society London B* 359:215–20.

Zink, R. M., D. F. Lott, and D. W. Anderson. 1987. Genetic variation, population structure, and evolution of California quail. *Condor* 89:395–405.

Index

Abadía-Cardoso, A., 126
Abbott, R., 82, 238
Abies sp., 40, 62
Abies bracteata, 53
Abies concolor, 53, 68, 74
Abies magnifica, 53, 61–62
Abies lasiocarpa, 68
Accipitridae, 48, 160, 169
Accipitriformes, 169
Accipitrinae, 169
Acciptridae, 169
accreted, 6, 8, 28, 30, 37
accretion, 22–25, 30–31, 91
Acer sp., 78, 237
Aceraceae, 78
Acinonyx jubatus, 180
Acipenser medirostris, 252
Ackerly, D., 48, 78, 92
Acritarchs, 21–22
Acrodonta, 147
Acrodus sp., 117
Acrotrichis xanthocera, 104, 214*table*
Actinemys marmorata, 151, 153, 217–19, 222, 224, 246
Adams, W., 53
Addicott, W., 120
Adenostoma fasiculatum, 98

admixture, 18, 106, 139, 155, 163, 170, 183, 186, 214, 218, 228, 248
advocacy, 249
Aegypiinae, 169
aeolian, 12
AFLP(s), 91, 110, 163, 199
Africa, 38, 49, 133, 149, 182, 197
African cheetah, 180
Agavaceae, 97
Agelaius phoenicecus, 163, 170
Agnarsson, I., 200
Agriades, 105
agriculture, 94, 185, 241, 246–47, 251–52
Aguilar, A., 123
Aguirre-Léon, G., 120
Aïnouche, K-A., 83
Aiolornis incrediblis, 60
Alameda County, 42, 117, 230
Alaska, 4, 33, 38, 57, 106–7, 119, 129–30, 132, 179, 182, 184, 189, 198–99, 202
Alaskan Gyre, 119*map*, 202
Albach, D., 89
albacore tuna, 128
Alberta, Canada, 210
Alcidae, 160–61, 169
alder, 40
Aleutian Islands, 131, 198, 229

Aleutian Low, 118
Alexander, M., 164, 213, 220, 224*table*
Alexander Valley, 44
algae, 5, 22, 24–25, 33, 63, 131, 133, 229
alkalinity, 126
allele(s), 15, 17–18, 94, 238
Allen, L., 9, 43,
Allendorf, F., 49, 240
Alliaceae, 85
alliance, conservation, 249
alligator lizard, southern, 60
Allison, E., 58
Allium sp., 85
allopatric, 6, 16, 35, 90, 104, 106, 126, 138,
    164–65, 223, 237–38
allopolyploidy, 99
allozyme(s), 69, 72–73, 105, 128, 136–37, 188,
    191, 198–99, 228
alluvial, 10–11
Alnus sp., 40, 62
alpine habitat, 19, 71, 73, 85, 88, 90–91, 96,
    104–9, 115, 139, 163, 175–76, 190, 220,
    229–30, 236–37
Alps, Trinity, 8
Altingiaceae, 79
Altithermal, 62
Álvarez-Castañeda, S., 190–91
Alvin, K., 70
Amador County, 40
Amaranthaceae, 82
Ambrosia dumosa, 97, 228
Ambystoma spp., 44, 139, 145,
Ambystoma californiense, 139, 145,
    217, 240
Ambystoma macrodactylum, 217
Ambystoma macrodactylum sigillatum, 139
Ambystomatidae, 44, 139
Amebeledon sp., 49
American pika, 105, 190
Americans, Native, 175
Americas, 105, 181
Ammerman, L., 178
ammocoete, 131
Ammonita, 28
Ammonoidea, 28
Ammospermophilus leucurus, 191
amoeboid, 23, 33
AMOVA, 18, 98, 114, 131, 139, 177, 182,
    186, 201
Amphibian(s), 4, 44, 59, 68, 135–6,
    143–45, 193, 217–18, 220, 230, 246,
    250, 253
Anacapa Island, 11

Anacardiacae, 48
Anacker, B., 84
anadromous, 4, 56, 120, 124, 126–27, 131,
    249, 256
Anas platyrhynchos, 165
Anas rubripes, 165
Anaspida, 147
Anatidae, 165, 167
Anaxyrus boreas, 143, 214
anchovy, 128
Anderson, J., 147
Anderson, K., 97
Anderson, P., 67, 183
Andreasen, K., 89
Aneides flavipunctatus, 217
Aneides lugbris, 143
Anemone sp., 82
anemones, sea, 25
angiosperm(s), 85, 88, 115
Anguidae, 60
Annieliidae, 60
Anniella pulchra, 60
anomalies, 115, 153
anomalous, 136
Anseriformes, 33, 165
Antarctica, 33, 39, 165
anteaters, 171
antelope ground squirrel, 191
Antelope Peak, 192, 223
Anthophyta, 24, 33–34, 77, 101, 209
anthropogenic, 121, 233
anthropomorphic, 233
Antilocapridae, 175
Antirrhineae, 82
ants, 24
Anttila, C., 238
Anuran(s), 135, 143
Aphelocoma californica californica 162,
    170, 215
Aphelocoma californica insularis, 162
Aphelocoma californica sumichrasti,
    162
Aphelocoma californica woodhouseii, 162,
    170, 215
Aphredoderidae, 117
Apiaceae, 90, 94
Apodidae, 160
apollo, small, 104
apomixis, 90, 99
aquatic, 24, 94, 96, 126, 165, 212, 220, 255,
    256
Aralia sp., 82
Araliaceae, 82

Aranda-Manteca, F., 48
Arbogast, B., 175, 183, 186, 188, 212, 227, 231
arboreal salamander, 143
Arbutoideae, 82
Arbutus sp. 48
Arbutus menziesii, 40
Archaea, 5, 21
Archaeocyatha, 23
archaeological, 180, 198, 211
archaic, 38, 117, 210
Archean, 22
Archie, J., 150–51
architects, 247
Archoplites interruptus, 252
Archosaurians, 38, 209
arctic, 6, 47, 55, 90, 103–5, 198
Arcto-Tertiary, 78, 101, 222
Arctostaphylos sp., 48, 98–99
Arctostaphylos viscida ssp. *mariposa*, 53
Argentina, 163
Argentine ant, 241
Argyrotaenia citrana, 111
Argyrotaenia franciscana, 111
Argyrotaenia isolatissima, 111, 239
arid habitat, 49, 126, 153, 155
aridity, increased, 79, 103, 141
Aristolochia californica, 110
Aristolochiaceae, 110
aristolochic acids, 110
Arizona, 97, 135, 156, 158, 188, 192, 227
armadillos, 171
Armargosa River, 195
arroyo chub, 34
Artemisiospiza belli, 161
arthropods, 5, 23, 221
Artiodactyla, 39, 41–42, 49, 175–176, 200, 210
Ascaphus truei, 214*table*
asexual, 112
Ashland, Oregon, 249
Asia, 38–39, 67, 69, 83, 103–5, 108, 115, 126, 149, 171, 175, 182, 184, 209
Asphalt Stork, 60
assemblages, geological, 31, 34, 215,
assemblages, species, 6, 8, 79, 154, 244, 252
assembly, continental, 21–22
associations, floristic, 6, 78, 101, 115
Asteraceae, 82, 90, 94, 97, 111
Atlantic Ocean, 55,
atmosphere, 21–23
atmospheric, 41, 204
ATPase, 136, 155

Atwater, T., 10
Aubry, K., 182–83
Auk(s), 53, 160
Auriparus flaviceps, 161
Austin, G., 219
Austinomis sp., 33
Australia, 133
Australis, 39
autopolyploidy, 99
autosomal, 15
Aves, 4, 38, 147, 170, 253
avifauna, 53, 118, 160, 169
Avise, J., 13–18, 34, 133, 163, 165, 167, 170, 184, 193, 201, 214, 233
avocado, 48
Axelrod, D., 7, 12, 34, 40, 43, 47–49, 52, 56, 59, 61, 67–68, 70, 78–80, 88, 90–91, 94, 97, 101, 155, 195, 222, 226, 228
Ayala, F., 34, 214
Azevedo, J., 4

bacteria, 5, 21, 63
Baeolophus inornatus, 162, 170, 215, 229
Baeolophus ridgwayi, 162, 170, 215
Bagley, M., 121
Bahia Santa Maria, 132
Bailey's pocket mouse, 192
Baja California, 5, 12, 33, 58*map*, 69, 71, 98, 118–20, 128, 130–32, 138, 143, 151, 154–58, 161, 164, 167, 169, 186–88, 191–93, 197–98, 201–2, 209, 222, 225–28, 242
Baker, C., 201
Baker, H., 7
Baker, M., 164
Balaenopteridae, 200
Baldwin, B., 7, 17, 80, 82, 85–86, 89, 91, 96, 214, 222, 224
baleen whales, 47, 200
Baltica, 24
Barbour, M., 251, 253–54
Barhoum, 164, 224–25
Barlow, M., 130
Barn Owls, 160
barnacles, 133
Barnes, A., 187
Barnhoum, D., 225
Barnosky, A., 183, 189, 211
barrier(s), gene flow, 44, 57–58, 72, 95, 110, 112, 118, 120, 123, 132–33, 137, 140–41, 145, 154, 156–58, 180, 193, 216–18, 221, 227
Barron, J., 62
Barrowclough, G., 168, 213

Barry, S., 241
Barth, J., 118–19, 129, 197, 256
Basilemys sp., 31, 149
Baskin, J., 179
bass, striped, 131
Bassett, I., 97
batfriendly, 247
bat, western mastiff, 253
batholiths, 9
Batrachoseps spp., 140, 143, 145, 195, 217–18, 224
Batrachoseps attenuatus, 140, 145, 217–18
Batrachoseps gabilanensis, 217
Batrachoseps luciae, 217
Batrachoseps nigriventrus, 224*table*
Batrachoseps stebbinsi, 222
Battus philenor hirsuta, 110
Baum, D., 82, 97
Bautista Formation, 68
beaked whales, 205
Beamish, R., 131
beavers, giant, 175
Becker, B., 24
beech, 79
Beerli, P., 15
bees, 24, 252
beetle(s), 24, 107
beetle, clown, 110
beetle, darkling, 110–12
beetle, feather-winged
beetle, pictured rove, 110
Beidleman, R., 6
Bengtsson, J., 246
Bennett, K., 35, 55
Bennettiales, 31, 77
Benson, L., 107
benthic, 131, 229
Benton, M., 25, 38, 171, 174
Bering Strait, 39, 54, 57, 77, 171, 202, 229–30
Beringia, 59, 71, 78–80, 82–83, 97, 99, 103, 105, 115, 136, 176, 179, 182, 193, 210–12, 231
Berkeley, University of California, 6, 23, 25, 31, 117, 149
Bermingham, E., 18, 125
Bernardi, G., 130–31, 229
Bernatchez, L, 185
Bertoldi, G., 10
Betancourt, J., 72
Betoideae, 82
Betulaceae, 40
Beukelaer, S., 204*map*
Bickham, J., 199, 201, 229

bighorn sheep, 49
billfish, 128
biodiversity
biodiversity, importance of, 241, 243, 245–46
biodiversity, threats to, 134, 235–36, 238, 254
biodiversity conservation of, 244, 247–49, 256
bipolar 83–84
birds, 5, 14, 33, 48, 59, 94, 159, 161, 162, 164, 170, 211, 220, 253
Bishop pine, 59
Bison sp., 61
Bison antiquus, 59, 60*fig*
Bison bison, 59
Bison latifrons, 59, 175
bison, giant, 59
Bivalvia, 28, 38
Blackbird, 163
black surfperch, 129
Blakey, R., 26–27, 29, 45–46
Blancan, 171, 175
boa(s), 147, 149, 156
bobcat, 180
Bodega Bay/Head, 56, 120, 187, 194*map*, 218
Boechera breweri, 89, 214*table*
Boechera hoehleri, 214*table*
Boechera koehleri, 89
Boechera sparsiflora, 88–89, 214*table*
Boechera stricta, 88
Boechereae, 88
Boidae, 149, 156
Bonneville, 56, 120, 227
Boobies, 48
Booth, D., 195
Boraginaceae, 91
boreal
boreal habitat, 90, 163, 186–87
boreal species, 170, 182–83, 187, 212
botany, importance of, 245
bottleneck(s), 14, 18, 58, 73, 69, 93–94, 109, 168–69, 188, 201, 215, 185
boundary(ies), biogeographic, 118, 130, 133, 138, 201, 204, 235
Bouse Embayment, 12, 58, 156–58, 227, 230
Bouteloua sp. 98
Bovidae, 49, 175
Bowen, B., 133
Bowerman, N., 108
Boykinia sp., 82
Brachiopoda, 23, 25
brackish, 10, 130
Bramble, D., 156

Brandegee, M., 6
Brassicaceae, 82, 88, 90
Brattstrom, B., 59
Breckenridge, 112
Brewer spruce, 74
Brewer, W., 6, 74
Briggs, C., 51, 133, 250
bristlecone pine, 53, 62, 73
British Columbia 73, 92, 94, 107, 117–19, 125, 137–38, 163, 180, 188, 201, 211, 242
Brito, P., 17
brittlebush, 97
broad-nosed weevil, 112
browsers, 53
Brubaker, L, 71
Brunelli, J., 126
Brunsfeld, S., 73
Brzuszek, R., 247
Bubo sinclairi, 60
Bufo boreas, 143
Bufonidae, 143
Bunn, D., 255
Burge, D. 79
Burgess Shale, 23
Burghardt, K., 247
Burnham, K., 140, 228
burro bush, 97
Burton, R., 57, 118
bush, burro, 97
Buteo sp., 48, 169,
Buteo lineatus elegans, 169
Buteo swainsoni, 253
Buteoninae, 169
Buteos, 169
Butte County, 25, 30–31, 33, 43, 117, 149, 159
Butte County meadowfoam, 95
butterfly (ies), 24, 104–5, 109–10, 115, 212, 241–42

caballine horses, 177
Cabras, Point, 131
caches, seed, 70–71
Cactus Wren, 161
caddisflies, 24
Cadotte, M., 242
Caecilians, 135
Caesar, R., 104, 248
Cahuilla, Lake, 191
Calandrina sp., 98
Calaveras County, 188
CALFED, 253
California Bight, 57, 118, 120, 130, 132
California Coastal Conservancy, 251

California Condor, 169
California Gnatcatcher, 161
California Gull, 167
California myotis, 178
California Floristic Province, 5, 13, 89, 216
California Quail, 167
California Rangeland Conservation Coalition, 251
California Thrasher, 164, 166*fig*, 223
Callipepla californica, 167
Calocedrus sp., 41, 62
Calocedrus decurrens, 68
Calochortus spp., 85, 92, 214
Calochortus tiburonensis, 85
Calochortus macrocarpus, 92
Calochortus umbellatus, 85
Calsbeek, R. 14, 43, 53, 112, 223
CalTrout, 243
Calycanthus sp., 82
Calystegia collina, 92–93
Camargo, A., 14
Cambria, California, 204*map*
Cambrian Period, 22–24, 26*map*
Camelidae, 42, 175
Camelops, 60
camels, 42
Campanula sp., 83
Campanula scabrella, 83
Campanula scouleri, 83
Campanulaceae, 83, 91
Campos, P., 176
Campylorhynchus brunneicapillus, 161
Canada, 103, 119, 163, 178, 182, 198
Cane, M., 52
Canidae, 175, 181
Canis sp., 60, 182–84, 239
Canis dirus, 60 184
Canis etruscus, 183
Canis latrans, 182–84, 239
Canis latrans orcutti, 184
Canis lepophagus, 183
Canis lupus, 182–84
Canyon, Monterey, 216
Canyon, Sequoia-Kings, 237
Canyon, Titus, 42
Cape Blanco, 119*map*
Cape Mendocino 75, 100, 118–19, 122, 124, 132, 144, 148, 194
Caprifoliaceae, 82
Caprifolium clade, *Lonicera*, 82
Carabidae, 107
Carangidae, 128

carbonate platforms, 22
carbonate soils, 84
Carboniferous, 24–25, 67, 74, 135
Carde, R., 24
Cardellina pusilla, 162–63, 170
Cardinalidae, 159
Cardinals, 159
Carex sp., 83
Carex whitneyi, 90
Caribbean Ocean, 55
Carnivora, 175, 178, 197–98
Carnivore(s), 38, 59, 61, 180, 210
Carpenteria fossil bed, 60
Carquinez Strait, 10
Carr, S., 176
Carreño, 155, 228
Carretta, A., 199
Carson/Sonora Pass, 108
Carstens, B., 15–16, 214*table*
cartilaginous fishes, 25
Caryophyllales, 77
Cascade(s) Range, 5, 8–9, 37, 43, 61–62, 71,
    73, 86, 90, 92–93, 104, 107–9, 115, 136,
    151, 153, 162, 168, 176, 179, 181–82, 185,
    189–90, 212–15, 219–20
Cassel, E., 9
Castanea sp., 79
Casteel, R., 133
Castilleja sp., 98
Castoe, T., 155, 226
Castoridae, 175
cat, scimitar, 180
Cataostomidae, 117
Caterino, M., 7, 96, 112–13, 223, 230
catfish, 117
Cathartidae, 53, 159, 169
Catharus ustulatus, 163, 170
Catostomidae, 117, 121
Catostomus microps, 125
cattails, 61
Caudata, 135, 143, 210
Cavender, T., 51
caviomorph, 174
Convention on Biological Diversity, 244
CCS (California Current System), 50, 57, 118,
    129, 160, 256
CDFG (California Department of Fish and
    Game), 149
CDFW (California Department of Fish and
    Wildlife), 243–44, 254, 256
Ceanothus spp., 48, 79
Ceanothus subgenus *Cerastes*, 79
cedar, incense 68

Cenozoic Era, 5, 8–9, 24, 31, 37–39, 41–42,
    55, 74, 91, 171–72, 215–16, 250
Centrarchidae, 44, 117, 121
CEQA (California Environmental Quality
    Act), 243
Crotalus atrox, 155, 180, 226
Crotalus cerastes, 155, 226
Crotalus mitchellii, 155–56, 226
Crotalus tigris, 156, 226
Cercis sp., 80, 82
Cercis canadensis, 82
Cercis occidentalis, 82
Cercocarpus sp., 48
Cervidae, 175
Cervus spp., 176
Cervus elaphus nannodes, 177
Cervus elaphus roosevelti, 176
Cervus elaphus nelsoni, 177
Cetacean(s), 47, 197, 200–202, 205
Cetartiodactyla, 200
Chabot, L., 132
Chaetodipus sp., 58
Chaetodipus. baileyi, 49, 192
Chaetodipus fallax, 191
Chaetodipus formosus, 49
Chaetodipus hispidus, 49, 154
Chaetodipus penicillatus, 192
Chaetodipus rudinoris, 193
Chamaea fasciata, 164, 170, 223–25
channelization, 250, 256
chaparral, 48, 61, 79, 92, 98, 149, 164, 187,
    191, 217, 236, 253
Charadriiformes, 167
Charina
Charina bottae, 149–50, 158, 219
Charina bottae bottae, 149, 158
Charina bottae umbratica, 149
Charina trivirgata, 156, 227
Chater, O., 83
Chatzimanolis, S., 96, 111–13, 222–24, 230
Chavez, A., 190
Cheatham, N., 253
checkerbloom, 89
Checkley, D., 118–19, 129, 197, 256
cheetah, African 180
Chelonia, 30–31
Chelonii, 31, 147, 149
Cheloniidae, 31, 149
Chelonioidea, 31
Chelonioididea, 149
Chen caerulescens, 167
Cheng, Y., 69
Chenopodiaceae, 82

Chert, Rhynie, 24
chestnut, 79
Chicago Botanical Garden, 245
Chickadee, Mountain, 162
Chickadees, 161
Chickens, 33
Chico Formation, 31, 33, 117, 149, 159
Chihuahuan Desert, 155, 157–58, 165, 226–27
Chilcote, M., 250
Childs, M., 237
Chile, 95, 133
China, 67
Chino Hills, 113
Chinook salmon, 56, 127*map*, 252
Chironomidae, 109
Chiroptera, 175, 177
Chiropterans, 178
Chivers, S., 201
chloroplast markers, 16, 70, 72, 94, 99
Chondrichthyes, 25, 117
Chordata, 24
Christiansen, P., 180
chub, arroyo, 34
chub, Gila, 56, 120, 227
chub, Klamath blue, 34
chuckwalla(s), 154, 158
Chytridiomycosis, 250
Cicero, C., 161, 229
Cichorieae, 82
Ciconia maltha, 60
Ciconiidae, 60
circulation, oceanic, 47, 57, 256
circumboreal, 70, 74, 83
circumpolar, 201
CITES, 244
cities, 234, 237, 244, 255
cladistic, 16
Cladogenesis, 13
Clark, P., 58, 71, 108, 165
Clarke, J., 33, 159
Claytonia megarhiza, 90
Clementz, M., 42
cliffs, 11
clinal, 137
cline, 182, 222
clonal, 70, 154
*clp*P, 114
Clupeidae, 128, 133
Cnidaria, 25, 38
coachwhip, 60
Coart, E., 237

Coastal dunes, 85
coastal marshes, 186, 252
coastal redwood, 6
coastal species, 25, 51, 57, 73, 93, 113, 118, 126, 127*map*, 129–30, 133, 136–37, 150, 153, 142, 163–64, 175, 179, 201–2, 213, 218, 222*table*, 243, 248, 251, 256
coastal sage scrub 191, 235–36, 239, 251
Coast Range, 30, 41, 52, 187
Coates, A., 50
Coccolithales, 33
Coccyxus americanus occidentalis, 253
Cockerham, C., 18
Cody, S., 50
Coelus ciliatus, 111
Coelus globosus, 111
Coelus gracilis, 111, 180
Coelus pacificus, 111
coho salmon, 127
COI, 16, 104–6, 108–14, 162–63, 165, 168
COII, 16, 106–8, 164, 168
COIII, 164, 192
Cole, C., 155
Cole, K., 97–98
Coleman, A., 17
Coleoptera, 24
Colfax, California, 40
Colias sp., 109, 220
Colias behrii, 109, 220
Colias meadii, 109
collection(s), importance of, 142, 189, 246
Collinsia spp., 84–86, 89, 91, 96, 214, 220, 222, 224, 231
Collinsia antonina, 85, 96
Collinsia bartsifolia, 86*fig*
Collinsia callosa, 86*fig*
Collinsia childii, 86*fig*
Collinsia concolor, 87*fig*, 96
Collinsia corymbosa, 85
Collinsia grandiflora, 86, 89
Collinsia greenei, 85, 87*fig*
Collinsia heterophylla, 96, 224
Collinsia latifolia, 86*fig*
Collinsia linearis, 86*fig*, 89*fig*, 214
Collinsia parryi, 96
Collinsia parviflora, 86*fig*, 89
Collinsia rattanii, 86, 91
Collinsia sparsiflora, 86*fig*
Collinsia tinctoria, 86*fig*
Collinsia torreyi, 86*fig*
Collinsia torreyi var. *brevicarinata*, 86*fig*
Collinsia verna, 89

Collinsia violacea, 86, 89
Collinsia wrightii, 86*fig*, 96, 224*table*
Colombia, 79
colonization, ancient, 117, 120, 179, 228
colonization, of California, 3, 5–6, 8
colonization, after glaciation, 89, 99, 120–21, 130, 163, 211
colonization, to high Sierra Nevada, 90, 107
colonization, island, 111
colonization, long-distance, 77, 80, 92, 105, 107, 115, 126, 165, 169, 223
Colorado Basin, 128, 228,
Colorado Desert, 5, 11, 122*map*, 124*map*, 127*map*, 144*map*, 148*map*, 149, 191, 237
Colorado Plateau Geosystems, 26*map*, 27*map*, 29*map*, 32*map*, 45–46*map*
Colorado River 12, 57–58, 98, 123, 126–28, 154–58, 192, 193, 195, 226–28
Coluber constrictor, 60
Colubridae, 60, 150–151, 157
Columbian black-tailed deer, 176
Colusa County, 235
Colusa grass, 95
communities, animal, 243, 246
communities, coastal, 243, 251
communities, fish, 126
communities, haplophytic, 4
communities, plant 4, 41, 61–62, 70, 98, 101, 243, 246–47, 249, 251, 254
communities, riparian, 186
competition, 5, 12, 38, 182, 233
composting, 247
Condor, California 169
conduit, migrational, 50, 120, 222, 229
Condylarthra, 38–39
conflagrations, 253
confluence, biotic, 213, 221
congener(s), 69, 161, 167, 231, 237–238
conifer(s), 4, 24–25, 40–41, 43, 53, 67–69, 72–74, 75*map*, 78, 82, 84, 104, 163, 179, 187, 209, 215, 217, 235, 237
Coniferophyta, 25, 30–31, 33, 40, 68, 209
Conkle, M., 57, 72–73
connection(s)
connection(s), desert, 12, 230, 255
connection, between North and South America, 230
connection, between Sierra Nevada and Coast Ranges, 181, 221–22, 230, 255
connections, between Siberia and Alaska 57, 171

connections, marine, 45, 51, 120, 140, 158, 216, 219, 229–30
connectivity, habitat, 236, 239, 249–50, 253
connectivity, forest 67, 71, 80, 215,
Conroy, C., 185–86
Conservancy, Sierra Nevada, 250
Conservancy, California Coastal, 251
conservation, 3–4, 6–8, 92, 95, 200, 233–36, 240–41, 242–51, 254–56
Constance, L., 7
conterminous U.S., 4, 182, 254
Contia tenuis, 158
Continental Divide, 155, 226
continental margin, 21–22, 25, 26*map*, 30, 34, 37
continental shelf, 5, 22, 28, 130
contraction, 91, 98, 107, 195, 223
Contra Costa County, 230
contractions, range, 34, 98, 184
Convention on the International Trade in Endangered Species (CITES), 244
convergent evolution, 41, 82
Convolvulaceae, 92
Cooper, J., 6
Copepoda, 130
coppers, great, 106
coral(s), 25, 30
Cordilleran arc, 29*map*, 32*map*,
Cordilleran ice sheet, 175, 183, 195
Coriolis Effect, 52
Cormorants, 48, 53
Cornaceae, 78
Cornus sp., 78, 82, 93, 220, 237
Cornus nuttallii, 93, 220
Correa, E., 79
corridor(s), habitat, 5, 34, 54, 59, 127, 185, 193, 233, 238–39, 244, 254
Cortéz, Sea of, 193
Corvallis, Oregon, 249
Corvidae, 160, 162, 165, 227
Corvus corax, 165, 227
Coryphaena hippurus, 131
Coryphaenidae, 131
cosmopolitan taxa, 131–32, 160, 169
Costa Rica, 129,
Cottidae, 56, 120, 128, 227
Cottoidae, 117
cottonwood, 40
Cottus pitensis, 125
cougar, 180
cows, sea, 47
Cox, B., 12

cox1, 140
coyote(s), 182–184
cpDNA, 69–73, 83, 93–94, 97–98, 114, 214
cpHS, 93
cpHT, 93
CR, mitochondrial, 16, 129–33, 143, 176, 179, 187–88, 191, 200–2
Craig, M., 7, 132
Crandall, K., 15
Cranioceras sp., 49
craton, North American, 24–25
Creodonta sp., 38–39
creosote, 97
Crespi, B., 58, 221
Cretaceous Period, 25, 30–31, 33–34, 37, 41, 54, 67–71, 74, 77, 79, 115, 117, 147, 149, 159, 165, 209–10, 215, 219
Cricetidae, 58, 175, 185, 187, 189
cricket, Mahogany Jerusalem, 113
Critchfield, R., 84
Crocodilia, 147
Croizat, L., 12
Crotalinae, 149
Crotalus sp., 149, 155
Crotalus atrox, 149, 155–156, 226
Crotalus cerastes, 226
Crotalus mitchellii, 156, 226
Crotalus ruber, 156, 226
Crotalus tigris, 156, 226
Crotaphytidae, 147
Crustaceans, 131, 229, 253
Cruz, B., 82
Cuckoo, Western Yellow-billed, 253
Culley, T., 237
cultivars, 237
Culver, M., 181
Cumberland, 152*fig*
Cunninghamia, 69
Cupressaceae, 6, 34, 41, 53, 62, 67–69
Cupressus, 41
Curculionidae, 112
currents, oceanic, 57
cutthroat trout, 56, 121, 124, 127
Cuyama River, 151, 224*table*
cyanobacteria, 21
Cycadaceae, 30, 40
Cycadophyta, 30–31, 33, 67
CYCLOIDEA, 89
Cyclopedidae, 171
Cyperaceae, 83, 90
Cypher, B., 183
cypress, bald, 69

cypress, Guadalupe, 69
cypress, Monterey, 59, 69
cypress, Tecate, 69
cypress, woodland, 48
Cyprinidae, 34, 44, 117–18, 121, 123, 125
Cypriniidae, 121
Cyprinodon spp., 126, 255
Cyprinodon diabolis, 128, 228
Cyprinodon fontinalis, 128
Cyprinodon macularius, 56, 120, 128, 227–28
Cyprinodon nevadensis amargosae, 128, 228
Cyprinodon nevadensis mionectes, 128
Cyprinodon nevadensis nevadensis, 128, 228
Cyprinodon radiosus, 128, 228
Cyprinodon salinus, 128, 228
Cyprinodontidae, 56, 118, 120, 227
Cyrtonyx cooki, 48
Cyt *b*, 16, 129–30, 132, 136, 138, 140–41, 143, 150, 153, 154, 163–65, 176, 178–79, 182, 185–88, 190–92, 199
cytoplasmic markers, 14, 16–17

Dakotas, 92
Dall's porpoise, 202, 204*map*
Daniels, M., 93
darkling beetle, 110–12
Darwin, C., 12
Dasypodidae, 171, 175
Datisca sp., 82
Datiscaceae, 82
Davidson, A., 119, 203
Davidson, C., 250
Davies, J., 242
Davis, C., 82,
Davis, E., 194*map*, 195,
Davis, F. 235–36, 240
Davis, M., 242
Davis, O., 120, 229
Davis, University of California, 7
dawn redwood, 34
Dawson, M., 57, 118, 130, 133
Debruyne, R., 59
DeChaine, E., 90, 107
deciduous, 43, 80, 82, 209
DeCourten, F., 9
deer
deer, ancestral, 49, 175–76
deer, black-tailed, 176
deer, mice, 58
deer, mule, 175–76
degradation, 185, 193, 245, 249–50, 252, 254
Delaney, K., 162, 170, 215, 245

Della Sala, K., 248
DeLong, J., 198
Delta, San Francisco Bay, 10, 62, 75*map*,
    100*map*, 121–23, 124*map*, 125, 127*map*,
    140, 144*map*, 148map, 151, 158, 181,
    187–88, 194*map*, 217–18, 221–22,
    252
Delta, Baja California, 33, 209
Delta, Colorado River, 128, 228
demographic(s), 18–19, 98, 109, 114, 140, 191,
    235, 240
Dempster, L., 93
Dendragapus fuliginosus, 167
Dendragapus obscurus, 167
Dephinidae, 205
deposition, soil, 10, 22, 29*map*, 38, 40, 51, 57,
    59, 218
Dermochelyid(s), 31, 149
Dermochelyidae, 31, 149
Descurainia sp., 82
Desmodus rotundus, 178
Desmostylia sp., 42
Desmostylians, 47
Devitt, T., 155, 157–58, 226
Devonian, 24, 135
Diablo Range, 187
Diadophis punctatus, 61, 150, 152*fig*, 153,
    158, 220
diamond turbot, 132
diatom, 62
diatomite, 44, 47
Díaz-Jaimes, P., 129, 131
Dicamptodon
Dicamptodon ensatus, 216–17
Dicamptodon tenebrosus, 137, 214
Dicamptodontidae, 137
Dickinson, W., 30, 91
Didelphidae, 171, 175
Dilleniaceae, 77
Dinoflagettes, 22
diorama, Pliestocene, 60
Dipodominae, 48
Dipsadinae, 156
Dipsosaurus dorsalis, 158
Diptera, 24
Dirca occidentalis, 92
Dire wolf, 184
disease, 145, 236
disjunct(s), 62, 74, 78–80, 83, 88, 91–92, 103,
    110, 156, 168, 182, 190, 215
disjunctions, 78, 82, 97, 99
dispersal events, 6, 12, 14, 71, 74, 83, 96, 105,
    107–8, 143, 150, 158, 216,

dispersal ability, 5–6, 14, 18, 34, 57, 78–81, 88,
    91–94, 96, 99, 103–4, 108, 111, 114–15,
    118, 128, 130–32, 139, 154–55, 179,
    181–82, 186, 193, 202, 216–17, 220,
    238–39, 241, 244
displaced, 45–46
disrupting, 191
disruption(s), 35, 62, 256
disrupts, 254
dissected, 9
distinctive, 128, 228
disturbance(s), 7, 19, 121, 181, 219, 231, 233–
    34, 236–37, 239, 241–42, 255
diverted, 10, 216
divisions, genetic, 70, 94–96, 108, 130, 133,
    136, 157–58, 163, 168, 176, 183, 190–91,
    193, 195, 201, 225
Dizon, A., 202
dogwood, 78, 93
dolphin(s), 203–205
dolphinfish, 131
Domengine-Ione Seaway, 40
Domning, D., 42, 47–48
Dong, J., 72
Donley, M., 4
Donoghue, M., 80, 82
Dorsey, R., 154, 226
Douglas, M., 6, 40, 75, 149, 156
Dowling, T., 237
downwelling, 118
Doyle, J., 33–34, 77, 133
drought, Eocene, 68
drought, Medieval, 62, 109
drought, summer, 85
drought, 14ka, 96
Drummond, A., 16
Dryas, Younger, 59
D_{ST}, 17
Ducea, M., 55
dugong(s), 47, 203
Dugongidae, 47
Dukes, J., 238
dunes, aeolian, 12
dunes, coastal, 85, 111
Dupré, W., 125, 217
Durham, J., 58
dusky-footed woodrat, 187
Dusky Grouse, 167
dusky-wing skipper, 242
Dutchman's pipe, 110
Duvernell, J., 128
Duxbury Reef, 47
Dyer, R., 238

Eagle Lake Field Station, 124
Eagles, 169
eared seals, 48, 198
earless seals, 48, 198
easement, conservation, 254
Echelle, A., 128
Echinodermata, 23, 25, 28, 38
Eckert, A., 15–16, 19, 68, 70–71, 73,
    84, 214
economic benefits, 246, 250
economic sacrifices, 248, 250,
economic viability, 256
ecosystem conservation, 244
ecosystem vulnerability, 47, 236, 241
ecosystem degradation, 111, 238,
    245, 254
ecosystem function, 51, 63, 236, 238, 241–42,
    245, 254, 256
ecosystem restoration, 240, 244, 247
edaphic endemism, 69, 78, 84–85, 90, 92,
    209, 220, 248
Eddy, Mount, 88, 119, 200
education, importance of, 243,
    245, 256
Edwards, S.V, 15, 17
Edwards, S.W., 48, 62, 68–69, 78–79, 98
Edwards, R., 237
Eel River, 92
Ehrlich, P., 58, 233
Eizirik, E., 180
Ekman Transport, 52
El Niño, 63, 199
Elam, D., 238
Elaninae, 169
elephant(s), 175, 198–99
Elephantidae, 38, 175
Elgaria multicarinata, 60, 158
McClintock, E., 7
Elk Creek, 138
elk, 138, 176
elk, Roosevelt, 177
elk, Tule, 177
Ellstrand, N., 237–38
embayment(s), 12, 58, 113, 140, 145, 153,
    156–58, 176, 217–18, 223, 225, 227,
    230
Emberizidae, 159, 161, 163
Embiotoca jacksoni, 129–31, 229
Embiotoca lateralis, 130–31, 229
Embiotocidae, 47, 121, 129
Embryophyta, 24
Emery, N., 95
Empetrum sp., 84

Emydidae, 151
enamel, tooth, 42
Enantiornithes, 33, 159
Encelia farinosa var. *phenicodonta*, 97
Encelia farinosa var. *radians*, 97
Enchodontidae, 117
Enchodus, 117
endemic, 53, 84–85, 90, 92, 99, 115,
    121, 134, 158, 194, 220, 214*table*, 227,
    233, 235
Endopterygota, 24
Engel, M., 24
Engraulidae, 128
Engraulis mordax, 128
Enhydra lutris, 197–98, 202
Enhydra lutris kenyoni, 198
Ensatina spp., 136–37, 195, 217–18
Ensatina eschscholtzii, 141, 143, 145
Ensatina eschscholtzii oregonensis, 137,
    141–43, 217
Ensatina eschscholtzii platensis, 137
Ensatina klauberi, 141
Ensenada, Mexico, 58, 157
entomologist(s), 6–7, 104
Enzel, Y., 63
Eocene Epoch, 37–41, 44, 68, 70, 79–80, 84,
    92, 117, 125, 135, 140, 143, 159, 171, 178,
    193, 209–10, 219
equator, 55
Equidae, 49, 53, 175
Equistaceae, 25, 209
Equiza, M., 34
Equus sp., 61
E. caballus, 177
Erethizontidae, 171, 175
Ericaceae, 40, 48, 53, 82–84, 98
Erickson, V., 240
Eriogonum sp., 84
Ernest, H., 181, 222
erosion, 31, 38, 40, 43, 53, 246, 251, 254
eruptions, volcanic, 9, 12
Erycinae, 147
Erynnis propertius, 242–43
Escobar-Garcia, P., 69
Escorza-Treviño, S., 202
Escudero, M., 83
Essig, E., 6
Estes, R., 135, 198
estuary(ies), 10, 44, 50–51, 56, 94,
    118, 120, 130, 132–34, 211–12,
    218, 229
estuarine habitat, 25, 38, 47, 120, 129, 140,
    218–19, 231, 252, 256

ETS, 89
Eucyclogobius newberryi, 47, 130
eudicots, 77
eukaryotes, 21–22
Eumetopias jubatus, 199–202, 229
Eumetopias jubata monteriensis, 199–200
Eumops perotis, 253
Euramerica, 24
Eurasia, 53, 70, 74, 78–80, 82, 105–7, 176, 178, 180, 182–83, 209–10, 212
Eureka, California, 62, 75*map*, 100*map*, 122*map*, 124*map*, 127*map*, 144*map*, 148*map*, 194*map*
Eureptilia, 147
Europeans, 6
Eurosid(s), 77
euryhaline, 128
even-toed ungulates, 39, 42, 200, 255
evergreen, 92, 236
Excoffier, L., 18
exotic, 30
expansion(s), range, 3, 6, 15, 41, 91, 93, 98, 105–8, 115, 133, 137, 140, 150–51, 154, 160–65, 169, 184, 188, 191–92, 195, 201, 220–23, 225–26, 253
extinction(s), 5, 25, 33, 35, 38–39, 44, 47, 54–56, 58–59, 103, 115, 118, 120, 134, 145, 159–60, 176, 199, 203, 212, 233–39, 241, 249–50, 252
extirpated, 4, 44, 57, 67, 73, 98, 121, 138, 150, 157, 159, 180, 184–185, 197, 212, 219, 252–53

Fabaceae, 83–84, 90
Fabidae, 77
faculty, 245
Fagaceae, 61, 79, 93
Fagus sp., 79
Falconidae, 168
Falconiformes, 168
false mermaid, 95
Farallon Islands, 129,
Farallon Plate, 30–31, 43
Farallones National Marine Sanctuary, 204*map*
farmland, 254
Farrar, C., 10
fauna, avi-, 53, 118, 160, 169,
fauna, California, 3, 24, 57, 118, 133, 243, 252
fauna, herpetological, 12, 143, 213, 216
fauna, ichthyo-, 210
fauna, mammal, 171, 193

fauna, marine, 38, 44, 51, 55, 118, 120–21, 134, 200
fauna, Pleistocene mega-, 59, 176, 212
fauna, paleo-, 23, 117, 175, 210
federally-listed taxa, 95, 139, 141, 234, 237, 254
Feduccia, A., 160
Fehlberg, S., 97
Feldman, C., 110, 112–13, 145, 158, 220, 223–25
Felidae, 49, 175, 180
Felids, 50, 180
Felinae, 50, 180, 210
felsic material, 22
Feng, Y., 82
ferns, 24–25
Ficus sp., 40
Fiedler, P., 92, 94
Fink, S., 51
fir, Chinese, 69, 75
fir, Douglas 40, 75*map*
fir, red 53, 75*map*
fir, Santa Lucia 53
fir, white, 53
F_{IS}, 17, 74
fisher, 179
fisheries, 118, 249, 256
fitness, impacts to evolutionary, 239–40, 243
Flagstad, O., 176
flatfish, 47
flat-tailed horned lizard, 157
Fleischer, R.,164
Flightless Auk, 53
flightless broad-nosed weevil, 112–13
Floerkea proserpinacoides, 95
flora, 3, 12, 84, 99, 101, 209, 251
flora, alpine, 90, 96, 243
flora, fossil, 33, 40, 43, 48, 53, 59, 61, 68, 77–80, 82, 84, 88, 98–99, 210
Florida, 152
floristic, 5, 13, 38, 48, 62, 77, 88–90, 216
floristically, 209
fluctuation(s), climatic, 5–6, 38–39, 44, 47, 50, 52, 54, 56, 61–63, 76, 88, 107, 109, 118, 120, 136, 175, 184, 188, 215, 218, 223, 243, 252
Flycatchers, Tyrant 160
Flynn, J., 198
Flyway, Pacific, 253
fog, 4, 47, 50, 118
Folsom, California, 40
Fong, J., 23*fig*
Fontanella, F., 152–53

foothill(s), 48, 110, 142, 153, 187–88, 218, 253
foothill yellow-legged frog, 138
Foraminifera, 38
Fordyce, J., 110
forest(s), 75*map*
forests, boreal, 163, 186
forests, conservation of, 235, 237–38,
    251–254
forests, deciduous, 82, 209
forests, kelp, 47
forest(s), as habitat, 49, 52, 93, 104, 149, 165,
    168–69, 179–80, 188, 217, 221, 248–49,
    252–254
forests, Paleogene/Neogene, 54, 62, 74,
    82–83
forests, Pliocene/Pleistocene, 67, 70, 83, 183,
    187–88, 215,
forest(s), Mesozoic, 4, 33
Forister, M., 107
Formaminifera, 33
formation, Bautista, 68
formation, of California 7–8, 22, 118
formation, Chico, 31, 33, 117, 149, 159
formation, Coast Range, 10, 14, 30, 141, 145,
    158, 191, 211, 230
formation, Gulf of California, 156, 191
formation, Kern River, 56
formation, Miocene diatomite, 44
formation, Hosselkus, 31, 117, 149, 209
formation, Monte de Oro, 30
formation, Monterey, 47
formation, Moreno, 31
formation, San Francisco Bay, 31, 38, 188
formation, Sierra Nevada, 219
fossils, Precambrian, 21–22
fossils, Cambrian, 23–24, 28
fossils, Carboniferous, 67, 74
fossils, Mesozoic, 30–34, 68, 70, 77, 79, 117,
    135, 149, 159, 210
fossils, Paleogene, 37–48, 73, 80, 147, 149,
    156–57, 172*fig*, 178–79, 198–99
fossils, Neogene, 53–54, 161
fossils, Pliocene/Pleistocene, 55, 57, 60–61,
    68, 73–74, 82, 97–98, 120–21, 133,
    143, 154–55, 169, 175, 180, 183, 189,
    191, 226
fossil records, general, 7–9, 10
fossilization, 54
fox, swift, 182
foxtail, 53, 73
fragmentation, habitat, 15, 74, 88, 98, 113–14,
    164, 181–82, 188, 227, 233, 238–39, 251,
    253–54

framework, 15, 78
Franciscan complex, 30, 32*map*, 45*map*,
    46*map*, 91, 140, 215–16
Frankham, R., 239–40
Fraser River, 119*map*
Fremont, J., 6
freshwaters, 121
Fresno County, 31, 62–63, 117, 149, 188
Fresno, California, 75*map*, 100*map*, 122*map*,
    124*map*, 127*map*, 144*map*, 148*map*, 187,
    194*map*
Friis, E.,77
Fritsch, P., 82
frog(s), 135, 138, 141–43
F_S, 140, 192, 223
F_{ST}, 17–18, 74, 93–95, 114, 131–32, 138–39,
    168–69, 177, 180–81, 183, 185, 199–202
F_T, 234
Fu, 140
fungi, 5, 23
Funk, V., 239

Gabilan Mountains, 10, 216–17, 251
Gahagan, L., 39
Galbreath, K., 190
Galeorhinus galeus, 132
Galium grayanum, 90
Gall, G., 121
Galliformes, 167
Gannet, 47
GAPDH, 151
gardener, 247
Garlock fault, 12
Garone, P., 253
garter, snake, 61, 150
Garza, J., 123
Gassel, M., 178
Gasterosteidae, 56, 117
Gasterosteus aculeatus, 56, 120
Gastropoda, 38, 133
Gauthier, J., 33
Gavia sp., 53
Gaviidae, 54
Ge, D., 171–172
Geese, 167
generalist(s), 39, 181, 238, 241
Geodercodes latipennis, 112
geoflora(s), 78–79
geomorphologic, 6
Geomyidae, 58, 175, 190
geophytes, 92
geosystems, 26–27, 29, 45–46
Gernandt, D., 70–72

Gila chub, 56, 120, 227
Gila spp., 123
Gila coerulea, 34, 214
Gila crassicauda, 252
Gila intermedia, 56, 120, 227
Gila orcuttii, 34
Gila River, 156, 227
Gilbert, C., 176
Gill, F., 161
Gillespie, A., 9
Gillett, G., 90
Gingerich, P., 200
Ginkgophyta, 30–31, 33, 67
Givnish, T. 85, 92, 214
glacial refugia, 14, 58, 67, 71–74, 89, 98, 105–
    106, 125, 130, 141, 163, 165, 183, 185–86,
    191, 212, 220–21
Gladenkov, A.Y. 47
Glaucomys spp., 175, 183, 212
Glaucomys sabrinus, 186
Glaucomys volans, 186
Glenn County, 235
glory, morning, 92
Glossopteridales, 77
Glyptodontidae, 171
Glyptostrobus spp., 69
Gnatcatcher, California, 161
gobies, 47, 55–56
Gobiidae, 47, 55, 120, 218
Godt, M.-J., 74
Goebel, M., 143, 214
Gold Lake, 43
Gompert, Z., 106, 214, 220
Gondwana, 28, 77
Good, D., 216
Goodman, D., 131
gopher, pocket, 13, 58, 190
gopher snake, 60
gopher tortoise, 156
Gopherus sp., 156–58, 255
Gopherus agassizii, 156, 158
Gopherus berlandieri, 156
Gopherus flavomarginatus, 156
Gopherus morafkai, 157
Gopherus polyphemus, 156
Gordon, S., 95
gorge, Pit River, 62
Gouday, C., 8
Govean, F., 59
government agencies, 159, 235, 243
GPS, 114
gradient(s), 39, 55, 84, 118, 240
Graham, A., 25, 53, 79, 92, 212

graminoids, 90
granite, Santa Lucia, 251
Granite Mountains, 187
granitic terranes, 9–11, 213, 215, 219
granitic soils 84,
grass, Colusa, 95
grasses, 41, 49, 246
grassland(s), 41–42, 49, 52, 92, 94–95,
    139, 149, 177, 210, 235–36, 241–42,
    252–53
Graves, S., 92, 241
grazers, 41, 49, 53, 246
great coppers, 106
Great Plains, 106, 152*fig*
grebelike, 33
Grebes, 54
green sturgeon, 252
Greene's tuctoria, 95
Greenland, 39, 159
Greya politella, 109–10, 214
Griffin, J., 62, 80, 84
Griffin, G., 254
Grimaldi, D., 24
Grinnell, J., 6, 58, 190
Griswold, T., 252
Grivet, D., 80, 84
grizzly bear, 185
Grizzly Island, 187
groundwater, 246, 255
Grouse, 160
Grouse, Sooty, 167
Grylloblatta spp., 104, 108, 214, 220
Grylloblatta chandleri, 108
Grylloblatta gurneyi, 108
Grylloblatta marmoreus, 104
Grylloblatta oregonensis, 104
Grylloblatta siskiyouensis, 104
Grylloblatta marmoreus, 104
Grylloblattidae, 104
G_{ST}, 17, 80, 168
Guadalupe cypress, 69
Gualala River, 123, 125, 216–17
Guatemala, 92
Gugger, P., 14, 19, 40, 73–74
Gulf of California, 12, 58*map*, 120–21, 132,
    153–56, 200, 204*map*, 226–27
Gulf of Mexico, 201
Gull(s), 60, 167
Gunnerales, 77
guppy, 242
Gymnogyps amplus, 169
Gymnogyps californianus, 169
Gymnophiona, 135

Gymnospermopsida, 31
Gyre, Alaskan, 119*map*,
Gyre, North Pacific, 118, 119*map*, 202

Habitat Conservation Planning Branch
    (HCPB), 244
Hadrotes crassus, 110
Hafner, D., 39, 44, 105, 190, 195, 212
Hagstrum, J., 30
Haig, S., 168
hake, 129
halibut, 132
Haller, R., 7
halophytic, 4
Hamrick, J., 68, 71, 74
hardhead, 125
Hardiman, N., 237
hardwood, 75, 235
Hardy-Weinberg equilibrium, 17
Harlin-Cognato, A., 201, 229
Harper, D., 38
Harris, J., 180
Harrison, S., 7, 138
Harvard University, 105
Harvey, H., 34
hatchery, fish, 123, 250
haulouts, 204
Hawaii, 4, 233
Hayhoe, K., 243
headwater(s), 109, 123, 221
heather, 189
hedgerows, 246
Hedin, M., 221
Helenes, J., 155, 228
Helm, R., 199
Hemizonia clementina, 111
hemlock, western, 53
Hennig, W. 13
herbivores, 22, 38, 154, 241
Herman, A., 34
herpetofauna, 143, 213, 216
Hesperia comma, 107
Hesperiidae, 107
Hesperocyparis forbesii, 69
Hesperocyparis guadalupensis, 69
Hesperocyparis macrocarpa, 59,
    69, 98
Hesperornis, 33, 159
heterogeneity
heterogeneity, habitat, 4, 30, 41, 52, 91, 104,
    109, 236
heterogeneity, rate, 179
Heteromyidae, 175, 190–192

Heterotrissocladius sp., 109
Hewitt, G., 13, 58, 162, 170, 213, 231
Hickerson, M., 14, 18
Hickey, L., 34
highways, 239, 247
Hilaria sp., 98
Hileman, L., 82
Hillebrand, H., 84
Hills, Chino, 113,
Hills, Panache, 117
Hills, Puente, 113
Hills, Red, 125, 216
Hilton, R., 30–31, 47, 149
Hippidion sp., 53
Hippopotamus sp., 200
Histeridae, 110
Hoelzel, A., 199
Hoffman, J., 21,
Hoffman, P., 186, 199
Hohmann, S., 82–83
holarctic, 59, 105, 107, 160, 165,
    182, 211
Holland, V., 253
Hollibaugh, J., 51
Hollingsworth, B., 154
Holman, J., 135, 147
Holocene Epoch, 9, 37–38, 56, 61–63, 71, 73,
    84, 93, 98, 105, 107–8, 114, 120, 164, 189–
    90, 215, 225
Holway, D., 241
homogenization, biotic, 71, 237, 241
homoploid species, 106
Homotherium serum, 180
Hood, W., 254
Hooper, D., 241
Hopkins, D., 182, 211
Horning, M., 199
horses, 49, 53, 177
horsetails, 25
Hosselkus formation, 31, 117, 209
hotspot(s), biodiversity, 5, 162, 194*map*,
    203*fig*, 204*map*, 213, 219, 222, 249
Hovanitz, W., 105
Howard, R., 9, 48, 60, 219
Howell, J., 70, 90
Howellia aquatilis, 91
Hudson, R., 14
Huelsenbeck, J., 16
Huenneke, L., 254
Hughes, G., 176
human disturbance, 121, 184, 198, 245
Humboldt County, 93
Hummingbirds, 159–60

Humpback whale, 200, 204
Hunter, R., 251
HVRI, 199
Hybodontidae, 117
hybridization(s), 6, 13, 62, 70, 72, 88–89, 93,
    97, 99, 106, 125, 137, 140–41, 162, 167–68,
    187, 213–15, 217, 219, 222, 230, 233,
    237–38
Hydrictis sp., 197
Hydrodamalis sp., 47
Hydrodamalis gigas, 202
Hydromantes brunus, 137–38, 145, 219
Hydromantes platycephalus, 109, 137–38,
    219–221
Hydromantes shastae, 145
Hylidae, 141, 143
Hymenoptera, 24
Hypocaccus lucidulus, 110
Hypomesus transpacificus, 252
Hypsiglena torquata, 156
Hypsopsetta guttulata, 132
Hyracotherium sp., 49
Hystricomorpha, 175

Icaricia, 105
icecrawlers, 108
ice sheets, 41, 53, 58, 177, 183
Ichthyornis, 33, 159
Ichthyosauria, 30
Ickert-Bond, S., 78
Ictaluridae, 117
Icteridae, 159, 163
icthyofauna, 121, 134, 210
Idaho, 22, 52, 74, 121, 143, 179, 185, 191
igneous rock, 11, 28
iguana, desert, 158
Iguania clade, 147
Iguanidae, 154
Iguaniidae, 147
Illinoian glaciation, 131, 182, 211
inbreeding, 17, 74, 95, 239
incense cedar, 68
Incredible Teratorn, 60
incursion(s), marine, 219, 239
indigenous, 38, 182
Indo-Pacific populations, 129
Ingersoll, R., 44, 118, 229
Ingles, L., 7, 12, 38
inner Coast Ranges 30, 153, 187, 215–16,
    218
Insecta, 25, 253
insect(s), 5, 24, 103–4, 114–15, 145, 209,
    242, 247

institutions, role of, 246
insular, 90, 154, 220, 239
Interagency, 252
interglacial, 56–57, 69, 94, 107, 201, 212
interlacustral, 190
Phenacomys intermedius, 189–190
intermountain, 90, 150
International Union for the Conservation of
    Nature (IUCN), 205, 240, 248, 250
interocean, 131
interpopulational, 255
intraspecific, 72, 88, 164, 219, 223
introduction(s) 3, 97, 123, 234, 238, 244
introgression, 238
intron(s), 17, 70, 73, 89, 114, 186
inundation, 57, 94, 111, 191, 227, 230, 243
invasive species, 7, 111, 145, 233–34, 236–39,
    241, 244, 246–47, 249–53, 255–56
invertebrate(s), 5, 24, 51, 131, 229–30, 255
Inyo County, 23, 88
Inyo Mountains, 11, 44, 211
Inyo Valley, 195
Irvingtonian age, 171, 175
Island Scrub-Jay, 162
Island, Anacapa, 11
Island, Grizzly, 187
Island, Guadalupe, 69
Island, San Clemente, 11, 55, 130
Island, San Miguel, 11, 111
Island, San Nicholas, 55, 111, 130
Island, Santa Barbara, 11, 130, 239
Island, Santa Catalina, 11, 130, 167, 186
Island, Santa Cruz, 11, 111, 150
Island, Santa Rosa, 11, 57,111
island tarweed, 111
Island, Vancouver, 119*map*
Islands, Aleutian, 131, 198, 229
Islands, Channel, 5, 11, 44, 52, 57, 110, 111,
    130–31, 154, 199, 211, 216, 226, 229–30,
    236, 252
Islands, Farallon, 129
isozyme, 72, 74
Isthmus of Panama, 50, 55, 210
Istiophoridae, 128
ITS, 17, 89
IUCNredlist, 200

Jaarola, M., 186
Jackman, T., 137
Jacobs, D., 11, 47, 50–52, 57, 111, 129–30,
    160–61, 200, 210
Jacobson, G., 85, 183, 212
Jaeger, E., 11, 195

Jagels, R., 34
jaguar, 180
jaguarundi, American, 180
Jahner, J., 241
Janis, C., 39, 49
Janzen, F., 150
Japan, 133, 197–99, 202
Jarvis, K., 108
Jay, Western Scrub, 162
Jeffery pine, 75
jellyfish, 25
Jepson, W., 6
Jerusalem cricket, 113
Jezkova, T., 192, 195
Jockusch, E., 222, 224
Johnson, J., 160–61
Johnson, N., 161
Johnson, W., 180
Jones, W., 123
Jorgensen, S., 71
Joshua tree, 97
Joshua tree National Park, 255
Juan de Fuca Strait, 119
Juglandaceae, 53
Juglans sp., 61
Juncaceae, 90
Junco myemalis, 163
Junco phaeonotus, 163, 170
Junco, Yellow-eyed, 163
Juniperus sp., 41, 96, 215
Juniperus californica, 98
Juniperus osteosperma, 96
Jurassic Period, 25, 30–31, 33, 38, 68,
    77, 135

Kadereit, J., 80, 82
Kafton, D., 69
kangaroo mouse, 48, 190
kangaroo rat, 48, 190
Karlstrom, K., 21–22, 34
Käss, E., 83
Katibah, E., 253
Katz, J., 249
Kaweah River, 110
Keator, G., 80
Keeler-Wolfe, T., 254
Keir, K., 94, 220
Kellogg, R., 48
kelp forests/beds, 47, 129
Kenagy, G., 190, 227
Kendall, A., 129
Kendrick, K., 55
Kennedy, M., 33

Kern County, 9, 110, 121, 186, 221, 235
Kern River, 56, 110, 121, 123, 221
Kern River rainbow trout, 124*map*
Khasa, D., 237
Kiefer, C., 88, 214
Kimsey, L., 103, 115
Kimura, M., 131, 163
Kings River, 123, 180, 188, 193, 221
kingsnake, California, 60,
kingsnake, California mountain, 60,
    150–51
Kirby, M., 50
Kishore, V., 95
kit fox, San Joaquin, 182
Kites, 169
Klamath blue chub, 34
Klamath Mountains, 27–28, 30–31, 32*map*,
    37, 43, 45*map*, 46*map*, 62, 73–74, 75*map*,
    86*fig*, 89, 93, 96, 100*map*, 104, 110,
    122*map*, 124*map*, 127*map*, 164, 185,
    194*map*
Klamath-Siskiyou Ranges, 5–6, 8–10, 25, 34,
    62, 68, 71, 76, 88–90, 92–93, 99, 104,
    106–7, 115, 135–37, 143, 144*map*, 145,
    148*map*, 168, 181, 213, 214*table*, 215, 230,
    235, 248–49,
Klamath River, 85*fig*, 89–90, 125, 137, 249
Kleppe, J., 109
Klicka, J., 165, 170
Kluge, A., 147
Knowles, L., 13, 15
Koch, M., 88, 214*table*
Kochia spp., 82
Koepfli, K.-P., 178–79, 197
Kornev, S., 198
Korveva, S., 198
Korth, W., 48
Kraft, N., 99, 216, 228, 235, 244
Kramer, A., 245
Kreissman, B., 4
Kruckeberg, A., 7, 84
Kuchta, S., 10, 136–37, 141–42, 214,
    217, 220
Kurtén, B., 183

La Brea Tar Pits, 59–61, 98, 143, 169, 180, 184
La Porte, 40
Labridae, 55
lacustrine, 10
Lagomorpha, 175
lagomorphs, 39
Lahontan Trough, 56, 120, 190, 227
Lake Cahuilla, 191

Lamb, T., 154, 157
Lamiaceae, 98
lamp shells, 23, 25
Lampetra tridentatus, 131, 249
lamprey, 131
Lampropeltis cf. getula, 60
Lampropeltis zonata, 150–51, 158, 195, 219–20, 224, 226
landforms, 5, 8, 11–12, 96
Landini, W., 55
landmass(es), 11, 25, 50, 111, 216
LaPointe, F., 230–231
La Porte, 40
larch, 40
Laridae, 60, 167
Larix sp., 40
Larrea tridentata, 97, 228
Larus californicus, 167
Lassen County, 9
Lassen Volcanic National Park, 61, 137, 150
Lasthenia sp., 94
Latta, R., 72
Lauraceae, 40, 48
Laurasia, 28
laurel, bay, 48
Laurentia, 22, 24
Laurentide, 183
lava, 9, 28
Lavinia exilicauda, 125, 217, 228
Lavinia symmetricus, 123, 125, 216–17, 228
lawns, 244
laws, 243, 250
Lawver, L., 39
LEA-like, 71
Leaché, A., 151, 155, 157, 176, 224, 228, 253
leadership, conservation, 247
leatherback turtles, 31, 149
leatherwood, 92
Leavitt, D., 153
LeBeouf, B., 7, 199
LeConte's Thrasher, 161, 164
Ledig, T., 57, 73–74, 214
Lee, S., 58, 73
LEED, 247
legislation, 243, 250
legless lizard, 60
Lehman, P., 252
Lepidoptera, 24, 105, 241
Leporidae, 175
Lesica, P., 240
Lessios, H., 125
Levine, J., 238

Lewis, H., 7
LGM (Last Glacial Maximum), 11, 56, 58–59, 62, 71, 73–74, 94, 96–97, 108, 111, 114, 163, 170, 176–77, 184, 192, 195, 212, 231
Lidicker, W., 197–98, 202
Lijtmaer, D., 160
lilac, 48
Lilaeopsis occidentalis, 94
Liliaceae, 85
lime, 26–27
limestone, 31, 28, 117
Limnanthaceae, 95
Limnanthes floccosa ssp. *californica*, 95
Lincoln, California, 40
Lind, A., 138, 216–17, 221
Lindberg, D. 55
Lindsay, S., 188
Linepithema humile, 241
Linnean, 200
lion, American, 180
Lipps, J., 55
Liquidambar sp. 79, 82
Liston, A., 68, 70–71, 82, 84, 97, 224
Little Willow Lake, 61
livestock, 246
Livezey, B., 165
lizard, 147, 154, 158
lizard, brush, 158
lizard, coast horned, 157
lizard, desert night, 153
lizard, fence, 60
lizard, flat-tailed horned, 157
lizard, legless, 60
lizard, Mojave fringe-toed, 255
lizard, side-blotched, 60
lizard, southern alligator, 60
lizard, spiny, 154
lizard, tree, 158
lizard, western fence, 150
lizards, 147, 154, 158
Loarie, S., 241–44
Lodgepole pine, 75
Lolita, 105
longfin smelt, 252
long-nosed snake, 61, 160
Longrich, N., 33, 159
Lonicera sp., 82
Lontra canadensis, 179
Loons, 54
Lophioprason clade, *Allium*, 85
Los Angeles/Los Angeles Basin, 11, 30, 44, 47, 52, 58, 61, 75, 100*map*, 118, 120, 122*map*, 124*map*, 127*map*, 129–30, 133, 142*map*,

144*map*, 148*map*, 151, 157, 191, 194*map*, 211, 219, 223–24, 229–30, 234
Los Padres National Forest, 254
Louisiana, 152*fig*
Lovette, I., 163
Lovich, J., 224
Loxolithax stocktoni, 47
Luikart, G., 49
lupine, 83
Lupinus sp., 83–84
Lupinus angustfiolius, 83
Lupinus luteus, 83
Lupinus mutabilis, 83
Lupinus nanus, 83
Lupinus polyphyllus, 83
Lutra sp., 179
Lycaeides idas, 105–6, 214, 220
Lycaeides melissa samuelis, 105
Lycaena dione, 106
Lycaena editha, 106
Lycaena xanthiodes, 106
Lycaenidae, 105–6
Lycanidae, 105–6
Lycophyta, 25, 209
Lynx rufus, 180, 239
Lyon, G., 199
lyre snakes, 225

Macdonald, D., 184
Macey, J., 136, 218
Machairodontinae, 50
mackerel, jack, 128
Mackintosh, N., 201
macrofossils, 97
Maddison, W., 15
Madrean-Tethyan hypothesis, 77–78, 80, 82
Madro-Tertiary flora, 78–79, 101
madrone, 40
Magnolia sp., 40
Magpie, Yellow-billed, 162
mahogany, mountain, 48
Mahoney, M., 136–37, 214*table*
Malamud-Roama, F., 63
Maldonado, J., 145, 187, 193, 218, 224, 229
Mallard, 165
Malus sp., 237
Malvaceae, 89
Malvidae, 77
Mammalia, 38, 49, 210
mammalogist, 7

mammal(s), 4–5, 12, 16, 33, 38–39, 42, 44, 49, 53, 59, 61, 94, 118, 145, 171, 175, 180, 182, 186, 193–94, 197–99, 202–4, 209–11, 217, 220, 230
mammals, placental, 38
mammoth(s), 38, 59, 175
Mammut sp., 38
Mammuthus sp., 38, 61
Mammuthus primigenius, 59,
Mammutidae, 38, 175
management, conservation, 200, 234–35, 239, 245–46, 248–51, 256
manatees, 47
Mancalla sp., 53
Manchester, S., 79, 82
Mann, K., 118
Manos, P., 80
manzanita, 48, 53
maple, 78
Marble Mountains, 8, 23, 106
Margeriella sp., 31
Marin, B., 24
Marin County, 24, 47, 150
Marincovich, L., 47
Marine Life Protection Act (MLPA), 256
Marine Protected Area, 203*fig*
Marlow, J., 52
marsh(es), 10, 126, 186, 218, 252
marsupials, 174
marten, American 161, 179
Martes americana, 179, 183, 212
Martes americana humboldtensis, 179
Martes americana sierrae, 179
Martes pennanti, 179–80, 193, 211, 221
Martin, A., 90, 107
Martín-Bravo, 83
Martínez-Solano, I., 140
Marty, J., 246
Mascarello, J., 58
Mason's Lilaeopsis, 94
Masticophis flagellum, 60
Masticophis lateralis, 60
mastodon(s), 38, 49, 175
Mastroguiseppe, J., 73
Mastroguiseppe, R., 73
*mat*K, 69–71, 89
Matocq, M., 187–89, 193, 217, 220, 222, 228
matrifocal, 201
matrilineal, 176, 201
Maurandya sp., 82
MaxEnt, 176, 195

Maxwell, A., 70
May-Collado, L., 200
Mayr, E., 160
McClenaghan, L., 69
McClintock, E., 7
McCloud arc, 27*map*
McCloud River, 123, 124*map*
McCollum, F., 197–98, 202
McCracken, G., 178
McKay, S., 126
McKittrick, 60
McMahan, L., 247
McNeal, D., 85
Mead, J., 59, 143
Medicine Lake, 9
Medieval Warm Period, 62
Mediterranean climate, 43, 47–48, 50, 61, 80, 82, 210, 229–30, 241
Mediterranean, species from the, 78, 83, 97, 125
Megachiroptera, 177
megafauna, 59, 176, 212
Megalonychidae, 171, 175
Megantereon sp., 180
Megaptera novaeangliae, 200–1
Megatheriidae, 171, 175
Melissa blue, 105
Mellish, J.-A.-E., 199
Mendocino County 138–39, 216,
Mendocino Frac Zone 45–46
Menispermaceae, 40
Merced River, 89, 137, 150–51
Merlucciidae, 129
Merluccius productus, 129
mermaid, false, 95
Merriam, C., 6
Merycodus sp., 49
Mesoamerica, 163
Mesohippus sp., 42
Mesozoic Era, 4, 7–9, 25, 28, 31, 33–34, 38, 68, 74, 91, 101, 115, 149, 209–10, 213, 215–16, 219
Messinger, O., 252
metamorphic, 8, 10–11
metamorphosed, 9
Metasequoia, 69
Metasequoia glyptostroboides, 34, 67
Metcalf, A., 112, 141–142, 219, 224–25
Mexican free-tailed bat, 178
Mexico, 34, 58*map*, 79, 103, 106, 110, 119*map*,128, 154–58, 162–63, 168, 170, 178, 188, 190–91, 201, 209, 225–28
Meyers, S., 97
mice, 49

mice, deer, 58
mice, pocket, 48, 58
Michel, F., 104
Microchiroptera, 178
microclimates, 251
Microdipodops megacephalus, 190
microevolutionary, 35
microhabitat, 107
microsatellite(s), 15, 17, 89, 93–95, 109, 123, 132, 139, 141, 165, 168–69, 176–77, 181, 183–84, 199, 201–2, 231
Microtus, 185–86, 188, 255
Microtus californicus, 186, 188
Microtus californicus mohavensis, 255
Microtus c. scirpensis, 255
Microtus longicaudus 185
middens, Native American, 133
middens, packrat, 72, 97
midge, nonbiting, 109
migrational corridor, 5, 38, 50, 59, 78, 80, 82–84, 99, 112, 115, 193, 204*map*, 210, 227, 239, 244, 254
migrational events, 14, 19, 38, 41, 50, 54, 78, 115, 180
migrational patterns, 5, 74, 161–62, 168, 184, 188, 215, 229, 242
migratory species, 5, 7, 54, 131, 160–63, 167, 170, 175, 178, 233
Milá, B., 163
Milankovitch Cycles, 34–35, 52
Millar, C., 40–41, 53, 70–71
Miller, 51, 70, 126, 191
Milne, R., 82–83
Milner, A., 126
mimics, 110
Mimidae, 161
minks, 197
minnow(s), 44, 123
minnow, Sacramento pike, 123, 125
Miocene Epoch, 6, 10–12, 34–35, 37–39, 41–50, 53, 56, 59, 68, 73–74, 78–80, 82–85, 97, 99, 111, 117–18, 120–21, 123, 126, 129–30, 135, 140, 145, 149, 151, 154–58, 160, 169, 171, 175–76, 179–80, 195, 198, 200, 210–11, 216–19, 225–29
Miogeocline, 26
Miohierax stocki, 48
Mirounga angustirostris, 198–99
Mississippi River Valley, 152*fig*
mitigation, 244
mitochondrial markers, 13, 109, 112, 121, 161, 163, 177, 201
Mitton, J., 72

Modesto, California 147
Modoc Plateau, 5, 9, 34, 75*map*, 100*map*, 122*map*, 124*map*, 127*map*, 144*map*, 148*map*, 181, 194*map*, 213
Modoc sucker, 125
Mohr, J., 93
Mojave Desert, 5–6, 8, 11–12, 23, 25, 30, 44, 49, 52, 62–63, 68, 75, 96–98, 115, 122, 124, 127*map*, 144*map*, 148*map*, 149, 151, 154, 157–58, 185, 187–88, 191–92, 194*map*, 195, 222–23, 224*table*, 226, 228, 230, 235, 237, 254–55
Mojave fringe-toed lizard, 255
Mojave ground squirrel, 192
Mojave National Preserve, 255
Mojave ragwort, 97
Mojave River, 192
Mojave River vole, 255
Mojavia, 155, 226
Moldenke, A., 104, 214
mole salamanders, 44
mollusc(s), 5, 23, 28, 57
Mollusca, 23, 25, 33
Molossidae, 178
Mönkkönen, M., 185
Mono Lake, 62, 68
Mono Lake kangaroo mouse, 190
monophagous, 110
monophyletic, 15, 71, 79, 85, 105, 108, 111, 147, 165, 177–78, 190, 198
monophyly, 138, 176, 192, 216
Montalvo, A., 241
Montana, 91, 143, 179, 185
montane, 25, 53, 90, 168, 182, 186, 215, 219, 242
Monte Carlo 16
Monte de Oro, 30
Monterey Bay, 56, 119*map*, 120, 125, 129, 133, 136, 204*map*, 217–19
Monterey cypress, 59, 69
Monterey Seaway, 10, 44, 140, 153, 158, 176, 188, 193, 195, 201, 216–19, 230
Monterey Formation, 47
Monterey pine, 59
Montiaceae, 90, 98
Mooney, H., 238
Moore, G., 33
Moore, M., 77
Moore, D., 83
Moraceae, 40
Moratto, M., 184
Moreno Formation, 31

Morgan, D., 4
Moritz, C., 18, 136, 218, 243
Morone saxatilis, 131
Moronidae, 131
Morro Bay, 44
Mortiz, C., 7, 200
Morus vagabundus, 47
mosaic, 6, 22, 61, 94, 213, 252
moth(s), 24, 239
moths, yucca, 98, 114
Mount Eddy, 88, 119, 200
Mount Lassen, 9, 71, 90, 219
Mount Shasta, 9, 75*map*, 88, 90, 100*map*, 122*map*, 124*map*, 127*map*, 144*map*, 148*map*, 194*map*, 219
Mount Whitney, 4, 75*map*, 100*map*, 122*map*, 124*map*, 127*map*, 144*map*, 148*map*, 194*map*
Mountain Chickadee, 162
Moyle, P., 4, 7, 118, 121, 125–26, 247, 249, 252, 256
MrBayes, 16, 86
mtDNA, 13, 16, 59, 72–73, 104–6, 108–14, 121, 125–26, 128–33, 135, 137–41, 143, 149–51, 154–58, 161–65, 167–69, 176, 179–82, 185–92, 195, 199–202, 217, 221, 225, 228
mud, 26–27, 30
Muhlfeld, C., 237–238
Muhs, D., 57
Mulcahy, D., 155–157, 224, 227–228
Mulch, A., 9
mule deer, 175–176
multicellularity, 22
multitaxonomic, 13
Murchey, B., 30
Murphy, R., 120, 157, 188, 193, 219
Murray, B., 54, 187
Murres, 160
Museum of Vertebrate Zoology, 6
Museum of Paleontology, University of California, 60
muskoxen, 175
Mustela sp., 179, 197
Mustela nigripes, 179
mustelid, 202
Mustelidae, 175, 178–79, 197, 210
Myers, N., 5, 234
Mylodontidae, 175
Mylopharodon conocephalus, 125, 228
Myodes rutilus, 183, 212
myotis, California, 178
Myotis californicus, 178

Myotis ciliolabrum, 178
Myotis leibii, 178
Myrmecophagidae, 171
Mysticeti, 47, 200

Nabokov, V., 105
NADH, 16
Naiman, R., 128, 228, 254
Nanhsiungchelyidae, 31, 149
Napa County, 92
Natal, 123
natal philopatry, 126, 163–64,
natal rookeries, 199
natives, 83, 238, 240–41
Natural Community Conservation Planning
    Act (CCPA), 243–44
Nature Conservancy, The, 255
Navarro River, 125, 216
ND1, 16, 114, 131, 136, 138, 140, 150–51, 153,
    156–57, 162, 165, 199
ndhB, 73
ndhE, 73
ndhF, 92
ndhH, 73
ndhI, 73
Nearctic, 157, 160, 182, 211
nearshore, 47, 57, 118, 129, 132, 256
Nebria, 107–9, 221
Nebria ovipennis, 108
Nebria spatulata, 108
Negrini, R., 63
Nei, M., 15, 17–18, 94
neighborjoining, 131
neoendemic(s), 6, 68, 84, 99, 107, 109, 194,
    216, 235, 228, 230
Neogene Period, 37, 43–44, 49, 52, 54,
    57, 78, 80, 83, 88, 99, 147, 193,
    210
Neognathae, 33
Neoproterozoic, 22
Neoptera, 24
Neornithes, 33
Neostapfia colusana, 95
Neotoma spp., 58, 72, 97
Neotoma fuscipes, 187–88, 189*fig*, 193, 217,
    221, 222*table*, 228
Neotropical, 110, 157, 160, 225
Neotropics, 160
Neovison, 179, 197
NEPA, 243
Nettel, A., 93, 214, 248
Neuwald, J., 186
New World Vultures, 53, 159, 169

New Zealand, 147, 203*map*
Nevada, 22, 28, 227
Nevadan, 43, 53, 109, 220
newt, California, 44, 135–36
newt, rough-skinned, 217
NGOs, 243
Nguyen, N., 85
Nicholas, 11, 55, 111, 130
Niebling, C. 72
Nielson, M., 214
Nigro, D., 161
Nilsson, T., 175
non-migratory, 161, 163, 169–70, 178
non-native, 241, 247
Norris, K., 245
North Atlantic Land Bridge 77–80, 209
North Atlantic right whale, 203*fig*
North Pacific Gyre, 118, 119*map*,
Noss, R., 248
Notholithocarpus densiflorus, 93, 214,
    248
Nothrotheriidae, 171, 175
Notophthalmus sp., 135
Novacek, R., 171
Nowak, R., 183–184
nucDNA, 17, 73, 83, 93, 97, 109, 112, 121, 140,
    151, 153, 155–58, 199
Nuna, 21
Nutcracker, Clark's 71
Nuthatch, Pygmy, 165
nutmeg, 69–70
Nyctoporis, 110, 112, 222, 224
Nyctoporis carinata, 110, 112, 222, 224
Nymphalidae, 104–5

oak woodlands, 53, 94, 139, 177,
    252–53
oak, live, 48, 53
oak, tan, 93
oak titmouse, 162
Obando, J., 50
O'Brien, S., 180
Ocean, Pacific, 5, 58*map*, 75*map*, 81*map*,
    100*map*, 118, 119*map*, 122*map*, 124*map*,
    126, 127*map*, 131–32, 140, 142*map*,
    144*map*, 148*map*, 153, 193,194*map*, 202,
    204*map*, 210, 230
Ocean, Panthalassa, 26–27*maps*
Ochotona spp., 172*fig*
Ochotona princeps, 105, 190, 212
Ochotonidae, 105, 172
Odocoileus sp., 175–76
Odocoileus hemionus, 175–76

Odocoileus hemionus columbianus, 176
O'Corry-Crowe, G., 201, 229
Odontoceti, 200
Odontophoridae, 159, 167
Oeneis chryxus, 104–5, 212
Oeneis chryxus ivallda, 105
Oeneis chryxus stanislaus, 105
offroad, 255
offshore, 11, 47, 129, 256
Okhotsk Sea, 202
Olcott Lake, 95
Olden, J., 237, 241
Oligocene, 10, 37–42, 44, 49, 68, 78–80, 117,
    149, 157, 180, 198, 210, 225
Oline, D., 73
Olivares, 129
Oliver, J., 106
Olsen, A., 132
Olson, S., 48, 255
Omland, K., 165, 227
Onagraceae, 90
one-toed grazing horse, 49
Oncorhynchus spp., 50–51, 120–21, 126, 221,
    237, 249, 252, 256
Oncorhynchus clarkii, 56, 120–21,
    126, 227
Oncorhynchus kisutch, 256
Oncorhynchus mykiss, 56, 121, 123, 126, 221,
    237, 249, 252, 256
Oncorhynchus mykiss aquabonita, 121, 221
Oncorhynchus rastrosus, 50, 51*fig*
Oncorhynchus tshawytscha, 56, 240, 252,
    256
Onychonycteris finneyi, 178
ophiolite(s), 8, 31, 91
opossums, 171
opportunistic, 125
Ordovician, 24, 28
Oregon, 5, 8, 51, 68, 71, 74, 89, 93–94, 106–8,
    110, 119, 129, 132–33, 136–39, 143, 151,
    162, 164, 167, 169, 179, 184, 186–91, 198,
    213, 242, 248–49, 251
Oregonian, 57, 118, 133
Oreobliton, 82
Oreodonta, 42
Orleans, California, 249
Orme, A., 68
ornamentals, 247
ornate shrew, 186
Ornduff, R., 7
ornithologist, 7
Ornithurae, 33
Orobanchaceae, 98

orogeny, 73, 139, 145
Oromerycidae, 42
Orr, R., 12, 199
Osprey, 169
Osteichthyes, 117
Otariidae, 48, 198–199
otter(s) 197–98, 203
otter, spotted-neck, 197
otters, New World river, 179, 197
otters, sea, 197–98
outbreeding, 240
outcrop(s), 92–93, 154
overfishing, 199, 134
overgrazing, 244, 246, 251, 255
overharvesting, 244
overhunting, 199
Ovibos moschatus, 175–176
Ovis canadensis, 49
Owen, J., 186
Owen, L., 59
Owens Valley, 128, 194*map*, 219, 228
Owls, 160
Owl, Great Gray, 168
Owl, Spotted, 168, 248
Owl, Sinclair, 60

Pacific Ocean, 5, 58*map*, 75*map*, 81*map*,
    100*map*, 118, 119*map*, 122*map*, 124*map*,
    126, 127*map*, 131–32, 140, 142*map*,
    144*map*, 148*map*, 153, 193,194*map*, 202,
    204*map*, 210, 230
Pacific sardine, 128
Paiute cutthroat trout, 124*map*
Pajaro River, 10
Palacios, D., 256
Palaeognathae, 33
Palearctic, 160
paleobotanical, 92
Paleocene, 34, 37–39, 79, 117, 135, 143, 159,
    209–10, 213, 219
paleoclimatic, 14, 105, 212
paleoendemics, 6, 68, 84
paleofauna, 23
paleofloras, 43
Paleogene Period, 9, 30–31, 33, 37–40, 42,
    54, 67, 69–71, 74, 78–79, 101, 103, 117, 147,
    159, 161, 169, 210, 213
paleomustelids, 179, 210
paleontology, 7, 13, 60
Paleoptera, 24
Paleozoic Era, 5–9, 22, 24–25, 28, 30, 115,
    209, 213, 250
Palos Verdes peninsula, 55

Palumbi, S., 15, 132, 201
palynological, 61–63, 72, 80
Pampatheriidae, 171
Panache Hills, 117
Panamint Mountains, 88
Pandionidae, 169
Pangaea, 25, 28, 30, 74, 135, 213
Panthalassa Ocean, 26*map*, 27
Panthera onca, 180
Pantheridae, 175
pantherines, 50
Pantropical, 160
Papaver sp., 82
Papaveraceae, 82
Papilionidae, 104, 110
Paralichthyidae, 132
Paralichthys californicus, 132
paraphyletic, 33, 111, 147, 138, 168
Parham, J., 31, 149
Paridae, 161–162
Parmesan, C. 242
Parnassius behrii, 104–5
Parnassius phoebus, 104–5, 212
Parnassius smintheus, 104–5
parthenogenetic, 111–12
Partington Point, 204*map*
Parulidae, 159–160, 162
Parus spp., 161, 211
Passeriformes, 54, 161, 168
Passerine, 161
pathogens, 199
Patterson, T., 85, 88, 92, 214map
Patton, J., 7, 16, 191, 195, 221–22, 224, 254
PAUP, 154, 230
Pavlik, B., 253
Peabody, F., 222
Peale's dolphin, 203*fig*
Pearse, D., 123
Pease, K., 176–77
Pecon-Slattery, J., 180
pelagic species, 128–29, 131–32, 202
Pelecaniformes, 160
Pelicans, 60, 160
Pelini, S., 242–43
Pellmyr, O., 98, 112, 223
Peninsular Ranges, 5, 11, 30, 43, 69, 71, 96–97, 122*map*, 124*map*, 127*map*, 144*map*, 148*map*, 152–53, 162, 164, 181, 193, 194*map*, 195, 216, 220, 224, 226, 230, 255
Pentapetalae, 77
perch, pirate, 118
perennials, 89–90

peripatric, 6, 16, 137–38, 141
Perissodactyla, 41–2, 175, 177
Permian, 24–25, 27, 115, 135
Perognathinae, 48, 58
Peromyscus, 58, 195, 229
Peromyscus californicus, 195, 229
Perrine, J., 183, 212
Persea sp., 48
pesticides, 250
Peterson, D., 62
Petren, K.,154
Petromyzontidae, 131
Péwé, T., 182, 211
Phalacrocoracidae, 48, 53, 60
Phasianidae, 160, 167
Phelan Peak, 192, 223
Φ_{CT}, 107
Φ_{SC}, 107
Φ_{ST}, 18, 91, 113–14, 132–33, 164, 168–69, 201–2
Phillips, C., 195, 199–200
Phillipsen, I., 112, 141–42, 219, 224–25
philopatric, 162, 190, 200
philopatry, 123, 126, 161, 163, 202, 212
Phipps, F., 91
Phoca vitulina, 201, 229
Phocidae, 48, 198
Phocoena phocoena, 201–2, 229
Phocoenidae, 47, 201–2
Phocoenoides dalli, 202
Phrynosoma mcallii, 157–58, 227
P. coronatum, 157
Phrynosomatidae, 60, 147, 150, 157–58
Phyllostomidae, 178
phylogeographers, 7
phytophagous, 24
Pica hudsonia, 162, 170, 215
Pica nuttalli, 162, 170, 215
Picea sp., 40, 68, 70, 183
Picea breweriana, 74, 214
Picea engelmannii, 68
Picidae, 160, 163
Picoides albolarvatus albolarvatus, 163–65, 213, 223, 224*table*
Picoides albolarvatus gravirostris, 163–64, 170, 223
Piedmont, 152
Piedras, 204
Pielou, 231
Pieridae, 109

pika, American, 105, 190
pike, Sacramento, 125
Pinaceae, 40, 53, 62, 70, 72
pine, Bishop, 59
pine, bristlecone, 53, 62, 68, 73–74
pine, Monterey, 59, 98
pine, pinyon, 48, 75
pine, ponderosa, 40, 68, 71–72, 165
pine, sugar, 68, 71, 84, 224
pine, Torrey, 57
Pinnacles, 251
Pinnipedia, 47, 197–98, 202, 204
Pinophyta, 67
Pinus spp., 40–41, 57, 61, 72, 90, 96, 165, 214–15
Pinus albicaulis, 68, 71, 84
Pinus balfouriana, 53, 73, 214
Pinus belgica, 70
Pinus coulteri, 71–72
Pinus flexilis, 68, 74
Pinus jeffreyi, 71–72
Pinus lambertiana, 68, 71, 84, 224
Pinus longaeva, 53, 62, 68, 73–74
Pinus monophylla, 96
Pinus monticola, 53, 68, 84, 90
Pinus muricata, 59, 98
Pinus ponderosa, 40, 68, 71–72, 165
Pinus ponderosa var. *washoensis*, 71–72
Pinus radiata, 59, 98
Pinus sabiniana, 62, 71–72, 98
Pinus torreyana, 57, 71–72
Pinus/Picea, 70
Pinyon pine, 48, 75
pioneering, 6–7
pipefish, 132
pipelines, 238
pipevine swallowtail, 110
pirate perches, 118
Pit River, 123, 125, 216
Pit River sculpin, 125
pit viper, 149
Pitkin, J., 234
Pitterman, J., 34, 41, 68
Pituophis catenifer, 60
Piute Mountains, 194*map*
Piute Pass, 105
Placer County, 40
plains, Great, 106, 152*fig*
plains, southern, 155
planet, 38, 252
plankton, 41, 47, 51, 118, 256

planktonic, 130
Plantaginaceae, 82, 84, 89, 97
Plantago section *Albicans*, 97
Plantago ovata, 97
plantain, 97
Platanaceae, 61, 82
Platanus sp., 61, 82, 237
Modoc Plateau, 5, 9, 26–27, 29, 34, 45–46, 75, 100*map*, 122*map*, 124*map*, 127*map*, 144*map*, 148*map*, 181, 194*map*, 213
Pleistocene Epoch, 6, 8, 11–12, 17, 37–38, 44, 48–53, 55–57, 59–61, 63, 67–69, 71–73, 80, 83–84, 88, 90–91, 93, 96–99, 103, 106–8, 110–11, 113–15, 117–18, 120–21, 126, 129–30, 137, 140–41, 143, 145, 150, 153–58, 160–61, 164–65, 167, 169–71, 175–81, 183–93, 195, 199–201, 210–12, 215, 217–21, 223, 225–27, 229, 231, 239
Plesiosauria, 30
Plethodon sp., 145
Plethodon elongatus, 136–37, 214*table*
Plethodon stormi, 136–37, 214*table*
Plethodontidae, 109, 136, 140, 143
Pleurodonta, 147
Pleuronectes spp., 47
Pleuronectidae, 47, 132
Pliocene Epoch, 9–11, 34, 37, 42–44, 46–56, 58, 68, 70, 73–74, 79–80, 82–83, 85, 96, 103, 106, 108, 111, 113, 118, 120–21, 125–26, 132, 140, 145, 153–58, 160, 164, 171, 175, 179, 188, 191–92, 195, 210–11, 215–16, 218, 220–21, 223, 225–27, 229
Pliohippus sp., 49
Plumas County, 28
pluvial, 107, 126, 190, 192
Poaceae, 90, 95, 98
pocket gophers, 13, 58, 190
Podicipedidae, 54
Podistera nevadensis, 90,
Poecile gambeli, 162, 220, 224
Pogonichthys macrolepidotus, 252
Point Año Nuevo, 204*map*
Point Cabras, 131
Point Conception, 57, 75*map*, 79, 100*map*, 118, 119*map*, 122*map*, 124*map*, 127*map*, 129, 132–33, 144*map*, 148*map*, 194*map*
Point Eugenia, 119*map*
Point Lobos, 248
Point Loma Formation, 33
Point Reyes, 75, 122, 124, 127, 140, 144, 148, 194, 216

Polemoniaceae, 91
Polihronakis, M., 112, 223
Polioptila californica, 161
Polioptilidae, 161
pollen, 40, 61–62, 67, 70, 72, 93, 212
pollinator(s), 93–95, 98, 109, 114, 239, 242, 245–46
pollution, 203, 250–52, 255–56
Polygonaceae, 84, 90
Polyomatinae, 105
Polyommatus sp., 105
polyploidization, 88
polyploidy, 90
Polystichum munitum, 90, 214
ponderosa pine, 40, 68, 71–72, 165
PopGraph, 81
Popp, M., 84
Populus sp., 40, 61, 237
porcupines, 171
Porinchu, D., 109
porpoise, 47
porpoise, Dall's, 202, 204*map*
porpoise, harbor, 201, 204*map*
Posasda, D., 15
postfire, 240
postglacial, 56, 63, 74, 109, 120, 123, 137, 161, 163
postrefugial, 168, 215
Potter, I., 131
Powell, 11, 111, 239
prairie(s), 235, 239, 251
Precambrian, 24–25
precipitation, 4, 19, 39, 44, 48, 56, 63, 230–31
predation, 5, 38–39, 119, , 129, 131, 198–99, 250
preserves, 233
Preston, R., 93
Prince William Sound, 132
Pringle, R., 233
Pritchard, J., 18
Proboscidea, 38, 175
Procellariiformes, 48, 160
Procyonidae, 175
Prodoxidae, 109, 114
Prodoxus weethumpi, 114
Prodoxus sordidus, 114
prokaryotic, 22
pronghorn(s), 49, 175
propagule(s), 5–6, 67, 210, 228, 240
Proposition 84, 250
Prosalirus bitis, 135

protection, 199, 233–34, 242–44, 246, 248–49, 251–52, 255–56
Proterozoic, 21–23, 30
Prothero, D., 169
protist(s), 22, 33
Protitanops sp., 42*map*
protozoa, ameboid, 23
province(s), biogeographic, 3
Province, California Floristic, 5, 13, 89, 216
province(s), faunal, 23
province, Klamath, 34
*psa*C, 73
*psb*A, 92, 98
Pseudacris cadaverina, 141–42, 145, 224–25
Pseudacris regilla, 143
Pseudaelurus sp., 50
Pseudotsuga sp., 40
Pseudotsuga macrocarpa, 68
Pseudotsuga menziesii, 62, 68, 73–74
Psittaciformes, 168
Psittacopasserae, 168
Pteridophyta, 25, 30–31, 33, 209
Pteridospermatophyta, 25, 209
Pterygota, 24
Ptiliidae, 104
Ptychocheilus grandis, 125, 228
Puente Hills, 113
Puffins, 160
Puget Sound, 119*map*
Puijila darwini, 198
Puma concolor, 180–81, 222
Puma yaguaroundi, 180
Punta Banda, Baja California, 131
Punta Eugenia, Baja California, 119*fig*
pupfish(es), 56, 120, 128, 227–28, 255
pups, 132
Pygmy Nuthatch, 165
Pyrus sp., 237

quail, ancestral, 48
Quail, California, 167
Quail(s), New World, 48, 159
quartzite, 85, 220
Quaternary Period, 37, 55, 61, 67, 69, 74, 83, 126, 130, 175, 187–88, 218, 229, 231
Quercus sp., 61–62, 80, 237, 242
Quercus agrifolia, 53, 242–43
Quercus garryana, 242–43
Quercus kelloggii, 242
Quercus lobata, 62, 80–81*map*, 84
Quesnel terrane, 27*map*
Quinones, R., 249

racer, 60
radiation, 6–7, 24, 33–34, 47, 62, 83, 89, 99, 109–10, 129–30, 145, 147, 159, 161, 167, 169, 175–76, 209–10, 219–20, 231
Radiolaria, 23
RAG-1, 140
ragwort, Mojave, 97
rainbow trout, 121, 124*map*, 126*map*
Rambaut, A., 16
Rana spp., 136, 141, 145, 218
Rana aurora aurora, 216
Rana aurora draytonii, 138, 143, 145, 216, 246
Rana boylii, 138, 143, 216–17, 221
Rana muscosa, 140, 143, 225, 250
Rancho La Brea, 59–60, 143, 169
Rancholabrean, 171, 175
rangeland, 251
Ranidae, 138, 140, 143
Ranker, T., 97
Ranunculaceae, 82, 90
rarity, 235–36
ratites, 33
rats, kangaroo, 48
rattlesnake
rattlesnake, red diamond, 156
rattlesnake, speckled, 155
rattlesnake, tiger, 156
rattlesnake, western diamondback, 155
*rbc*L, 70–71, 83, 92
rDNA, 73, 132, 143
recolonization, 120, 131, 150, 163, 200, 219, 239
recommendations, 233, 240, 251
reconstruction(s), 85, 105, 212
recovery, 240
recreation(al), 234, 256
Recuero, E., 143
recycling, 247
redbacked vole, northern, 183, 212
red diamond rattlesnake, 156
redband trout, 124*map*
Redding, California 75*map*, 100*map*, 122*map*, 124*map*, 127*map*, 144*map*, 148*map*, 194*map*
Redenbach, Z., 125
red-legged frog, 138
red sidewinder, 226
redlist, IUCN, 205
redwood, 75*map*
redwood, coastal, 6
redwood, dawn, 34
Reed, D., 239–240

reef(s), 25, 30, 47
refugia, 5, 9, 14, 19, 40–41, 56–58, 71–74, 83, 88–89, 94, 97–99, 105–6, 112, 120, 130, 137, 143 145, 155, 160–1, 165, 170, 176, 179, 182–83, 185, 192, 195, 201, 211–14, 220, 223, 225–26, 229–31
reintroductions, 240, 245
Reiter, J., 199
relictual, 9, 34, 62, 111, 186–87, 215, 222
RELN, 151
Remington, C., 155, 219, 222
Remizidae, 161
remnant(s), 80, 126, 128, 136, 157, 218, 227–28, 251
Rensberger, J., 189
Repenning, C., 189, 200
reptile(s), 4, 28, 30–31, 59, 38, 143, 145, 147–49, 153, 158, 220, 230, 253
Resh, V., 24
resilience, 240
resistance, 236, 248
restoration, 19, 95, 240, 244–45, 247–52
resubmerged, 11, 111, 216
Reticulitermes
Reticulitermes flavipes, 114
Reticulitermes hesperus, 114
Reticulitermes tibialis, 114–15
Rey Benayas, J., 245
Rhamnaceae, 48, 79, 98
Rhamnus californica, 98
rhinoceros, short-legged, 49
Rhinocheilus lecontei, 61
Rhinotermitidae, 114
Rhus sp., 48
Rhynchocephalia, 147
Rhynie Chert, 24
Rhyniognatha hirsti, 24
Rick, T., 198
Riddle, B., 7
Ridge Basin, 151
Riedman, M., 198
Rieseberg, L., 238
Riley, S., 239, 250
ring-necked snake, 61
Rio Santo Domingo, 131
Rios, E., 191
Riparia riparia, 253
riparian, 34, 61, 92, 94, 138, 163, 169, 186, 193, 218, 237–38, 246, 250, 252–54
Rissler, L., 136, 213, 216, 230–31, 236
riverine, 50, 56, 120, 126
river(s), 10, 40, 50, 62, 94, 110, 123, 130, 134, 156, 188, 227, 243, 256

River, Gualala, 123
River, Gila, 156, 227
River, King, 188
River, Mad, 62
River, McCloud, 123, 124*map*
River, Pajaro, 10
River, Pit, 123
River, Rouge, 249
River, Russian, 125, 217
River, Sacramento, 10, 123, 125
River, Salinas, 10
River, Santa Ana, 60–61, 143
River, San Joaquin, 110, 188
River, Santa Clara, 113, 145, 225
River, Smith, 249
River, Van Duzen, 62
Riverside County, 68
roach, 123
roads, 238
Robbins, C., 167
Robinson, M., 155
Rocha-Olivares, F., 129
rockfish, 129
Rocky Mountains, 49, 72–73, 85, 90, 92,
    104–9, 162, 167, 177, 190, 212, 219–20
rodent(s), 70, 174, 185
Rodentia, 39, 175, 185
Roeder, M., 61
Roderick, G., 7, 104–5, 108, 212, 220, 233
Rodinia, 22
Rodríguez-Robles, J., 113, 145, 149–51, 193,
    195, 219–20, 225, 227
Rodriguez, R., 178
Roed, K., 176
Rogers, D., 71, 241
Ronquist, F., 16
rookeries, 198–99
Rooney, T., 237
Roosevelt elk, 177
Rosaceae, 48, 90, 98
Rosenberg, N., 72
Rosidae, 77
Rosinae, 77
rosy boa, 156
Rouge River, 249
rove beetle, 110
Rovito, S., 109, 138, 221
rpl20, 70
rpl16, 92
rpoB, 93
rps7, 73
rps15, 73
rps16, 83

rps18, 70
R_{ST}, 17–18, 80, 93–94, 181, 183
rubber boa, 147, 149
Rubicaceae, 90
Rubinoff, D., 11, 111, 239
Ruegg, K., 163
Rumelhart, P., 44, 118, 229
ruminating ungulates, 42
Runck, A., 183, 212
Rundel, P., 85, 88, 90–91, 96
Russian River, 125, 217

saber-toothed cat, 50–51, 180, 198
saber-toothed salmon, 50, 51*fig*
Sabinianae, 71
Sabo, J., 253–254
Sacramento, California, 75*map*, 100*map*,
    122*map*, 124*map*, 127*map*, 144*map*,
    148*map*, 194*map*, 234
Sacramento County 40, 139
Sacramento River, 10, 123, 125
Sacramento-San Joaquin Delta, 121, 125, 138,
    140, 187–88, 217–18, 221
Sacramento pike minnow, 125
Sacramento splittail, 252
Sacramento Valley, 10, 75*map*, 95, 100*map*,
    122*map*, 124*map*, 127*map*, 144*map*,
    148*map*, 194*map*, 253
Sage, R., 185
sage, coastal 191, 239, 251
Sage Sparrow, 161
sagebrush stepe, 61
Sahney, S., 25
Sala, O., 237, 248
salamander(s), 109, 137, 139–40, 143, 217
salamander, arboreal, 143
salamander, black, 217
salamander, coastal giant, 137
salamander, slender, 140
salamander, California slender, 140
salamander, Gabilan Mountains slender, 217
salamander, Mount Lyell, 109
salamander, Santa Lucia slender, 217
salamander, southern long-toed, 139
salamander, California tiger, 139
salamanders, mole, 44
salamanders, woodland, 136
Salamandridae, 44, 135
Salamonids, 127
Salicaceae, 40, 53, 90
Salinas River, 10, 217
saline, 44
Salinian Block, 140, 215–16

Salinian terrane, 10
salinity, 126, 252
Salix sp. 40, 61
Salix nivalis, 90
Salmonidae, 50, 117–18, 121, 125
salmonid(s), 125–27, 249–50
Salmoninae, 125
Salomon, A., 256
Salton Trough, 12, 58*map,* 157, 191, 226–27, 230
Salvia sp., 98
San Andreas, 31, 43, 52, 140, 218, 251
San Bernardino Mountains, 11, 45*map,* 46*map,* 55, 99, 113, 141, 164, 192, 194*map,* 195, 222, 224–25, 235, 255
Sanctuary, National Marine, 204*map,* 236
Sanderson, M., 91
San Diego, 11, 58*map,* 75*map,* 92, 100, 118, 119*map,* 122*map,* 127*map,* 127*map,* 133, 144, 148*map,* 151, 194*map*
San Diego County, 33, 48, 57, 133, 157, 163, 194
San Diego pocket mouse, 191
San Dimas, 236
sandstones, 28
San Clemente Island, 11, 55, 111, 130
San Emigdio-Tehachapi Mountains, 221
San Gabriel Mountains, 11–12, 45–46, 96, 112–13, 141–42, 157, 163–64, 192, 222–25, 255
San Gorgonio Mountains, 11, 88, 153, 222, 226
San Jacinto Mountains, 11, 55, 68, 112–13, 141, 153, 194, 222, 225
San Joaquin kit fox, 182
San Joaquin River, 108–10, 121, 123, 139, 188, 193, 195, 221
San Joaquin Valley, 10, 44, 48, 75*map,*100*map,* 122*map,* 124*map,* 127*map,* 138–40, 144*map,* 148*map,* 150–51, 153, 183, 188, 193, 194*map,* 217–19, 236, 254
San Luis Obispo County, 48, 139, 151, 215, 230
San Francisco, 10, 31, 38, 51, 56–57, 62–63, 75*map,* 92–93, 100*map,* 119*map,* 122*map,* 123, 124*map,* 125, 127*map,* 130, 136–37, 142, 144*map,* 145, 148*map,* 150–51, 158, 175, 181, 184, 187, 191, 194*map,* 195, 201, 204*map,* 215–18, 251–52, 256
Sanmartín, I., 54, 103
San Mauro, 135
San Miguel Island, 11, 111

Santa Ana Mountains, 113
Santa Ana River, 60–61, 143
Santa Barbara, 48, 131, 140, 217
Santa Barbara County, 10, 93, 118, 139, 151, 186, 230, 240
Santa Barbara Island, 11, 111, 130, 239
Santa Catalina, 11, 111, 130, 167, 186–87
Santa Clara River, 113, 145, 225
Santa Clara Valley, 44
Santa Cruz, 111, 150, 204*fig,* 251
Santa Cruz Island, 11
Santa Cruz Mountains, 194*fig*
Santa Lucia Mountains, 10, 55, 187, 194*fig,* 215–17, 235, 251, 254
Santa Lucia fir, 53
Santa Maria, Bahia, 132
Santa Maria Valley, 44
Santa Monica Mountains, 44
Santa Lucia Mountains, 112
Santa Lucia Range, 55, 194*fig,* 254
Santa Rosa Island, 11, 57, 111
Santa Rosa Mountains, 11, 222*fig*
Santa Ynez Mountains, 10, 43–44, 113, 224
Santa Ynez Valley, 44
Santo Domingo, Baja California, 131
sardine, 133
sardine, Pacific, 128
Sardinops ocellatus, 133
Sardinops ocellatus caeruleus, 133
Sardinops sagax, 128, 133
Sarna-Wojcicki, A., 10
Sauromalus spp., 154
Sauromalus ater, 158
Sauromalus australis, 154
Sauromalus obesus, 154
Savage, W., 7, 12, 121, 139, 222
savannahs, 39, 253
Savolainen, O., 19
Sawyer, J., 8–9, 74, 88, 215
Saxifraga sp., 83
Saxifragaceae, 82–83, 109
Scaphopoda, 38
Sceloporus sp., 60, 158,
Sceloporus magister, 154–55 224*table,* 227–28
Sceloporus occidentalis becki, 150
Sceloporus occidentalis biseriatus, 150
Sceloperus occidentalis bocourtii, 150
Sceloporus occidentalis longipes, 150
Sceloporus occidentalis taylori, 150–51
Sceloporus orcutti, 155, 228
Scher, S., 70
Schierenbeck, K., 91, 237

Schinske, J., 132
schist, 85, 220
Schluter, D., 170
Schoenherr, A., 9, 121, 123, 126, 128
Schorn, H., 40, 43
Schoville, S., 104–5, 107–9, 141, 212, 214, 220–21, 225*table*
Schrader, J., 92
Schramm, Y., 200
Schwartz, M., 182–83
Schweiger, O., 242
scimitar cat, 180
Sciuridae, 175, 186, 191–92, 209
Sciurus sp., 175
sclerophyllous, 48, 77, 79, 236
Scomber japonicus, 128
Scombridae, 128–29
Scott, R., 8, 90
sculpin(s), 56, 117, 120, 125, 227
sea anemones, 25
Sea, Bering, 202, 229
seabird(s), 33, 48, 53, 159
sea cow(s), 47
sea cow, Steller's 202
Sea of Cortez, 193
sea lion, California 200, 204
sea lion, Steller's, 199, 203*fig*
seal, northern elephant, 198–99
Sea, Okhotsk, 202
seals, eared, 48, 198–99
seals, earless, 48, 198
sea otters, 197–198
sea stars, 25
sea turtles
sea turtles, green, 31
sea turtles, leatherback 31
sea urchins, 23
Sea, Salton, 128, 191, 228
seas, inland 10, 43, 52, 151, 219, 226–27
seaway, 10, 40, 120, 143, 161, 191, 193, 216, 219, 227
seaway, Domengine-Ione, 40
Searles Lake, 52
Sebastes spp, 129, 219
Sebastes mystinus, 129
Sebastes subgenus *Sebastomus*, 129
Sebastidae, 129, 210
sediment(s), 10, 22, 28, 30–31, 33, 38, 40, 61, 63, 107, 131, 133, 209, 218
sedimentation, 11, 22, 30, 39, 43, 50, 53, 56, 91, 210–11, 216, 219

seep(s), 91, 107
Segraves, K., 112, 223
semiaquatic, 95
semiarid, 155
semideserts, 126
Senecio mohavensis ssp. *mohavensis*, 97
Sepedophilus castaneus, 112–13, 224–25
Sequoia sempervirens, 6, 34, 41, 48, 53, 62, 67–69, 84, 98, 236, 251
Sequoiadendron giganteum, 34, 67–69, 84
Sequoioid, 34
serpentine, 31, 84–85, 91–92, 236
serpentinite, 91
sewage, 247
Sgariglia, K., 165, 220, 225
Shafer, A., 213
Shaffer, B., 7, 113, 138–40, 151, 153, 216–19, 222, 224–25, 240
shale(s) 23, 28, 85
Shapiro, A., 7, 105–6, 212, 241
Sharsmith, C., 85
Shasta County, 23, 25, 28, 31, 117, 137, 149, 209
Shaver Lake, 188
Shaw, R., 242
sheep, bighorn 49
shells, lamp, 23, 25
Shepard, J., 242
Sherwin Glacial Maximum, 73–74
Shimamura, M., 200
shoals, 57, 118
shorebird(s), 33, 159–60, 169
shoreline, 30, 57, 58, 252
short-legged rhinoceros, 49
shovel-tusked mastodon, 49
shrew, ornate, 186
shrew, vagrant, 187
Siberia, 57, 189
Sidalcea hickmanii, 89
Sidalcea glaucescens, 89
Sidalcea malachroides, 89
Sidalcea oregana, 89
Sidalcea stipularis, 89
Siddiqui, A., 237
sidewinder, 155
sidewinder, red, 226
Sierra de la Laguna, 186–87
Sierra Nevada, 5, 9–12, 14, 25, 27–28, 30, 34, 37, 39–41, 43–44, 48, 50, 52–53, 56, 59, 61–63, 67, 71–73, 76, 80, 84–86, 88–93, 96, 99, 104–10, 112–13, 115, 136–42, 145, 150–51, 153, 158, 162, 164, 167–68,

179–82, 185–91, 193, 195, 210–15, 218–22,
    224–25, 230, 235, 243, 248, 254–55
Sierra Nevada Conservancy, 250
Sierra Pelona, 11, 96, 112–113, 142*map*, 151,
    224*table*, 225, 255
Sierra San Pedro Mártir, 188
Sigler, W., 117–118
Signor, P., 23–24
silicious, 85
Simmons, K., 178
Sinclair owl, 60
Sinclair, E., 153
Sinervo, B. 14
Sirenia, 47
Siskiyou County, 53, 123
Siskiyou Mountains, 5–6, 8–9, 62, 68, 71,
    76, 84, 88–89, 92, 99, 104, 106, 115,
    135–37, 143, 145, 181, 213–15, 230, 235,
    248–49
Sistrurus sp., 149
Sitta pygmaea, 165
Sittidae, 165
Skinner, M., 248, 253
skipper, dusky-winged, 242
skipper, holarctic, 107
Slack, K., 33, 159
Slatkin, M., 14, 17
Slauson, K., 179
slender salamander, 140, 217
Sloop, C., 95
sloths, ground, 171
sloughs, 252
Smilodon sp., 50, 60, 180
Smilodon fatalis, 180
Smith, C., 98, 114,
Smith, G., 11
Smith, J., 88
Smith, M., 16, 191, 195, 224*table*, 229, 254
Smith, S.A., 80, 82
Smith, S.G., 56, 120, 125, 227
Smith, S.V., 51
Smith, W., 136
Smith, T., 163
Smith River, 249
snowfall, 85, 109
Snyder, M., 256
Solano County, 95, 139
Soltis, P. and D., 9, 88, 90, 186, 213–14
Soltz, D., 128, 228
songbird(s), 162
Sonoma County, 92, 139, 217, 240
Sonoran Desert, 5, 11–12, 97–98, 100*map*,
    154–58, 192, 195, 225–26

Sonora Pass, 108
Sonoyta Basin, 128, 228
Sooty Grouse, 167
Sorensen, F., 72
Sorex ornatus, 186–88, 193, 217–18,
    222*table,* 224, 229
Sorex sinuosus, 222
Sorex vagrans, 187–88, 218
Soricidae, 175, 186
Soricomorpha, 175
Sork, V., 7, 80–81, 238
South Dakota, 210
southern long-toed salamander, 139
southeastern North America, 13, 58, 136, 156,
    179, 209
southwestern North America, 26*map,*
    27*map,* 29, 45–46, 79, 88, 92, 94, 97,
    108–9, 112, 115, 136, 143, 154, 156–57, 168,
    179, 182, 188, 191, 226
sparrow, sage, 161
Spaulding, W., 97, 156, 195, 227–28
specialist(s), 84, 98, 104, 107, 193, 241
specialization, 5, 187
speciation, 6–7, 13, 51, 55, 83, 106–7, 110, 125,
    137–38, 141, 156, 160–61, 163, 165, 167,
    170, 178, 211, 223, 228, 231
specificity, habitat, 129, 236
specificity, genotype, 239–40
speckled rattlesnake, 155
Spellman, G., 162, 165
Spermophilus sp., 175
Spermophyta, 31
Spicer, G., 7, 110, 112–13, 145, 158, 220,
    223–25
Spinks, P., 113, 151, 217–19, 222,
    224–25
spiny lizard, 154
Spirinchus thaleichthys, 252
splittail, Sacramento, 252
Sporobolus sp., 98
spruce, 40
spruce, Brewer, 74
Squamata, 60, 147, 149, 158
squirrel(s), 175
squirrel, antelope ground, 191
squirrel, Mohave ground, 192
squirrels, New World flying, 186
squirrels, New World tree, 188
stakeholders, 236
Stanislaus County, 51, 230, 237
Staphylinidae, 110, 112
Starrett, J., 221
Stearley, R., 125

Stebbins, G., 34, 52, 82, 84, 94, 97, 99, 222, 254
Stebbins, R., 137–138, 141
Steele, C. 137, 214*table*
steelhead trout, 50, 56, 124*map*, 126–27, 243
Steinhoff, R., 68, 84, 90
Steller's sea lion, 199, 202, 203*map*
Steneck, R., 47
Stenopelmatidae, 113
Stenopelmatus "mahagani"complex, 113, 224*table*
steppe, shrub, 61, 215
Stewart, J., 22
Stewart, B., 198
stickleback, three-spined, 56
Stidham, T., 31, 149
stilt-legged horses, 53
Stine, S., 62
Stockey, R., 41
Stone, K., 183, 212
Storer, T., 9
Storfer, A., 13, 137, 214
Stork, Asphalt, 60
Strait, Bering, 39, 171
Strait, Carquinez, 10
stratigraphic, 23, 37, 109
stratovolcano, 9
Strauss, S., 72
streamside, 253
Streptophyta, 24
Strigidae, 60, 160, 168
Strigiformes, 168
Strix sp., 168, 213–15, 248
Strix occidentalis, 168, 213, 248
Strix occidentalis caurina, 248
Strix nebulosa, 168, 214–15
stromatolites, 21, 23
strontium, 42
sturgeon, green, 252
Styracaceae, 82
Styrax sp. 82
Suarez, A., 241
subalpine, 53, 75, 88, 107, 109, 183, 214, 220, 243
subduct(ed), 30–31, 34, 213
subduction, 22, 28, 30–31, 43
submergence, 248
sub-Saharan Africa, 197
substrate(s), 8, 91, 99, 131, 213, 216, 229, 250
subsurface, 119
subterranean, 114
subtropic(al), 39–40, 48, 79

suburban, 244
suburbanization, 253
successional, 179, 248
sucker, Modoc, 125
suckers, 117
Suczek, C., 22
Suh, A., 168
Sulidae, 48
Sullivan, R., 105, 212
sunfishes, 44
Superasteridae, 77
supercontinent, 21–22, 24–25, 28, 43
Superrosids, 77
Sur-Obispo terrane, 45*map*, 46*map*
surfperch, 47
surfperch, black, 129
surfperch, striped, 130
Susanville, California, 40
Sutcliffe, A., 171
suture zone, 155, 219, 230
Swainson's Thrush, 163
swallowtail, pipevine, 110
Sweeney, B., 238
sweetgum, 79
Sweetwater Mountains, 90
Swenson, N., 219
swift fox, 182
Swifts, 160
sycamore, 61
Sylviidae, 164
sympatry, 6, 106, 130, 137, 141, 161, 211
Syngnathidae, 132
Syngnathus leptorhynchus, 132
Syring, J., 15, 70–72
Syverson, V., 169

Tadarida brasiliensis mexicana, 178
Taguchi, M., 201
Tahoe, Lake, 108, 139
Tajima, F., 18
Tallamy, D., 247
Talpidae, 175
talus, 85
Tamaulipan Plain, Mexico, 155
Tamias sp., 175
Tamiasciurus spp., 175, 183, 188, 212
Tamiasciurus douglasii, 188
Tamiasciurus hudsonicus, 188
Tamiasciurus mearnsi, 188
Tan, A.-M., 10, 136, 195, 214*table*, 217–18, 220
Tapiridae, 175
Taricha sp., 135–36, 145, 218

Taricha granulosa, 214, 217
Taricha torosa, 44, 135, 145, 195, 214*table*, 217–18,
Taricha torosa sierrae, 136, 214, 220
tarweed, island, 111
Taxaceae, 53, 69
Taxodium sp., 69
Taxus sp., 40
Taxus brevifolia, 69
Tayassuidae, 175
Taylor, E., 125–26, 248
Taylor, T., 24
Tecate cypress, 69
tectonic(s), 8, 12, 22, 26–27, 29, 45–46, 50, 57, 74, 103, 171, 239
Tedford, R. 50, 179
teeth, 41, 49–50, 61
Tegeticula antithetica, 114
Tegeticula synthetica, 114
Tehachapi Mountains, 10–11, 48, 112, 151, 181, 185, 187, 194*map*, 195, 221, 222*table*, 230, 235, 254
Tehama County, 235
Tejon Pass, 186
Tejon Ranch, 254
Teleoceras, 49
Temblor Range, 10
Templeton, A., 15
temporal, 6, 9, 14, 47, 101, 231, 240
Tenebrionidae, 110, 111–12
Teratorn, Incredible, 60
Teratornithidae, 60
termites, 114
terrace, 57
terrain, 93, 130, 219
terrane(s), 10, 27, 30, 34, 37, 45–46
terrapins, 147
Tertiary Period, 37
Tertoris sp., 60
Testudinidae, 156
Tetrameryx irvingtonensis, 175
Texas, 85, 152, 156, 178, 184
Thalictrum alpinum, 90
Thamnophis sirtalis, 150
Thamnophis sp., 61
thermal barrier, 132
thermal pollution, 256
Thinopinus pictus, 110
Thomomys sp., 58
Thomomys bottae, 190–91, 195, 224*table*
Thomomys laticeps, 191
Thomomys umbrinus, 190
Thompson, C., 120, 254

Thompson, R., 56, 59, 88, 97
Thornburgh, D., 8, 74
Thrasher, California, 164, 166*fig*, 223
Thrasher, LeConte's, 161, 164,
threat(s), 145, 235, 251, 253–55
threatened, 95, 159, 192, 199, 205, 234, 237, 243, 249, 256
three-spined stickleback, 56
three-toed horse, 42*fig*
Thrush, Swainson's, 163
Thunnus alalunga, 128
Thymelaecaceae, 92
Tiarella sp., 82
Tiffney, B., 54, 79, 82–83
tiger salamander, California, 139
tiger rattlesnake, 156
Tigriopus sp., 130
timber, 245
tinamous, 33
Tioga Crest, 108–9
titanotheres, large, 42
titmouse, oak, 162
Titus Canyon, 42
T_{MRCA}, 105, 108–9
toads, western, 143
Todisco, V., 105
Tolmiea menziesii, 214*table*
Tomales Bay, 125, 216
Tonella floribunda, 86*fig*
toothed birds, 33
toothed cetacean, 47
toothed whales, 200
Topatopa Mountains, 151
tope, 132
topographically, 256
topography, 4, 11
topology(ies), 15, 16, 71, 192
torix, 111
Torrey pine, 57
Torreya sp., 69–70
Torreya californica, 69
tortoise(s), 147, 156, 255
Tortricidae, 111
Townsend, T., 147
Toxochelyidae, 31, 149
Toxodontidae, 171
Toxostoma redivivum, 164–65, 170, 223–25, 228
Toxostoma lecontei, 161
Trachurus symmetricus, 128
transgression(s), 40, 58, 227
transient, 125
transitional, 33, 210

Transverse Ranges, 5, 11–12, 14, 34, 37, 43,
    45*map*, 46*map*, 50, 52, 55, 71, 75*map*, 80,
    86*fig*, 87*fig*, 96, 99, 100*map*, 106–7, 110,
    112–13, 122, 127, 141, 144–45, 148, 151,
    158, 162, 164–66, 170, 185–86, 191–92,
    194–95, 211, 215–16, 220–25, 230, 235,
    243, 254–55
Trautvetteria sp., 82
TreeAnnotator, 86
treeline(s), 183, 211
Triakidae, 132
Triassic Period, 28–31, 33, 117, 149, 209
tributary(ies), 123
Trichoptera, 24
Trilobita, 23
Trimorphodon sp., 225
Trimorphodon biscutatus, 157
Trinity Mountains, 8, 104, 213
Trinity River, 137
Tripodi, A., 115
tRNA, 156–57, 164, 179
tRNA-GLU, 141
tRNA-lysine, 164
*trn*C, 93
*trn*D, 16
*trn*F, 92
*trn*G, 71
*trn*H, 92–93, 98
*trn*K, 89, 93
*trn*L, 71, 73, 114
*trn*T, 16, 92, 114
*trn*V, 70
Trochilidae, 159–160
Troglodytidae, 160–61
tropic(s), 24, 34
trout, 124*map*, 243
trout, California golden, 124*map*
trout, cutthroat, 121
trout, Eagle Lake rainbow
trout, golden, 121, 221
trout, Goose Lake, 124*map*
trout, invasive, 250
trout, Lahontan cutthroat, 124*map*
trout, Little Kern golden, 124*map*
trout, Kern River rainbow, 124*map*
trout, McCloud River redband, 124*map*
trout, Paiute cutthroat, 124*map*
trout, rainbow, 121, 124*map*, 126, 221, 243
trout, steelhead, 124*map*
trout, Warner Valley redband, 124*map*
Truckee-Donner Land Trust, 250
Truesdale, H., 69
Trzyna, T., 251

Tsuga heterophylla, 53,
Tube-nosed Seabirds, 48, 160
Tuctoria greenei, 95
tuctoria, Greene's, 95
Tulare County, 139
Tulare Lake, 10, 63, 68, 139, 221
Tule elk, 177
Tule Lake, 53
tuna, albacore, 128
Tuolumne County, 88, 230, 237
Tuolumne River, 125, 150–51
turbidity, 134
turbot, diamond, 132
Turdidae, 163
Turner B., 128, 228
turtle(s), 147, 149
turtles, green sea, 31
turtles, leatherback sea, 31, 147, 149
turtles, sea, 30
turtle, western pond, 151
Two-toed sloth, 171
Typha sp., 61
Typhaceae, 61
Tyrannidae, 159–60
Tyrant Flycatchers, 159–60
Tytonidae, 160

ucmpdb, 23, 25, 31, 117, 149
Ulmaceae, 53
ultramafic, 213
Uma scoparia, 158, 255
Umbellularia, 48
undercurrent, 119
undergraduate, 245
underprotected, 235
underwater, 6, 8, 25, 38, 48, 209–10
undisturbed, 94, 134
UNESCO, 244
unglaciated, 58, 71, 74
Ungulata, 176–177
ungulates, 42, 171, 200
universities, role of, 245
University of California Museum of Paleon-
    tology (UCMP), 60
UPGMA, 154, 156
uplift, 6, 9–10, 25, 31, 40, 43, 50, 52, 55–56,
    78, 96, 118, 136, 139–40, 145, 154, 191–93,
    210–11, 213, 215–16, 219, 223, 226, 229
Upton, D., 193
upwelling, 47, 50, 52, 118, 129, 133, 161, 169,
    210, 219, 256
urbanization, 241, 245, 251, 253, 256
urchins, sea, 23

Urosaurus graciosus, 158
Urosaurus lahtelai, 158
Urosaurus ornatus, 158
Ursidae, 175, 184
Ursus americanus, 175, 183, 212, 222
*Ursus arctos californicus,*185
Usinger, R., 9
Uta stansburiana, 60
Utah, 191

Vaccininna sp., 105
Vacquier, V., 17
vagile species, 107, 188, 222
vagility, 104, 107–8, 113, 115, 153,
    212–13
vagrant shrew, 187
Valensise, G., 55
Valmonte Formation, 47
vampire bat, 178
Van Devender, T., 59, 97–98, 143, 155–56, 191,
    195, 227–28
Van Duzen River, 62
van Frank, 135
Vandergast, A., 113–14, 224, 251, 255
vasculature, 67
Vegavis sp. 33, 165
vegetation, 39, 41–43, 48, 53, 59, 61, 63,
    78–80, 92, 94, 96–97, 155, 195, 228, 247,
    250–51, 253–54
vegetative, 93, 249
vehicle(s), 255
Ventana Wilderness Area, 254
Ventura County, 151, 186, 254
Veracruz, Mexico, 190
Verdin, 161
Vermeij, G., 51
vernal pools, 94–95, 140, 218, 239,
    253–54
vertebrate(s), 6, 51, 57, 135, 174, 187, 193, 195,
    251, 255
Vespertilionidae, 175, 178
vicariant, 5–6, 12–14, 18–19, 57–58, 78, 96,
    115, 139, 150, 154, 158, 185, 191, 193, 221,
    226, 228–29, 231, 240
Vignen, 83
Vila, R., 105
Vilà, C., 184, 195, 217, 238
Villa-Lobos, J., 248
Viperidae, 149, 155
viper, pit, 149
Vireonidae, 159
Vireos, 159
Viro, 185

Vizarino, 191
volcanically, 9, 150, 219
volcanism, 30, 43
volcano(es), 9, 25, 28, 121, 213, 219, 121, 221,
    251
vole, 183, 185, 189, 212, 255
vole, Armargosa River, 195
von Humboldt, 12
vulnerable taxa, 115, 236, 238, 241, 243
Vulpes cascadensis, 182
Vulpes macrotis mutica, 182–183
Vulpes necator, 182
Vulpes velox, 182
Vulpes vulpes, 182–183, 211
Vultures, New World, 53, 159, 169

Wagner, D., 156
Wahlert, J., 49
Walcott, C., 23
Walker, D., 15, 163, 170
walnut, 61
walrus, 203*fig*
Waltari, E., 212
Wang, I., 13
Waples, R., 50
Warblers, New World, 159–60, 162
Warbler, Wilson's, 162
Warheit, K., 161
Warner Mountains, 86*fig*, 96, 124*map*, 164,
    213, 215
Warner Springs, 255
Warren, M., 242
Wasatch Line, 26
Washington, 52, 74, 91, 112, 119, 129,
    132, 139, 142–43, 150–51, 179, 186,
    188–89
wasps, 24
Wassuk Range, 190, 227
watershed(s), 62, 86, 139, 141, 249
waterways, 44, 118, 126, 131, 256
weasels, 197
Webber Lake, 43
Webster, D., 3
weevil, broad-nosed, 112
Wegener, A., 12
Weinstock, J., 49, 53, 177
Weir, B., 18
Weir, J., 170
Weiss, S., 246
Wen, J., 78
Wendling, B., 83
Westergaard, K., 83
western hemlock, 53

western mastiff bat, 253
western white pine, 53
Westlake, R., 201, 229
wetland(s), 56, 91, 95, 145, 186–87, 193, 218–
    19, 231, 246, 252–53
white fire, 53
White-headed Woodpecker, 163, 166fig
White, M., 254
White Mountains, 23fig, 44, 63, 73, 88, 90,
    108, 190, 211, 227
Whiting, M., 108
Whitmeyer, S., 21–22
Whitney, Mount, 4, 75map, 100map,
    122map, 124map, 127map, 144map,
    148map, 194map
Whittaker, R., 244, 248
Whorley, J., 191
wilderness, 249
wilderness, Ventana, 254
wildland, 244, 249
wildlife, 4, 122, 124, 127, 144, 148–49, 243,
    246–47
Wiley, E., 13,
willow, 40
Willyard, A., 70–72
Wilson's Warbler, 162
Wilson, A., 132
Wilson, C., 185
wingless, 107
wingless insects, 24
Wingo, S., 246
wingspan, 60
winter(s), 33, 50, 61, 85, 97–98, 118, 126, 130,
    162–63, 198, 201, 253
wintering, 161, 167
Wisconsin, 106
Wisconsin glaciation, 59, 153, 175, 182–83,
    211, 215
Wisely, S., 180
Wolf, A., 92
wolf, gray, 182
wolf, Dire, 184
Wolfe, J., 39–41, 43
Wolff, O., 185
wolves, 183
woodland(s), 48, 53, 70, 80, 92, 94, 136, 139,
    149, 163, 177, 217, 187, 235–36, 252–53
Woodpecker(s), 160, 163, 166

woodrat(s), 58
woodrat, dusky-footed, 187
Woolfenden, W., 68
Woolly mammoth, 59
wrasses, 55
Wren, Cactus, 161
Wrens, 160
Wrentit, 164, 166
Wright, S., 17
Wright, T., 52
Wyoming, 210

Xantusia henshawi, 224table
Xantusia riversiana, 154, 226
Xantusia vigilis, 153–54, 226
Xantusiidae, 153
xenarthrans, 174
Xerospermophilus mohavensis, 192, 223–24,
    255
Xerospermophilus tereticaudus, 192
xerotherm, 73
Xiang, J., 54, 82
Xiphiidae, 128
Xue, M., 9, 43
Xylococcus sp., 98

Yanev, K., 10, 195, 230
Yellow-eyed Junco, 163
yew, 40, 69
Yollo Bolly Mountains, 8
Yolo County, 95
Yosemite, 150, 237
Yuba River, 40
Yucca brevifolia, 97–98, 114, 153
yucca moths, 114
Yucca Valley, 153

Zacherl, D., 242
Zalophus californianus, 200, 202
Zamia sp., 40
Zanazzi, A., 41
Zehfuss, P., 9
Zhanxiang, Q., 171
Zielinski, W., 179
Zink, R., 7, 48, 160–61, 167, 170
Ziphiidae, 205
Zusi, R. 165
Zygophyllaceae, 97

Milton Keynes UK
Ingram Content Group UK Ltd.
UKHW051822140624
443920UK00008B/82